Oc[illegible]

With Love

Ron

# SONOLUMINESCENCE

The sonoluminescence bubble is the light blue dot in the middle of the jar. The acoustic resonator consists of two transducers separated by a thin glass cylinder. Graduate student Sean Cordy watches the sonoluminescence. Photograph graciously supplied by Professor Larry Crum, University of Washington, Seattle. (Crum LA Sept 1994 *Physics Today* 47, 22.)

# SONOLUMINESCENCE

**F. Ronald Young**

Newlands
Brookshill Drive
Harrow Weald
Middx UK

newlands@tinyworld.co.uk

020-8954-1950

CRC PRESS

Boca Raton London New York Washington, D.C.

**Library of Congress Cataloging-in-Publication Data**

Young, F. Ronald.
Sonoluminescence / F. Ronald Young.
p. cm.
Includes bibliographical references and index.
ISBN 0-8493-2439-4 (alk. paper)
1.Sonoluminescence. I. Title.

QC480.2.Y68 2004
535'.35--dc22 2004049731



Direct all inquiries to CRC Press LLC, 2000 N.W. Corporate Blvd., Boca Raton, Florida 33431.


International Standard Book Number 0-8493-2439-4
Library of Congress Card Number 2004049731
Printed in the United States of America 1 2 3 4 5 6 7 8 9 0
Printed on acid-free paper

**to Nancy**

# Contents

# Foreword

Twelve years after the discovery of single bubble sonoluminescence by Felipe Gaitan the publication of Dr. Young's book on this beautiful phenomenon signals that its understanding has come to some maturity. Dr. Young has brought together materials from various papers and proceeding contributions, highlightening both the experimental results (Chapters 2 and 3) and the theoretical understanding (Chapter 4). For those without previous knowledge on single bubble sonoluminescence, the book gives a thorough introduction to the field, and for the experts it is an encyclopedic summary of the progress achieved in the last decade.

In his search for single bubble sonoluminescence, Gaitan, then a graduate student of Larry Crum at the University of Mississippi, at some point worked in a regime with a moderate acoustic forcing pressure around 1.2–1.4 atm and with the water degassed to around twenty percent of its saturated concentration of air. He observed that "as the pressure was increased, the degassing action of the sound field was reducing the number of bubbles, causing the cavitation streamers to become very thin until only a single bubble remained. The remaining bubble was approximately 20 μm in radius and was remarkably stable in position and shape, remained constant in size and seemed to be pulsating in a purely radial mode. With the room lights dimmed, a greenish luminous spot the size of a pinpoint could be seen with the unaided eye, near the bubble's position in the liquid".

What Gaitan discovered can be viewed as the hydrogen atom of cavitation physics, as a single spherical cavitation bubble is the simplest building block of a sound driven bubbly fluid, just as hydrogen is for more complicated atoms, molecules, or solid states.

It is astounding how many sub-disciplines of physics and chemistry are necessary to understand this conceptionally simple building block: a single bubble oscillating in a sound field. These sub-disciplines range from acoustics, to fluid dynamics, to plasma physics, thermodynamics, atomic physics, spectroscopy, chemistry, dynamical system theory, and applied mathematics in general. Correspondingly, a text book on single bubble sonoluminescence has to address all these aspects, and Dr. Young's book does so. In 1860, Michael Faraday wrote the famous book "The chemical history of a candle", in which he addressed and discusses various aspects of a burning candle. He states: "So abundant is the interest that attaches itself to the subject, so wonderful are the varieties of outlet which it offers into the various departments of philosophy. There is not a law under which any part of this universe is governed which does not come into play and is touched upon in these phenomena. There is no better, there is no more open door by which you can enter into the study of

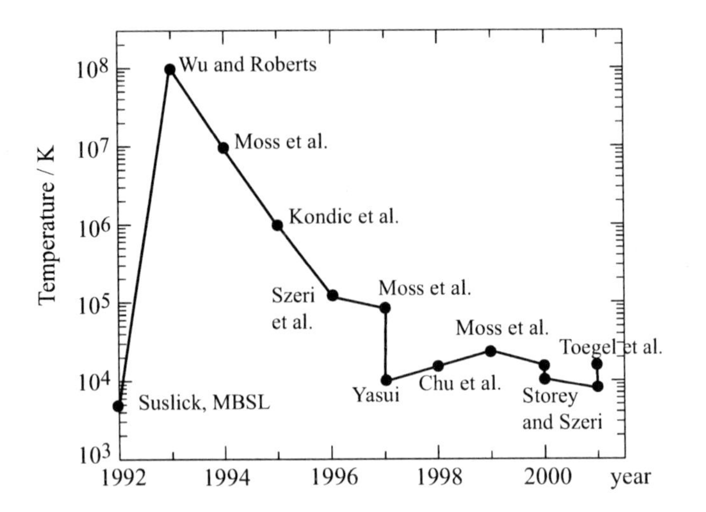


Figure 1 Theoretical predictions for the temperature of a typical sonoluminescing bubble over the years. Every reduction by an order of magnitude signaled that yet an additional physical effect had been added to the original model, namely thermal damping, viscous effects, the consideration of the cooling water vapor inside the bubble, and finally even chemical reactions of that water vapor. Meanwhile there seems to be "convergence" to temperature values between 10,000 K and 20,000 K. The references for the respective papers are given in the Chapters 3 and 4. The first value, 5,000 K, refers to Suslick's *measurements* of the temperature in multi bubble sonoluminescence, MBSL.

natural philosophy than by considering the physical phenomena of a candle. I trust, therefore, I shall not disappoint you in choosing this for my subject rather than any newer topic, which could not be better, were it even so good." It may be overdoing it, but in some sense a book on the light emission from a sound driven bubble can be viewed as the modern version of Faraday's approach.

The backbone of the understanding of single bubble sonoluminescence is the now classical theory of bubble dynamics, as developed by Lord Rayleigh, Milton Plesset, Andrea Prosperetti, and others. Dr. Young has already described this theory in his book "Cavitation" (1989), and is reiterating it here. The crucial part of (nonlinear) bubble dynamics is the so-called Rayleigh collapse, the collapse of a spherical void in a fluid. It is this dramatic event which leads to the adiabatic heating of the gas inside the bubble, its partial ionization, and consequently to the light emission on recombination of the electrons and ions. The temperature achieved at bubble collapse determines both intensity, length, and spectral properties of the light pulse. In the early years of single bubble sonoluminescence, temperatures as high as $10^8$ K had been suggested. Meanwhile the field has converged to more "modest" values between 10,000 K and 20,000 K, see Figure 1. Every reduction of the original theoretical temperature estimate by an order of magnitude signaled that yet an additional physical effect had been added to the original model, namely thermal damping, viscous effects, the consideration of the cooling water vapor inside the bubble, and finally even chemical reactions of that water vapor. However, the complication still remains that the temperature cannot be directly measured. It has to be deduced either from the bubble dynamics, or from the properties of the light, and modeling assumptions enter on both sides.

It is also the bubble dynamics which determines the conditions under which stable single bubble sonoluminescence can occur. As we have derived in our 1996 *Physics of Fluids* paper (Vol. 8, pages 2808–2826) these are (i) the threshold for the Rayleigh collapse to occur, (ii) the shape stability of the bubble, (iii) its diffusive stability, and (iv) the chemical stability of its ingredients.

The chemical activity inside the bubble (discussed in Section 2.2.7) is a consequence of the high temperatures achieved. In fact, the sonoluminescing bubble can be viewed as a high temperature, high pressure microlaboratory or reaction chamber, which can be controlled through the external parameters such as forcing pressure, frequency, or dissolved gas concentration. When the bubble is expanding, gas dissolves in the liquid and liquid vapor enters the bubble. At the adiabatic collapse, these gases are partly trapped inside the hot bubble, and react. E.g. nitrogen molecules will first dissociate to nitrogen radicals and later react to NH, NO, etc., which as highly soluble gases all dissolve in water when the bubble cools down and reexpands. Subsequently, the next reaction cycle starts.

Another milestone in the understanding of single bubble sonoluminescence was Gompf's precise measurements of the length of the light pulse as a function of the control parameters (1997, here discussed in Section 2.3.9). These measurements showed that the light pulse was much longer than previously thought, and there was no need for exotic theories predicting ultrashort light pulses of less than a picosecond. In fact, all experimental and theoretical information we have today is consistent with the thermal bremsstrahlung model of the sonoluminescing bubble. It is remarkable that Dr. Young had foreseen this in his book "Cavitation" back in 1990, when writing in the conclusions: "My own view is that sonoluminescence results from the heating of the cavity contents during the compression phase of the oscillating bubble. This causes excitation of the gas in the cavity, thereby promoting the formation and subsequent recombination of excited species, with the emission of light."

Where will the field go? With the detailed understanding of the hydrogen atom, atomic physics was not over, but only began to flourish. In analogy, now there is a basic understanding of single bubble sonoluminescence, I expect a similar florishing for cavitation physics. Most importantly, a better understanding of acoustically driven bubbly fluids must be achieved. Nearly all applications of collapsing bubbles will have to rely on more than one bubble (often in fact millions), and therefore bubble–bubble interactions become crucial. The nature of these applications are outlined in Chapters 5–8 of Dr. Young's earlier book on cavitation. Note that an application of the light emission itself seems unlikely, given that only a fraction of $\sim 10^{-4}$ of the energy in a collapsing bubble ends up as visual photons.

From my point of view the most important applications of bubble dynamics and cavitation are sonochemistry (on which Suslick and others have revealed astounding new insight in the last twelve years), ultrasound cleaning, (the prevention of) cavitation damage, and ultrasound diagnostics. A particular promising direction is the use of bubbles as vehicles for therapeutic applications. It has recently been demonstrated that the directed delivery of drugs as well as the transmission of genes through the wall of living cells is dramatically enhanced in the presence of ultrasound and microbubbles. It seems that here one has made use of cavitation damage in order to render the cell wall permeable for drugs and genes, but without permanently damaging the cell. The exact mechanism at work remains to be uncovered.

To achieve progress for these problems, a sound understanding of bubble dynamics, on which we learnt so much from the research on single bubble sonoluminescence, will be crucial – and therefore this present book can serve as a basis for this future progress.

*Detlef Lohse,*
*University of Twente, Enschede, The Netherlands*

# Preface

This book is intended as a source book on sonoluminescence, either as an introduction to those new to the subject, a text for the researcher, or as a place where an obscure reference with title and all the authors can be obtained. There will be things in the book that the reader will disagree with, as many aspects of sonoluminescence are not yet established. It has sometimes been difficult, and sound emission from a bubble is an example, whether to include it in Chapter 1 *Introduction*, Chapters 2 and 3 *Experiments* or Chapter 4 *Theories*. There is also the complication of multibubble sonoluminescence up to 1990, and single bubble sonoluminescence from 1990. I have done my best to structure the book logically but in many cases, a compromise is necessary.

Many people have stumbled across sonoluminescence by accident and I was no exception. Forty-two years ago, at an Open Day at Imperial College, London, I read a poster by Peter Jarman describing his work on sonoluminescence (Chapter 2, Jarman (1959a,b)). It seemed an intriguing puzzle, and I have not been disappointed in following its many twists and turns during the last 40 years when I have met many of the former giants of sonoluminiscence such as Milton Plesset, Hugh Flynn and Ernie Neppiras. The writing of this book has occupied me over the last five years, often in rather unlikely places such as Alaska and East Africa. I have spent many hours in the British Library which, fortunately, is only 15 miles from my house. Many researchers all over the world have been readily helpful in supplying offprints, diagrams and photographs, granting permissions and giving advice. I should like to thank Robert Apfel, Leif and Irina Bjørnø, David Crighton, Larry Crum, Anne Hartley, Ho-Young Kwak, Werner Lauterborn, Willy Moss, Wes Nyborg, Rainer Pecha, Andrea Prosperetti, Joan Ralfs, Brian Storey, Kenneth Suslick, Ola Törnkvist, Kyuichi Yasui and others.

Professor Detlef Lohse at the University of Twente has been immensely helpful in reading the manuscript, dispensing advice from his store of scholarship and writing the foreword. His wise and prompt counsel was a constant inspiration to a lone author.

I am very much indebted to Kathleen Killick who typed the manuscript several times with great expertise.

Finally, I would like to express my gratitude to my editors: Janie Wardle at Taylor and Francis for her encouragement and diplomatic suggestions, Jay Margolis at CRC Press for his professional advice, and Victor Selivanov in Moscow for type-setting the manuscript.

# List of Symbols

| | |
|---|---|
| $c$ | velocity of sound in liquid; gas concentration in liquid |
| $c_g$ | velocity of sound in gas |
| $c_0$ | saturation concentration of dissolved gas in liquid |
| $c_\infty$ | ambient concentration of dissolved gas in liquid far from bubble |
| $c_\infty/c_0$ | relative gas saturation in liquid |
| $f$ | frequency |
| $f_r$ | resonance frequency |
| $h$ | van der Waals hard core radius |
| $k$ | angular wave number = $2\pi/\lambda$; thermal conductivity of the gas; Boltzmann's constant |
| $k_\lambda$ | photon absorption coefficient |
| $p$ | dimensionless pressure $P_A/P_0$; pressure in liquid |
| $p_{max}$ | maximum liquid pressure |
| $q$ | heat flux (energy flow per unit mass per unit time); degree of ionization |
| $r$ | distance from bubble centre to point in liquid |
| $u$ | velocity of liquid at radius $r$ |
| $C$ | velocity of sound at bubble wall; concentration of dissolved gas in liquid |
| $C_p$ | specific heat at constant pressure |
| $C_v$ | specific heat at constant volume |
| $D$ | diffusivity of gas (mass or thermal) = $K/\rho C_p$ |
| $E$ | compressibility (bulk modulus of elasticity) |
| $K$ | thermal conductivity of the liquid |
| $M$ | acoustic Mach number = $\dot{R}/c$ |
| $N$ | number density of atoms |
| $P$ | pressure = $p(r, t)$ = local pressure in liquid at radius distance $r$ |
| $P_A$ | acoustic pressure amplitude |
| $P_e$ | Péclet number |
| $P_g$ | pressure of gas in the bubble |
| $P_L$ | pressure in liquid just outside bubble wall |
| $P_{max}$ | maximum pressure in collapsing bubble |
| $P_m$ | liquid pressure at transient collapse |
| $P_0$ | ambient pressure in liquid |
| $P_\infty$ | pressure in liquid far from bubble |

| | |
|---|---|
| $Q$ | initial pressure of gas in the bubble |
| $R$ | radius of bubble boundary |
| $R_0^c$ | threshold radius for Blake nucleation (Blake radius) |
| $R_e$ | radius of surface of emission of a bubble (Chapter 5) |
| $R_g$ | gas constant |
| $R_0$ | initial radius of bubble |
| $S_v$ | specific heat capacity at constant volume of liquid |
| $T$ | temperature |
| $T_A$ | period of sound field |
| $T_g$ | temperature of gas in bubble |
| $T_0$ | initial temperature of gas in the bubble |
| $V$ | gas volume |
| $Z$ | compression ratio = $R_0^3/R^3$ |
| $\alpha$ | phase angle |
| $\gamma$ | ratio of specific heats of the gas |
| $\lambda$ | acoustic wavelength |
| $\mu$, $\eta$ | dynamic (shear) viscosity |
| $v$ | kinematic viscosity = $\mu/\rho$ |
| $\nu$ | frequency |
| $\xi_0$ | amplitude of radial oscillations |
| $\xi$ | displacement of radial oscillations |
| $\rho$ | density of liquid |
| $\rho_g$ | density of gas |
| $\sigma$ | surface tension of liquid |
| $\phi$ | velocity potential |
| $\chi$ | thermal diffusivity |
| $\omega$ | angular velocity |
| $\Gamma$ | polytropic index |
| $T$ | time of collapse of bubble |

Suffix 0 refers to equilibrium value
g refers to gas
v refers to vapor
T refers to total gas content

# Useful Dimensionless Parameters

Reynolds number $R = \dfrac{\rho \upsilon d}{\mu} = \dfrac{\upsilon d}{\nu}$

where $\rho$ = density
$\upsilon$ = velocity
$d$ = typical linear dimension, e.g. diameter
$\mu$ = viscosity
$\nu$ = kinematic viscosity = $\dfrac{\mu}{\rho}$

(The higher $R$ is, the less is the importance of viscosity)

Froude number $F = \dfrac{\upsilon^2}{gd}$,

where $g$ = acceleration due to gravity

(The higher $F$ is, the less is the importance of gravity)

Strouhal number $S = \dfrac{fd}{\upsilon}$

where $f$ = frequency

(The higher $S$ is, the greater the frequency – of, say, a flat plate of length $d$ advancing in still air at speed $\upsilon$ and oscillating about some mean attitude at frequency $f$)

Weber number $W = \dfrac{\Delta P \times R}{\sigma}$

where $\Delta P$ = pressure difference between the two sides of the interface
$R$ = radius of interface
$\sigma$ = surface tension

(The higher $W$ is, the less is the influence of surface tension)

Péclet number $P_e = \dfrac{2\upsilon R}{D}$

$D$ = diffusivity in $m = -D\dfrac{\delta n}{\partial x}$ where $m$ is mass of solution passing across unit area per sec and $\dfrac{\partial n}{\partial x}$ is the rate at which the concentration of dissolved substance diminishes normal to the unit area

(The higher $P$ is, the less is the influence of diffusivity)

Deborah number $D_e = \frac{\upsilon T^*}{R}$

(The higher $D_e$ is, the greater is the importance of the fluid relaxation time)

$T^*$ = fluid relaxation time $= \mu/G$ where $G$ is the shear modulus

Prandtl number $P_r = \frac{v}{D}$

where $v$ = kinematic viscosity $= \frac{\mu}{\rho}$ where $\mu$ = viscosity, $\rho$ = density and where $D$ = thermal diffusivity $= \frac{K}{\rho C_p}$ where $K$ = thermal conductivity, $C_p$ = specific heat at constant pressure

Mach number $M = \frac{\dot{R}}{c}$

where $\dot{R}$ = velocity of bubble surface; $c$ = velocity of sound

Courant number $C = \upsilon \frac{\Delta t}{\Delta x}$

where $\upsilon$ is constant velocity in $x$ direction and $\Delta t$, $\Delta x$ are steps of time and distance in finite difference calculations

Other numbers can be found in *Measures in Science and Engineering. Their Expression, Relation and Interpretation,* by B.S. Massey, published by Ellis Horwood, 1986.

# Note on Units of Viscosity

Dynamic viscosity $\mu = \text{kg m}^{-1}\ \text{s}^{-1}$ (or $\text{N s m}^{-2}$)

Kinematic viscosity $\nu = \dfrac{\mu}{\rho} = \text{m}^2\ \text{s}^{-1}$

The Poise is the cgs unit of dynamic viscosity

The Pa s (Pascal second) is the MKS unit of dynamic viscosity

The Stoke is the cgs unit of kinematic viscosity

Dynamic viscosity of water at 20°C $= 1.00 \times 10^{-3}\ \text{N s m}^{-2}$

$= 1.00 \times 10^{-2}\ \text{gm s cm}^{-1}$

$= 1.00 \times 10^{-2}$ Poise

# Note on Units of Pressure

I have used both bars and atmospheres as units of pressure in this book. This is because sometimes one unit seems appropriate, as in discussing the tensile strength of liquids, and sometimes the other unit. Also, authors of papers use different units and as far as possible I have used in the text the same unit of pressure as used by the author in a referenced paper.

By definition 1 bar = $10^5$ N m$^{-2}$.

Since by definition 1 Pascal is the name given to 1 N m$^{-2}$, then

$$\boxed{1 \text{ bar} = 10^5 \text{ Pa.}}$$

1 atmosphere is defined as 101325 Pa.

Therefore, approximately

$$\boxed{1 \text{ bar} = 1 \text{ atmosphere.}}$$

Also, 1 atmosphere is equivalent (but not exactly equal) to the pressure exerted by 760 mm of mercury.

Thus, 1 atmosphere = $0.760 \times 1.36 \times 10^4 \times 9.81 = 1.013 \times 10^5$ N m$^{-2}$.

Therefore 1 atmosphere = $1.013 \times 10^5$ Pa, which nearly agrees with the statement above.

CHAPTER ONE

# Introduction

It is not the critic who counts, not the man who points out how the strong man stumbled or where the doer of deeds could have done better. The credit belongs to the man who is actually in the arena; whose face is marred by dust and sweat and blood; who strives valiantly; who errs and comes short again and again; who knows the great enthusiasms, the great devotions, and spends himself in a worthy cause; who, at the best, knows in the end the triumph of high achievement; and who, at the worst, if he fails, at least fails while daring greatly, so that his place shall never be with those cold and timid souls who know neither victory nor defeat.

*Theodore Roosevelt*

Universes that drift like bubbles in the foam upon the River of Time

*Arthur C. Clarke*

The Wall of Darkness, in *Super Science Stories*, collected in *The Other Side of the Sky*, Signet, New York, 1959, Chap 4

## 1.1 HOW THE BOOK IS ORGANIZED

We start with the discovery of sonoluminescence in 1933, and then the independent discoveries of single bubble sonoluminescence in 1962 in Japan and in 1970 and 1990 in the USA.

Any treatise on sonoluminescence must be built on bubble dynamics and an account of this is included in Chapter 1. This leads on to the important basic subjects of Bjerknes forces, rectified diffusion, and sound emission from a bubble.

We are now in a position to consider sonoluminescence itself. Chapter 2 deals with *multibubble sonoluminescence*, and also the light from hydrodynamic cavitation, agitated mercury, and collapsing glass spheres. This is mainly up to 1990.

Chapter 3 describes *single bubble sonoluminescence*, when from 1990 a whole series of systematic experiments were performed. With a stable single bubble, the gas content of the bubble could be controlled. And with a specified driving frequency and pressure we are well on the way to controlling the bubble parameters. Many research groups investigated how the sonoluminescence depended on these parameters.

Chapter 4 describes the theories of sonoluminescence. Early theories, now largely discontinued, to explain multibubble sonoluminescence, are discussed. With the advent of single bubble sonoluminescence in 1990, theoreticians all over the world have been engaged in explaining where the light comes from. Some of these later ideas are now discounted. There have been over 15 theories of the origin of sonoluminescence put forward. No one theory has yet gained universal acceptance.

Chapter 5 contains some conclusions, uses of sonoluminescence, and suggestions for future work.

## 1.2 HISTORY OF SONOLUMINESCENCE

When studying the action of ultrasonic waves on the development of a photographic plate it was accidentally discovered by Marinesco and Trillat (1933) that fogging of the photographic plate sometimes occurred. They explained the action as due to the ultrasonic waves accelerating the processes of reduction which take place in the sensitive plate by the violent mixing of the reactants, but Frenzel and Schultes (1934) in Cologne discovered that the fogging was accompanied by a faint luminescence. This light fogged the plate. Chambers (1937) obtained sonoluminescence from 14 liquids using the eye as a detector.

Sonoluminescence is always preceded by cavitation, and can easily be seen by the naked eye in a dark room if glycerine is cavitated by a velocity transformer attached to a 20 W, 20 kHz transducer. It appears as a bluish-white light.

Usually the sonoluminescence is so weak that photomultiplier tubes are used to detect it. This led to the discovery that the light appears as discrete flashes which are periodic with the sound field. Work then proceeded to find out at what phase of the sound field, or volume of the cavitating bubbles, the flashes occurred. The definitive experiment on this was performed by Meyer and Kuttruff (1959). They produced cavitation bubbles on the end face of a nickel rod magnetostrictively excited at 2.5 kHz and obtained a series of photographs showing the life cycle of the cavitation bubbles. The photographs showed that the bubbles started to appear halfway through the sound period, grew to a maximum and collapsed rapidly. The sonoluminescent flash occurred at the end of the collapse.

Günther et al. (1957, 1959) trained a photomultiplier tube on each of the pressure antinodes of a 30 kHz standing wave and found that the sonoluminescence flashes were periodic and occurred shortly before the end of the compression part of the cycle. The frequency of occurrence of these flashes matched the frequency of the sound source.

Negishi (1960, 1961) also showed correspondence of sonoluminescence with the sound pressure cycle.

Leighton et al. (1988) and Hatanaka et al. (1999) showed experimentally that sonoluminescence occurred at the pressure antinodes of a standing-wave sound field in water.

Work proceeded during the next thirty years in determining the dependence of sonoluminescence (SL) on hydrostatic pressure, sound pressure amplitude, frequency of the sound field, temperature of the liquid, nature of the solvent, and the role of the dissolved gas. The spectra of SL was also studied. All these dependencies are described by Young (1989, 1999), Finch (1963) and Walton and Reynolds (1984).

In 1962, Yosioka and Omura discovered single bubble sonoluminescence, and in 1970 Temple independently discovered it, but in both cases the work was not confirmed (see §2.3.1).

This carries the story of SL up to 1990 when Gaitan (1990, 1992) and his PhD supervisor Crum trapped a single bubble in a standing sound wave. This enabled a whole host of systematic experiments to be performed (Chapter 2). Sonoluminescence became a theoretician's delight to explain how the light occurred. There are over 15 different theories (Chapter 4).

Recent introductions to cavitations and sonoluminescence have been provided by Lepoint and Lepoint-Mullie (1998, 1999).

## 1.3 BUBBLE DYNAMICS

### *1.3.1 Introduction*

A bubble in a liquid is a free-boundary problem in which the mechanical and thermal behavior of the fluids – liquid and gas – is described by the usual conservation equations coupled by the suitable conditions at the gas–liquid interface (Prosperetti (1999)).

### *1.3.2 Bubble nuclei: Blake threshold*

Consider a small free spherical bubble in a liquid. Such a bubble will slowly float to the liquid surface. A bubble of radius 10 μm in water would rise at the rate of about 0.3 mm $s^{-1}$ (Epstein and Plesset (1950)). Bubbles will grow during the ascent as shown in Fig. 1.1, from Liger-Belair and Jeandet (2002). Also, as the distance between two successive bubbles increases and since bubbles are released from the nucleation site with clockwork regularity, it can be deduced that a bubble accelerates when rising through the liquid. An entertaining account of the rise of bubbles in a glass of beer is given in Shafer and Zare (1991).

Also, gas will diffuse out of the bubble into the liquid. Epstein and Plesset (1950) estimate that a 10 μm radius air bubble in air saturated water will take about 7 seconds to dissolve. A favored stabilization mechanism was put forward by Harvey et al. (1944) which supposes that a pocket of gas is trapped in a small-angled crevice or crack in the container surface or in an imperfectly wetted dirt particle. See Young (1999) pages 40–42.

There are four ways in which a bubble is prevented from being stable: it will

- Rise due to buoyancy
- Dissolve due to diffusion of gas out of the bubble
- Contract due to surface tension
- Grow due to gas pressure

For a note on the Blake threshold pressure, see Young (1999) page 11.

### *1.3.3 General equations of bubble dynamics*

Let us apply the laws of conservation of mass, momentum and energy to a spherical bubble in an infinite liquid. From Neppiras (1980) and Landau and Lifshitz (1987) we have

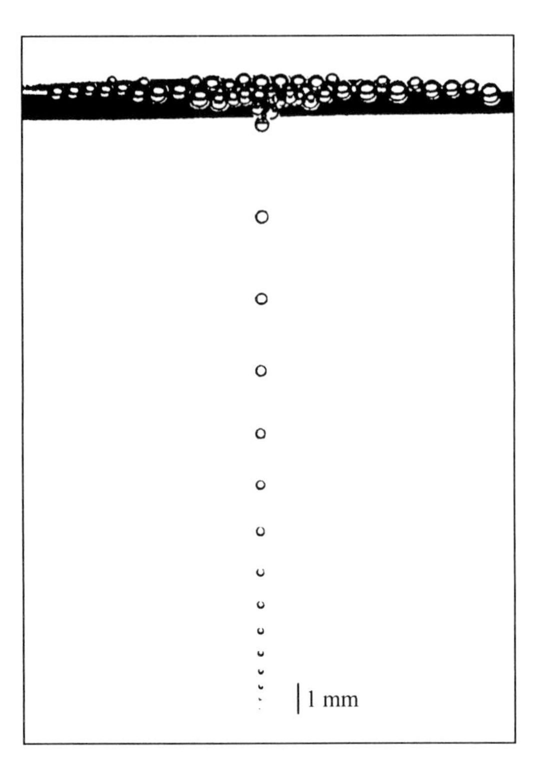


Figure 1.1 Typical photograph of a regular bubble train. The dark line running horizontally through the picture is due to the liquid meniscus between the free surface and the glass wall. (Liger-Belair and Jeandet (2002).)

$$\text{Continuity (liquid): } \frac{\partial \rho}{\partial t} + \frac{\partial}{\partial t}(\rho u) + \frac{2\rho u}{r} = 0 \tag{1.1}$$

$$\text{Momentum (liquid): } \frac{\partial u}{\partial t} + u\frac{\partial u}{\partial r} = -\frac{1}{\rho}\frac{\partial P}{\partial r} \tag{1.2}$$

$$\text{Energy (liquid): } \rho S_{\mathrm{v}}\left(\frac{\partial T}{\partial r} + u\frac{\partial T}{\partial r}\right) = -P\left(\frac{\partial u}{\partial r} + \frac{2u}{r}\right) + \frac{4\mu}{3}$$

$$\times\left(\frac{\partial u}{\partial r} - \frac{u}{r}\right)^2 + K\left(\frac{\partial^2 T}{\partial r^2} + \frac{2}{r}\frac{\partial T}{\partial r}\right) + \rho q \tag{1.3}$$

where $\rho$ = density of liquid
$u$ = particle velocity of liquid
$P$ = particle pressure of liquid
$T$ = temperature in liquid
$S_{\mathrm{v}}$ = specific heat capacity at constant volume of liquid
$q$ = heat flux (energy flow per unit mass per unit time)

$\mu$ = viscosity of liquid
$K$ = thermal conductivity of liquid.

For gas diffusion in the liquid we have

$$\frac{\partial C}{\partial t} + u\frac{\partial C}{\partial r} = D\left(\frac{\partial^2 C}{\partial r^2} + \frac{2}{r}\frac{\partial C}{\partial r}\right) \tag{1.4}$$

where $C$ is the concentration of dissolved gas in the liquid and $D$ is diffusivity of the gas. The boundary conditions at the bubble wall, where $r = R(r)$, are

$$D\frac{\partial C}{\partial r} = \frac{1}{4\pi R^2}\frac{\mathrm{d}}{\mathrm{d}t}\left(\frac{4}{3}\pi R^3 \rho_g\right) \tag{1.5}$$

$$P + \frac{2\sigma}{R} = P_T + \frac{4\mu}{3}\left(\frac{\partial u}{\partial r} - \frac{u}{r}\right) \tag{1.6}$$

where $\rho_g$ = density of gas in bubble
$\sigma$ = surface tension of liquid
$P_T$ = pressure of total gas content in bubble.

Neppiras (1980) gives the laws of mass and thermal diffusion in the liquid as

$$K\frac{\partial T}{\partial r} = \frac{1}{4\pi R^2}\frac{\mathrm{d}}{\mathrm{d}t}\left(\frac{4}{3}\pi R^3 \rho_T\right)\left[L + \frac{4\mu}{3\rho}\left(\frac{\partial u}{\partial r} - \frac{u}{r}\right)\right] + \rho_T\frac{RC_V}{3}\frac{\partial T}{\partial t} + P_T u \tag{1.7}$$

where $\rho_T$ is the density of the total gas content, and $C_v$ is the specific heat of gas at constant volume.

For the case of an acoustic driving pressure, at $r = \infty$,

$$P = P_0(t) = P_0 - P_A \sin \omega t \tag{1.8}$$

We need the equation of state for the liquid. The simplest one is

$$\rho = \text{a constant} \tag{1.9}$$

i.e. the liquid is incompressible.

We also need the equation of state for the gas. The perfect gas equation is used in the simplest case.

$$PV = R_g T \tag{1.10}$$

These 10 equations are the fundamental equations of bubble dynamics.

### 1.3.4 *Dynamical equation of a spherical collapsing bubble*

#### 1.3.4.1 Empty bubble

Suppose the empty bubble of radius $R$ in a liquid is expanding or contracting. $\dot{R}$ will be the radial velocity. Let $\dot{r}$ be the radial velocity in the liquid at any distance $r$.

Then

$$\dot{r} = \frac{R^2\dot{R}}{r^2}$$

The velocity potential is (Lamb (1879), Batchelor (1967))

$$\phi = -\int_r^\infty \dot{r}\,dr = -\frac{R^2\dot{R}}{r}$$

The equation of motion of the liquid is now given by Bernoulli's equation as

$$\frac{P - P_\infty}{\rho} = -\frac{\partial\phi}{\partial t} - \frac{1}{2}u^2 = \frac{2R\dot{R}^2 + R^2\ddot{R}}{r} - \frac{1}{2}\frac{R^4\dot{R}^2}{r^4}$$

where $P$ is the pressure in the liquid of density $\rho$ at distance $r$, and $P_\infty$ is the pressure in the liquid at infinity.

For the motion of the bubble wall, put $r = R$ and deduce

$$\boxed{\frac{P_\mathrm{L} - P_\infty}{\rho} = R\ddot{R} + \frac{3}{2}\dot{R}^2} \quad (1.11)$$

where $P_\mathrm{L} = P(t)$ = pressure in the liquid at the bubble wall.

#### 1.3.4.2 *Gas* bubble

The gas will act as a cushion absorbing the energy of the liquid collapsing inwards and will eventually completely stop the inward motion and then reverse it. Suppose the gas obeys the gas equation

$$P_\mathrm{g}\left(\frac{4\pi R^3}{3}\right) = R_\mathrm{g}T$$

If we assume adiabatic changes, $P_\mathrm{g}(4\pi R^3/3)^\gamma$ = a constant, where $\gamma$ is the ratio of the specific heats of the gas. Suppose the initial gas content (at $R = R_0$) gives a gas pressure in the liquid of $(P_0 + 2\sigma/R_0)$, where $P_0$ is the ambient pressure in the liquid and $\sigma$ is the surface tension. If the radius changes from $R_0$ to $R$ at constant temperature, the gas pressure becomes

$$P_i = \left(P_0 + \frac{2\sigma}{R_0}\right)\left(\frac{R_0}{R}\right)^{3\gamma} \tag{1.12}$$

Equation (1.11) then becomes (Neppiras and Noltingk (1951), Lamb (1932) and Poritsky (1952))

$$R\ddot{R} + \frac{3}{2}\dot{R}^2 = \frac{1}{\rho}\left[\left(P_0 + \frac{2\sigma}{R_0}\right)\left(\frac{R_0}{R}\right)^{3\gamma} - \frac{2\sigma}{R} - P_\infty\right] \tag{1.13}$$

This equation was first derived and explored in many ways by Noltingk and Neppiras in their famous pair of papers of 1950 and 1951 (Noltingk and Neppiras (1950) and Neppiras and Noltingk (1951)).

In 1952, Poritsky added a term to allow for the effect of viscosity of the liquid. If $\mu$ is the viscosity of the liquid we now have

$$R\ddot{R} + \frac{3}{2}\dot{R}^2 = \frac{1}{\rho}\left[\left(P_0 + \frac{2\sigma}{R_0}\right)\left(\frac{R_0}{R}\right)^{3\gamma} - \frac{2\sigma}{R} - \frac{4\mu\dot{R}}{R} + P_\infty\right] \tag{1.14}$$

Equations (1.11), (1.13) and (1.14) are often called the *Rayleigh–Plesset equation.*

### *1.3.5 Rayleigh analysis of a cavity*

#### 1.3.5.1 Rayleigh analysis of an empty *cavity*

Rayleigh (1917) derived an elegant solution for the bubble wall velocity and the time of complete collapse $\tau$.

If $R$ is the radius and $\dot{R}$ is the velocity of the bubble boundary at time $t$, and $\dot{r}$ is the simultaneous velocity at any distance $r$ (greater than $R$) from the center, then

$$\frac{\dot{r}}{\dot{R}} = \frac{R^2}{r^2}$$

and if $\rho$ is the density of the liquid, the whole K.E. of the liquid is

$$\frac{1}{2}\rho\int_R^\infty \dot{r}^2 4\pi r^2 \mathrm{d}r = 2\pi\rho\dot{R}^2R^3$$

If $P_0$ is the ambient liquid pressure and $R_0$ is the initial value of $R$, then the work done by the hydrostatic pressure is $\frac{4}{3}\pi P_0(R_0^3 - R^3)$, assuming isothermal compression. When we equate these expressions we get

$$2\pi\rho\dot{R}^2R^3 = \frac{4}{3}\pi P_0(R_0^3 - R^3)$$

$$\dot{R}^2 = \frac{2P_0}{3\rho}\left(\frac{R_0^3}{R^3} - 1\right) \tag{1.15}$$

expressing the velocity of the boundary in terms of the radius. Since

$$\dot{R} = \frac{\mathrm{d}R}{\mathrm{d}t}$$

then

$$\frac{\mathrm{d}R}{\mathrm{d}t} = \sqrt{\frac{2P_0}{3\rho}\left(\frac{R_0^3}{R^3} - 1\right)}$$

therefore,

$$\mathrm{d}t = \mathrm{d}R\sqrt{\frac{3\rho}{2P_0}\left(\frac{R^3}{R_0^3 - R^3}\right)}$$

and

$$t = \sqrt{\frac{3\rho}{2P_0}}\int_R^{R_0}\frac{R^{3/2}\mathrm{d}R}{R_0^3 - R^3} = R_0\sqrt{\frac{3\rho}{2P_0}}\int_\beta^1\frac{\beta^{3/2}\mathrm{d}\beta}{(1 - \beta^3)^{1/2}}$$

where $\beta = R/R_0$.

The time of collapse to a given fraction of the original radius is proportional to $R_0\rho^{1/2}P_0^{-1/2}$. The time $\tau$ of *complete* collapse is obtained by putting $\beta = 0$. Writing $\beta^3 = Z$, we have

$$\int_0^1\frac{\beta^{3/2}\mathrm{d}\beta}{(1 - \beta^3)^{1/2}} = \frac{1}{3}\int_0^1 Z^{-1/6}(1 - Z)^{-1/2}\mathrm{d}Z$$

which may be expressed by means of Γ functions. Thus,

$$\tau = R_0\sqrt{\frac{\rho}{6P_0}}\frac{\Gamma\left(\frac{5}{6}\right)\Gamma\left(\frac{1}{2}\right)}{\Gamma\left(\frac{4}{3}\right)}$$

therefore,

$$\tau = 0.91 R_0 \sqrt{\frac{\rho}{P_0}} \tag{1.16}$$

In the case of a bubble of radius 0.1 mm in water of density $10^3$ kg m$^{-3}$ and with $P_0 = 10^5$ N m$^{-2}$,

$$\tau \approx 10^{-5}\ \mathrm{s}$$

1.3.5.2 The adiabatic collapse of a *gas-filled* cavity

For an empty cavity, equation (1.15) shows that as $R$ decreases to 0, the bubble wall velocity $\dot{R}$ increases to infinity. To avoid this, Rayleigh filled the empty cavity with a gas that was compressed *isothermally*, though in most practical cases the gas is compressed adiabatically.

In the *isothermal* case the external work done on the system is equal to the sum of the work done in compressing the gas, and the kinetic energy of the liquid; that is

$$\frac{4\pi P_0}{3}(R_{\max}^3 - R^3) = 2\pi\rho\dot{R}^2 R^3 + 4\pi Q R_{\max}^3 \ln(R_{\max}/R) \tag{1.17}$$

where $Q$ is the initial pressure of the gas, $R_{\max}$ is the maximum radius of the bubble which is the initial radius for a *collapsing* bubble and $\dot{R} = 0$ when

$$P(1 - Z) + Q \log Z = 0$$

where $Z = \left(\frac{R_{\max}}{R}\right)^3$.

Rayleigh (1917) says: "Whatever be the (positive) value of $Q$, $\dot{R}$ comes again to zero before complete collapse. The boundary oscillates between two points, one of which is the initial".

Thus Rayleigh predicted oscillations in 1917.

Equation (1.17) gives, for the bubble wall velocity,

$$\dot{R}^2 = \frac{2P_0}{3\rho}\left[\left(\frac{R_{\max}}{R}\right)^3 - 1\right] - \frac{2Q}{\rho}\left(\frac{R_{\max}}{R}\right)\ln\left(\frac{R_{\max}}{R}\right)$$

In view of the very short collapse time, an adiabatic rather than an isothermal collapse is more realistic. The bubble wall motion is now given by equation (1.13), which, if we neglect surface tension and if we replace the liquid pressure at infinity $P$ by $P_{\mathrm{m}}$ (the liquid pressure at transient collapse), becomes

$$R\ddot{R} + \frac{3}{2}\dot{R}^2 = \frac{1}{\rho}\left[Q\left(\frac{R_{\max}}{R}\right)^{3\gamma} - P_{\mathrm{m}}\right] \tag{1.18a}$$

Under adiabatic compression we easily find the maximum pressure $P_{max}$ and maximum temperature $T_{max}$ reached by the gas:

$$P_{max} \approx Q\left[\frac{P_m(\gamma-1)}{Q}\right]^{\gamma/(\gamma-1)} = QZ^{\gamma}$$

$$T_{max} \approx T_0\frac{P_m(\gamma-1)}{Q} = T_0 Z^{\gamma-1} \tag{1.18b}$$

where $Q$ is the initial pressure, $T_0$ is the initial temperature of the gas in the bubble and $Z = \frac{V_{max}}{V}$.

### *1.3.6 Bubble dynamics accounts*

There are accounts of bubble dynamics by Young (1999), Hsieh (1965) with a very good introduction, Flynn (1964, 1975a,b), Lauterborn (1976), Prosperetti (1984a,b; 1999), Walton and Reynolds (1984), Gaitan et al. (1992), Leighton (1994), Wu and Roberts (1994), Brenner (1995), Cheeke (1997), Barber et al. (1997), Putterman and Weninger (2000) and Brenner et al. (2002).

### *1 3. 7 Bjerknes forces*

Let us first consider the acoustic radiation pressure. This is the pressure exerted on an object by a sound wave. An early work by King (1934) calculates the acoustic radiation pressure acting on small spheres. Although his theory agreed well with experiments for rigid spheres, there was some discrepancy for soft spheres. Yosioka and Kawasima (1955) showed that the incorporation of compressibility of the spheres solved the problem. A further paper by Yosioka et al. (1955) takes heat conduction into account and gives good agreement between theory and experiment for bubbles. Mitome (2001) summarizes Yosioka and Kawasima's 1955 paper as follows. The acoustic radiation force $\langle P\rangle$ acting on a sphere whose radius $a$ is much smaller than the acoustic wavelength is obtained by integrating the radiation pressure over the surface of the sphere and taking the time average as follows:

$$\langle P\rangle = \pi a^2 4ka\bar{E}\sin 2khF(\lambda, \sigma) \tag{1.19}$$

where $k$ is the wave number of the medium, $\bar{E}$ is the mean total energy density in the standing-wave field, $h$ is a distance of the center of sphere from the anti-nodal plane of sound pressure, and $\langle\ \rangle$ stands for time-averaging operation. $F$ is a function of density ratio $\lambda = \rho^*/\rho_0$ and sound-speed ratio $\sigma = c^*/c_0$ and is given by:

$$F(\lambda, \sigma) = \frac{\lambda + [2(\lambda-1)/3]}{1+2\lambda} - \frac{1}{3\lambda\sigma^2}$$

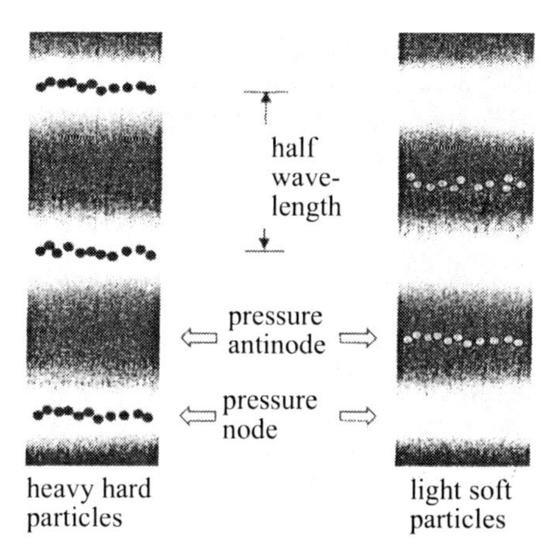


Figure 1.2 Acoustic radiation force acting on particles in a standing-wave field. (Mitome (2001).)

where $F$ is given by

$$F(\lambda, \sigma, k^*a) = \frac{\sigma(k^*a)[3\lambda - (k^*a)^2]}{\sigma^2(k^*a)^6 + [3\lambda - (k^*a)^2]^2}$$

where * means values pertinent to the sphere and 0 to the medium. The sign of $F$ determines the motion of spheres in the standing-wave field; if $F$ is positive, they move towards the pressure node, and if negative towards the antinode. Since heavy and hard particles have positive $F$, they move towards the pressure nodes. On the other hand, light and soft particles have negative $F$ and they move towards the pressure antinodes, as shown in Fig. 1.2. If a suspension is poured into the standing-wave field, small particles agglomerate every half wavelength and this phenomenon can be used for separation or filtration of particles. Since the radiation force depends on particle size and properties, classification of particles is possible based on this principle.

One may think, from this argument, that small bubbles are light and soft and move towards the pressure antinodes. However, equation (1.19) is applicable on the condition that $\lambda = 0$ (1) which does not hold for bubbles in water. Yosioka and Kawasima (1955) show that the following radiation force is applicable on the condition that $\lambda = 0$ $[(ka)^2]$:

$$\langle P \rangle = -\frac{4\pi}{k^2}\bar{E}\sin 2kh F(\lambda, \sigma, k^*a) \qquad (1.20)$$

where $F$ is given by

$$F(\lambda, \sigma, k^*a) = \frac{\sigma(k^*a)[3\lambda - (k^*a)^2]}{\sigma^2(k^*a)^6 + [3\lambda - (k^*a)^2]^2}$$

Since equation (1.20) has a negative sign, the radiation force pushes bubbles towards a pressure antinode if $F$ is positive and towards a pressure node if $F$ is negative. The sign

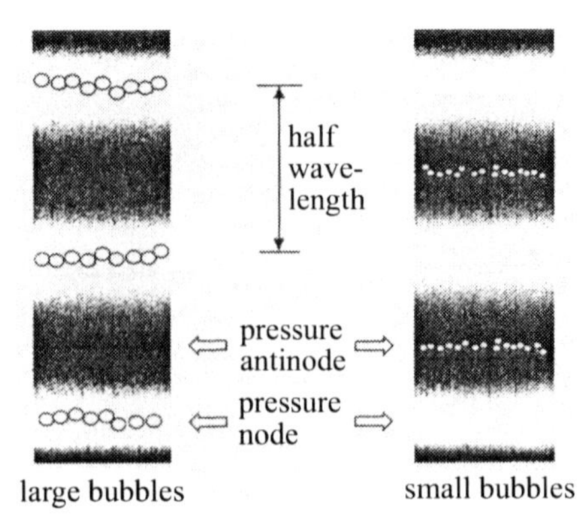


Figure 1.3 Acoustic radiation force acting on bubbles in a standing-wave field. (Mitome (2001).)

of $F$ changes at $k^*a/\sqrt{3\lambda} = 1$, which corresponds to resonant bubbles. In other words, bubbles smaller than the resonant bubble size move toward a pressure antinode and larger ones move towards a pressure node, as shown in Fig. 1.3 (Eller (1968), Henglein (1987)). The resonant frequency $f_r$ of a bubble of radius $a$ is

$$f_r = \frac{1}{2\pi a}\sqrt{\frac{3\gamma^* p_0}{\rho_0}} \qquad (1.21)$$

where $\gamma^*$ is the ratio of the specific heats of the gas in the bubble and $p_0$ is the pressure in the liquid (Young (1999)). For air bubbles in water, equation (1.21) reduces to

$$f_r a \approx 3 \ \ [\mathrm{m\ s^{-1}}]$$

where $f_r$ is in Hz and $a$ is in meters. The resonant radius of an air bubble is 300 μm at 10 kHz and 30 μm at 100 kHz.

To understand the bubble behavior, the *Bjerknes force* or radiation force on a bubble is helpful and will now be considered.

A body in an inhomogeneous pressure field experiences a force in the direction of lower pressure. In a gravity field, for instance, this leads to the buoyancy force. We will consider a bubble in a sound field. In this case, forces arise due to the sound pressure gradient and, moreover, due to the oscillations of the bubble volume, which also induce pressure gradients.

Bjerknes forces are separated into primary and secondary forces, depending on the origin of the pressure gradient as described by Crum (1975). *Primary Bjerknes forces* relate to the primary sound field originally causing the bubble to oscillate. *Secondary Bjerknes forces* are due to the sound oscillations emitted from other bubbles on the given bubble. The primary Bjerknes forces thus act on a bubble by the external sound field, while the secondary Bjerknes forces act between oscillating bubbles.

The sign and magnitude of the Bjerknes forces depend on whether the bubbles are oscillating harmonically or strongly nonlinearly. We will first consider the harmonic case.

Consider the force on a bubble in a stationary wave. In a field where a pressure gradient $\Delta P$ exists, a body of volume $V$ is acted on by a force $-V\nabla P$. Since $V$ and $P$ vary with time, it is necessary to consider a time-averaged force

$$F = -\langle V \nabla P \rangle \tag{1.22}$$

where the bracket $\langle \ \rangle$ indicates the average over a cycle. This force is called the primary *Bjerknes* force described in his book *Fields of Force* (1906). For a stationary wave the amplitude is due to an incident wave *and* a reflected wave. If the wave is reflected at a rigid boundary, the reflected wave amplitude is equal to the incident amplitude and the resultant amplitude is twice the incident amplitude i.e. $2P_A$.

If $2P_A << P_0$, the bubble radius $R(t)$ will oscillate linearly as

$$R(t) = R_0 - \xi \cos(\omega t + \alpha) \tag{1.23}$$

where the phase angle $\alpha$ is zero for bubbles smaller than resonance, and equals $\pi$ for bubbles larger than resonance (Leighton et al. 1990), and where $\xi$ is the amplitude of the radial oscillation given by

$$\xi = \xi_0 \sin kx \tag{1.24}$$

where $k$ is the circular wave number ($2\pi/\lambda_A$) of the sound field, and $x$ is a position in the sound field.

The negative sign is taken since a positive acoustic pressure causes a reduction in bubble volume when the two are in phase.

The total varying pressure in the liquid is

$$P = P_0 + 2P_A \cos(\omega t + \alpha)$$

We must now consider a spatial dimension (given by $x$) in the standing-wave field.

$$P = P_0 + 2P_A \sin kx \cos(\omega t + \alpha) \tag{1.25}$$

Substituting (1.24) in (1.23)

$$R_t = R_0 - \xi_0 \sin kx \cos(\omega t + \alpha) \tag{1.26}$$

Using (1.26), the bubble volume $V = \frac{4}{3}\pi R^3$ may now be written as

$$V = \frac{4}{3}\pi[R_0 - \xi_0 \sin kx \cos(\omega t + \alpha)]^3$$

$$V = \frac{4}{3}\pi[R_0^3 - 3R_0^2 \xi_0 \sin kx \cos(\omega t + \alpha)]$$

neglecting higher terms

$$V = \frac{4}{3}\pi R_0^3\left[1 - \frac{3\xi_0}{R_0}\sin kx\cos(\omega t + \alpha)\right]$$

$$V = V_0\left[1 - \frac{3\xi}{R_0}\sin kx\cos(\omega t + \alpha)\right] \tag{1.27}$$

where $V_0 = \frac{4}{3}\pi R_0^3$ is the equilibrium volume of the bubble.

Substituting Eqs. (1.25) and (1.27) into Eq. (1.22) and giving the steps leads to

$$F = -\langle V\nabla P\rangle$$

$$F = -V_0\left[1 - \frac{3\xi_0}{R_0}\sin kx\langle\cos(\omega t + \alpha)\rangle\right][2P_A k\cos kx\langle\cos\omega t\rangle]$$

$$F = -V_0\left[2P_A k\cos kx\langle\cos\omega t\rangle - 3\times 2\frac{\xi_0 P_A k}{R_0}\sin kx\cos kx\langle\cos(\omega t + \alpha)\cos\omega t\rangle\right]$$

The time averages are $\langle\cos\omega t\rangle = 0$ and $\langle\cos(\omega t + \alpha)\cos\omega t\rangle = \langle\cos^2\omega t\rangle\cos\alpha + \langle\sin\omega t\cos\omega t\rangle\sin\alpha$, the second term of which is zero.

Since $\sin kx\cos kx = \sin 2kx$ and $\langle\cos^2\omega t\rangle = \frac{1}{2}$, the *primary Bjerknes force* on the bubble is

$$\boxed{F = \frac{3\xi_0 P_A k V_0\sin 2kx}{2R_0}\cos\alpha} \tag{1.28}$$

For bubbles smaller than resonance, $\alpha = 0$ and

$$F = \frac{3\xi_0 P_A k V_0\sin 2kx}{2R_0} \tag{1.29}$$

For bubbles larger than resonance, $\alpha = \pi$ and

$$F = -\frac{3\xi_0 P_A k V_0\sin 2kx}{2R_0} \tag{1.30}$$

From Eqs. (1.25), (1.29) and (1.30), Leighton et al. (1990), and Leighton (1994) on pages 346–348, show that if the bubble is smaller than resonant size it will tend to move to the pressure antinode, and if the bubble is larger than resonant size it will tend to move to the pressure node.

Matula considers the Bjerknes (or acoustic radiation) force required to *levitate* a bubble in a standing wave field and this section is taken from Matula (1999). This force

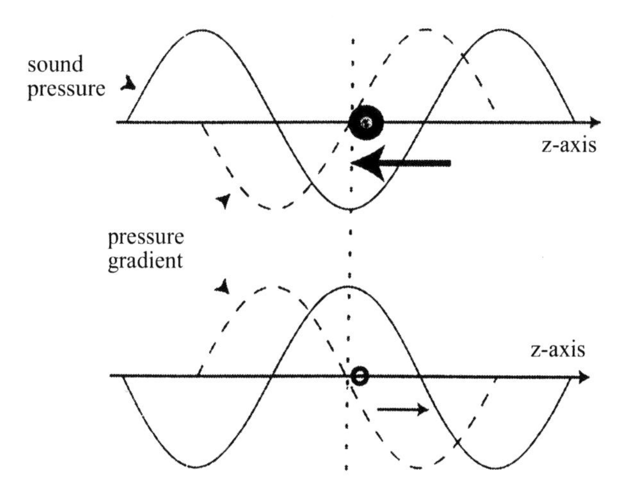


Figure 1.4 Levitation of a small bubble at a pressure antinode. When the bubble is large (during the tensile phase of the sound field) the force is directed towards the antinode. When the bubble is small (during the compressive phase of the sound field), the force is directed away from the antinode. (Matula et al. (1997).)

arises from the pressure difference (gradient) across the bubble (Eller (1968)). The force on the bubble is time dependent and varies according to the phase of the sinusoidal pressure field. Figure 1.4 shows this force for the case of *small* driving pressures (and for drive frequencies below the bubble's natural resonance frequency (Matula et al. (1997)). During the negative part of the sound field the bubble grows. There is a pressure force on the bubble due to the slight difference in pressure exerted on either side of the bubble's surface. This force directs the bubble towards the pressure antinode. During the compressive phase of the sound field the bubble is small and the force is directed away from the pressure antinode; however, since the corresponding volume is smaller, this force is smaller, and hence, over an acoustic cycle, the average Bjerknes (or average radiation) force directs the bubble towards the antinode. (This argument on the direction of the force applies only to a bubble that is driven below its natural frequency. For bubbles driven above their natural resonance frequency, a different phase response requires them to be forced away from the pressure antinode and towards a node.)

Though valid at lower drive-pressure amplitudes, this description of the Bjerknes force must be modified at higher driving pressures. Under moderately large pressure amplitudes, the bubble may continue to expand after the pressure turns positive, before finally undergoing a violent inertially dominated collapse. As the drive-pressure amplitude increases, the phase (time) of collapse continues to increase. Pedagogically, one can envision that as the drive pressure is increased further and further, the phase of collapse will continue to increase until the next acoustic cycle interferes with the bubble's motion. What actually occurs is more subtle. Consider that the average acoustic radiation force must balance the average buoyancy force in order to levitate a bubble. This relation can be expressed as

$$\frac{\rho g}{T}\int_0^T V(t)\,\mathrm{d}t = \frac{k_z P_\mathrm{a}}{T}\sin(k_z z)\int_0^T V(t)\sin(\omega t)\,\mathrm{d}t \tag{1.31}$$

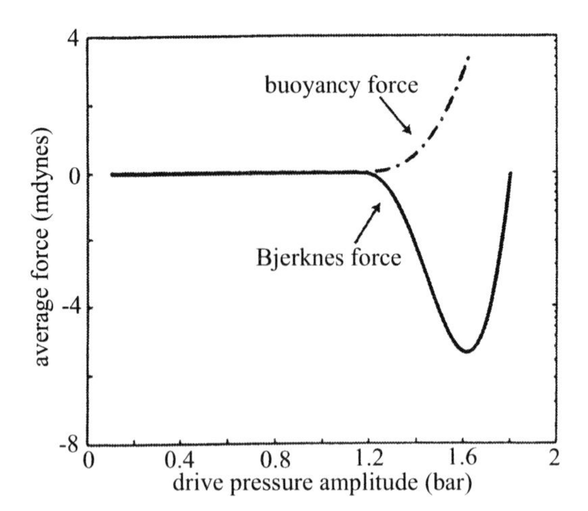


Figure 1.5 The Bjerknes and buoyancy forces on a small bubble (from equation (3.1)). These calculations assume that the size of the bubble does not change as the drive-pressure amplitude is increased. Although this assumption is not valid, there is only a slight dependence on bubble size over the region occupied by SBSL. (Matula (1999).)

where $\rho$ is the fluid density, $g$ is the acceleration due to gravity, $V(t) = \frac{4}{3}\pi R^3(t)$ is the bubble volume, $k_z = 2\pi/\lambda_z$ is the vertical wave number, $T$ is the acoustic period, $z$ is the average equilibrium position of the bubble above the pressure antinode (where $z = 0$), and $P_a$ is the applied drive-pressure amplitude. Note that the bubble is assumed to be spherical, and the Bjerknes force is parallel to the gravitational force. With no body forces in the horizontal plane, the symmetry of the sound field should preclude a horizontal component of the acoustic radiation force.

Figure 1.5 plots the average acoustic radiation or Bjerknes force (the right-hand side of Eq. (1.39)), as well as the average buoyancy force (the left-hand side of Eq. (1.31)) using experimental parameters (Matula et al. (1999)). Note that the Bjerknes force is negative for drive-pressure amplitudes below about 1.8 atm. Above this value, the average force actually pushes the bubble *away* from the pressure antinode, thus precluding bubble levitation at these high drive-pressure amplitudes.

From Eq. (1.31), a simple expression for the average equilibrium levitation position of the bubble can be obtained, provided one assumes the bubble is located near the pressure antinode. Then $\sin(k_z z) \approx k_z z$ and

$$z \approx \frac{\rho g}{k_z^2 P_a} \frac{\Lambda_1}{\Lambda_2}$$

where

$$\Lambda_1 = \int V(t)\mathrm{d}t \text{ and } \Lambda_2 = \int V(t)\sin(\omega t)\mathrm{d}t$$

Matula (2003) extends his (1997) paper by adding the stable parameter space $[P_a, R_o]$, the added mass force and the drag. The results are compared with a simple force balance that equates the Bjerknes force to the buoyancy force. Under normal sonoluminescence conditions, the comparison is quite favorable. A more complete accounting of the forces shows that a stabling levitated bubble undergoes periodic translational motion.

Lauterborn et al. (1999) discuss *secondary Bjerknes forces* referring to Oguz and Prosperetti (1990), Mettin et al. (1997) and Pelekasis and Tsamopoulos (1993). The force of an oscillating bubble "1" on a neighboring bubble "2" is given, to some approximation, by

$$\mathbf{F}_{B2} = -\frac{\rho}{4\pi}(\dot{V}_1\dot{V}_2)\frac{\mathbf{x}_2 - \mathbf{x}_1}{\|\mathbf{x}_2 - \mathbf{x}_1\|^3}$$

where $\mathbf{x}_1$ and $\mathbf{x}_2$ denote the locations of the interacting bubbles. For harmonic bubble oscillations, we obtain

$$\mathbf{F}_{B2} = -\frac{\rho\omega^2}{8\pi}(V_{1A}V_{2A})\cos(\varphi_1 - \varphi_2)\frac{\mathbf{x}_2 - \mathbf{x}_1}{\|\mathbf{x}_2 - \mathbf{x}_1\|^3}$$

where $V_{1A}$, $V_{2A}$ and $\varphi_1$, $\varphi_2$ are the amplitudes and the phases of the volume oscillations $V_i(t) = V_{i0} + V_{iA}\cos(\omega t + \varphi_i)\,(i = 1, 2)$, respectively. According to this result, a bubble smaller than the linear resonance radius and a bubble larger than the linear resonance radius repel each other, while pairs of smaller or larger bubbles are subject to an attracting secondary Bjerknes force.

Hatanaka et al. (1999) give a good account of the relationship between a standing wave field and a sonoluminescing field. Each field was measured experimentally. The sound field energy density was calculated and the sound pressure distribution was observed optically. It became clear that the sonoluminescence occurred at *pressure antinodes* of the standing wave field.

Hatanaka et al. (2001) points out that excessive growth of the bubbles under intense sound driving pressure leads to repulsion of the bubbles from the pressure antinodes and results in the sonoluminescence being quenched.

In the above discussion, it has been assumed that the bubbles oscillate sinusoidally. Actually, the bubbles oscillate with strong nonlinearity at large sound amplitudes (Tian et al. (1996)). Bubble motion can be described with the Rayleigh–Plesset equation as the zeroth-order approximation or the Keller equation taking into account the viscosity and compressibility of the liquid. These nonlinear oscillations of a bubble influence the Bjerknes forces and the force balance in the field as showed by Matula et al. (1997). Anyway, sonoluminescing bubbles move toward the pressure antinodes.

Other references on Bjerknes forces are Watanabe and Kukita (1993) on the transitional and radial motions of a bubble in an acoustic standing wave field, Akhatov et al. (1997) on the Bjerknes force threshold, Mettin et al. (1997) on the Bjerknes forces between small cavitation bubbles in a strong acoustic field, and very scholarly papers by Doinikov (2000, 2001, 2002a,b).

### *1.3.8 Sonoluminescence concentrates energy by 12 orders of magnitude*

This can be justified as follows.

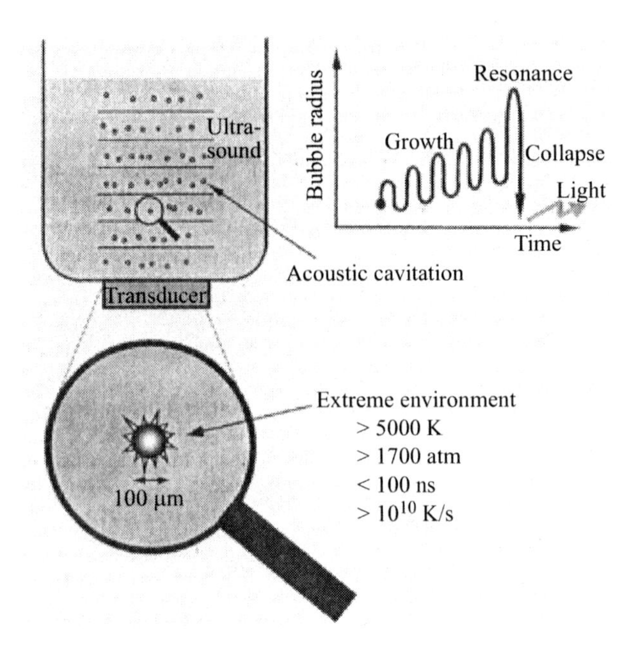


Figure 1.6 Extreme environment generated by collapse of micro bubbles. (Mitome (2001).)

The average acoustic energy given to an atom of the liquid by the sound field is

$$(\rho v^2) \times (\text{volume/atom})\ (\text{Leighton 1994 p. 18}) \approx 4 \times 10^{-12}\ eV/atom$$

where $\rho$ is the density of water and v is the velocity amplitude produced by the sound wave.

If the wavelength of the sonoluminescence is 200 nm,

Energy of a photon $= h\nu$

$$= 6.6 \times 10^{-34} \times \frac{3 \times 10^8}{2 \times 10^{-7}}$$

$$= 1.09 \times 10^{-18}\ \text{J}$$

As 1 eV $= 1.60 \times 10^{-19}$ J

*Energy of a photon = 6.8 eV*

∴ *Energy concentration ~ 12 orders.*

### *1.3.9 Rectified diffusion*

Lauterborn et al. (1999), from which this section is taken, point out that the amount of non-condensable gas inside a bubble is usually not constant in time as diffusion processes from and to the liquid lead to a change in the gas content of the bubble. The saturation concentration $c_0$ of a gas in a liquid is given by Henry's law, which relates the concentration linearly to the partial pressure $p_0$ of the gas above the solution, $p_0 = kc_0$, where $k$ is Henry's constant. The surface tension $\sigma$ of the liquid will always cause a pressure $2\sigma/R$ that adds to the partial gas pressure inside the bubble in pressure equilibrium. Consequently, spherical non-oscillating bubbles in a saturated liquid under pressure $p_0$ will always tend to *dissolve*.

For oscillating bubbles, however, the situation differs, as a phenomenon called *rectified diffusion* sets in. This was first recognised by Harvey et al. (1944). Contributions were then made by Blake (1949), Strasberg (1961), Hsieh and Plesset (1961), Plesset and Hsieh (1961) relating domestic difficulties, Eller and Flynn (1965), Safar (1968), Eller (1969), Gould (1974), Eller (1975), Neppiras (1980), Crum (1980, 1984), Apfel (1981a,b), Church (1988a,b), Atchley and Crum (1988), Brennen (1995), Roberts and Wu (1998), Young (1999), Brenner et al. (2002), Louisnard and Gomez (2003).

Then, the following factors contribute to a net *growth* of an oscillating bubble. During the oscillation, the bubble volume, bubble surface, and pressure inside the bubble change cyclicly and cause, via Henry's law, a variable gas concentration in the liquid layer at the bubble wall. Now the *surface area* during expansion is larger than the surface area during contraction. Hence the inward mass flow is greater than the outward mass flow. Also, the nonlinearity of the radial oscillation leads to a longer fraction of the cycle during which the bubble is large (and gas diffuses inward) and only short intervals during which the bubble is small (and gas is pushed out). Furthermore, the radial fluid motion near the bubble alternately steepens and shallows the concentration gradient in the liquid at the bubble wall. Since the mass transport through the bubble wall is also proportional to this concentration gradient, this *shell effect* also contributes to diffusion rates and may intensify the bubble's growth. When the bubble grows to a sufficiently large size, it becomes unstable and collapses adiabatically in a very short time, generating a bubble temperature exceeding 5,000 K and 1,700 atm within 10 ns as shown in Fig. 1.6 from Mitome (2001).

In a recent detailed analytical work, Fyrillas and Szeri (1994) treat the problem of diffusion for a spherically oscillating bubble in a liquid of arbitrary initial saturation; that is, the gas content in the liquid far from the bubble, $c_\infty$, may differ from the saturation concentration $c_0$ under normal pressure $p_0$. Fyrillas and Szeri (1994) take into account all the mechanisms already mentioned and consider the case where the (slow) time scale of diffusion can be separated from the (fast) time scale of the bubble oscillation (i.e. for a large Péclet number $P_e = R\omega/D >> 1$, where $D$ is the diffusivity). One important result is that the mass flow is governed by the quantity

$$\langle p[R(t)]\rangle_\tau = \frac{1}{\int_0^T R^4(t)\mathrm{d}t}\int_0^T R^4(t)p[R(t)]\mathrm{d}t$$

where $p[R(t)]$ denotes the pressure inside the bubble at time $t$, depending on the instantaneous bubble radius $R(t)$, $\langle ...\rangle_t$, abbreviates the indicated time averaging weighted by $R^4(t)$, and $T$ is the period of oscillation. In particular, a diffusional equilibrium of the oscillating bubble in the bubble occurs for

$$\frac{\langle p[R(t)]\rangle_\tau}{p_0} = \frac{c_\infty}{c_0}$$

Whenever $\langle p[R(t)]\rangle_t > p_0 c_\infty/c_0$, the bubble shrinks, and for $\langle p[R(t)]\rangle_t < p_0 c_\infty/c_0$ it grows. Thus, one recognizes again that a static bubble with $p[R(t)] = p[R_n] = (p_0 + 2\sigma/R_n)$ can be in equilibrium only with oversaturated liquid, $c_\infty > c_0$. With respect to the stability of a diffusional equilibrium, it has to be ensured that a growing bubble leads to a *larger* $\langle p[R(t)]\rangle_t$, and vice versa.

In Fig. 1.7 from Lauterborn et al. (1994), curves of equilibrium radii are shown for different ratios $c_\infty/c_0$ in the $R_n - \hat{p}_a$ plane ($p_0 = 100$ kPa). In this figure, a positive slope of a curve indicates stable equilibria, while a negative slope indicates unstable equilibria. Liquid saturated or oversaturated with gas ($c_\infty/c_0 \geq 1$) provides only unstable equilibrium points, since the curves are monotonically decreasing. However, reducing the gas content with respect to normal saturation can lead to arcs with positive slope, emerging near non-linear resonances in the undulating portions of the curves (Brenner et al. (1996)) and in a pronounced manner for very low content, high pressure amplitude, and small bubbles in the region of the large response (Akhatov et al. (1997)). Two examples are shown in Fig. 1.7 to show how to read the graph. For saturated liquid, $c_\infty/c_0 = 1$, and an excitation pressure amplitude of 60 kPa, we find one unstable equilibrium radius at $R_n \approx 8$ μm, denoted by a circle. The arrows indicate the direction of bubble growth or shrinkage. For reduced gas content, $c_\infty/c_0 = 10^{-4}$, and a pressure amplitude of 150 kPa, there are two equilibrium points, at $R_n \approx 1$ μm and $R_n \approx 5$ μm. Only the $R_n \approx 5$ μm point is diffusionally stable, denoted by a cross. *It is these conditions that allow for stable single bubble sonoluminescence in degassed water.*

Hao and Prosperetti (2002), in a scholarly paper, show that rectified heat transfer will cause a vapor bubble to grow even in a subcooled liquid. They study how translation, and the ensuing convective effects, influence this process.

### 1.3.10 Sound emission

Barber et al. (1997) show a detailed diagram of bubble collapse in Fig. 3.5 from Chapter 3. Trace (*b*) clearly shows a sizable sound emission spike at 26 μs. This sound emission is very important as it damps the bubble's motion. Equation (3.3) from Chapter 3 shows the sound emission term $\frac{R}{c}\frac{\mathrm{d}}{\mathrm{d}t}(p_g)$, where $c$ is the velocity of sound in the liquid and $p_g$ is the pressure of the gas in the bubble.

$$R\ddot{R} + \frac{3}{2}\dot{R}^2 = \frac{1}{\rho}\left[(P_g - P_0 - P(t)) - 4\eta\frac{\dot{R}}{R} - \frac{2\sigma}{R} + \frac{R}{c}\frac{\mathrm{d}}{\mathrm{d}x}(p_g)\right] \qquad \text{((3.3) from Chapter 3)}$$

Sound emission is discussed in detail in Chapter 3 in Secs. 3.5.5 and 3.5.6.

### 1.3.11 Bubble clustering

Hatanaka et al. (2002) studied the influence of the clustering of cavitation bubbles on multi-bubble sonoluminescence (MBSL) in standing wave fields with a photomultiplier tube, and observed the bubble behavior with a high speed video camera and an intensified charge-coupled device camera. If the acoustic power is steadily increased, the MBSL is suddenly quenched at a function generator output of 435 $\mathrm{mV_{p\text{-}p}}$. At this point, the bubble behavior clearly changes; the bubbles which form dendritic branches (branching figures like plants) of filaments change into clusters due to the secondary Bjerknes forces. A cluster is formed of several large bubbles surrounded by many tiny bubbles, in which bubbles repeatedly

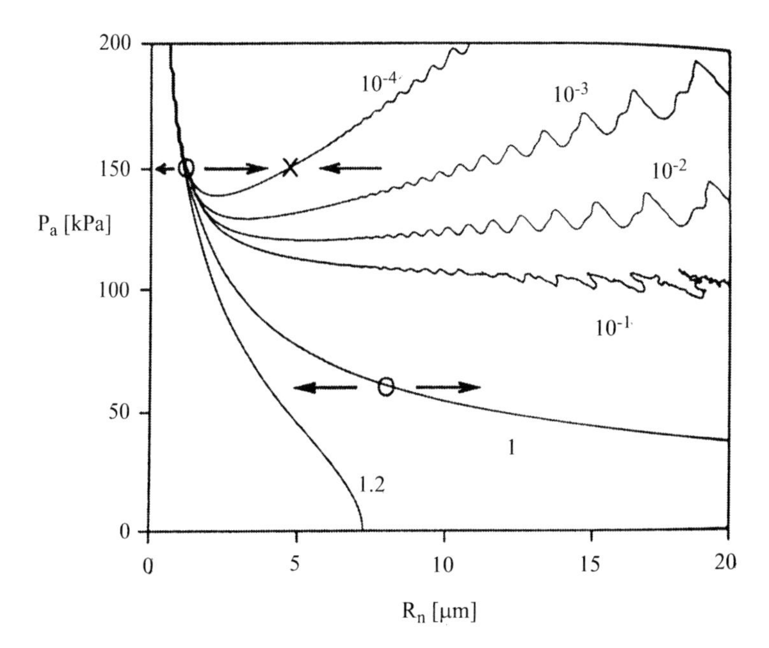


Figure 1.7 Curves of diffusional equilibrium in a plane of rest radius $R_n$ and driving pressure amplitude $P_a$ ($= \hat{p}_a$). The labels indicate $c_\infty/c_0$, the gas concentration in the liquid far from the bubble, normalized with respect to the saturation concentration under normal air pressure. Only equilibrium points on curve segments with positive slope are stable. As examples, the arrows give the direction of change of the rest radius for a constant pressure amplitude $P_a = 60$ kPa (one unstable equilibrium, denoted by an open circle) and $P_a = 150$ kPa (one unstable and one stable equilibrium, the latter denoted by a cross). The undulation of the curves in the upper right corner is caused by nonlinear resonances and chaotic oscillations of the bubble. (The undulations in the upper left corner are due to a plotting artefact of coarse graining.) (Lauterborn et al. (1999).)

coalesce and fragment, and run away from the pressure antinodes. When the clusters are broken up by forced fluid motion, the quenching of MBSL is suppressed.

Hatanaka et al. (2002) list some useful references on bubble clustering.

## 1.4 ACOUSTIC CAVITATION

For an account of acoustic cavitation see the author's sister volume, Young (1999).

## REFERENCES

Akhatov I, Mettin R, Ohl CD, Parlitz U and Lauterborn W 1997 Bjerknes force threshold for stable single bubble sonoluminescence, *Phys Rev E* **55**, 3747.

Apfel RE 1981a Acoustic cavitation prediction, *J Acoust Soc Am* **69**, 1624.

Apfel RE 1981b *Methods of Experimental Physics*, Vol 19, Ultrasonics, Academic Press p 355.

Atchley AA and Crum LA 1988 *Ultrasound, Its Physical, Chemical and Biological Effects*, Suslick KS (ed) VCH, Weinheim, page 1.
Barber BP, Hiller RA, Löfstedt R, Putterman SJ and Weninger KR 1997 Defining the unknowns of sonoluminescence, *Phys Rep* **281**, 65.
Bjerknes VFJ 1906 *Fields of Force*, Columbia University Press, New York.
Blake FG 1949 Technical Memo 12, Acoustic Research Laboratory Harvard University.
Brennen CE 1995 *Cavitation and Bubble Dynamics*, Oxford University Press.
Brenner MP, Hilgenfeldt S and Lohse D 2002 Single bubble sonoluminescence, *Rev Mod Phys* **74**, 425.
Brenner MP, Lohse D, Oxtoby D and Dupont TF 1996 Mechanisms for stable single bubble sonoluminescence, *Phys Rev Lett* **76**, 1158.
Chambers LA 1937 The emission of visible light from cavitated liquids, *J Chem Phys* **5**, 290.
Cheeke JDN 1997 Single-bubble sonoluminescence: "bubble, bubble, toil and trouble", *Can J Phys* **75**, 77.
Church CC 1988a Prediction of rectified diffusion during nonlinear bubble pulsations at biomedical frequencies, *J Acoust Soc Am* **83**, 2210.
Church CC 1988b A method to account for acoustic microstreaming when predicting bubble growth rates produced by rectified diffusion, *J Acoust Soc Am* **84**, 1758.
Crum LA 1975 Bjerknes forces on bubbles in a stationary sound field, *J Acoust Soc Am* **57**, 1363.
Crum LA 1980 Measurements of the growth of air bubbles by rectified diffusion, *J Acoust Soc Am* **68**, 203.
Crum LA 1984 Rectified diffusion, *Ultrasonics* **22**, 215.
Doinikov AA 2000 Influence of neighbouring bubbles on the primary Bjerknes force acting on a small cavitation bubble in a strong acoustic field, *Phys Rev E* **62**, 7516.
Doinikov AA 2001 Acoustic radiation, interparticle forces in a compressible fluid, *J Fluid Mech* **444**, 1.
Doinikov AA 2002a Viscous effects on the interaction force between two small gas bubbles in a weak acoustic field, *J Acoust Soc Am* **111**, 1602.
Doinikov AA 2002b Translational motion of a spherical bubble in an acoustic standing wave of high intensity, *Phys Fluids* **14**, 1420.
Eller A 1968 Force on a bubble in a standing acoustic wave, *J Acoust Soc Am* **43**, 170.
Eller A 1969 Growth of bubbles by rectified diffusion, *J Acoust Soc Am* **46**, 1246.
Eller A 1975 Effects of diffusion on gaseous cavitation bubbles, *J Acoust Soc Am*, **57**, 1374.
Eller A and Plynn HG 1965 Rectified diffusion during nonlinear pulsations of cavitation bubbles, *J Acoust Soc Am* **37**, 493.
Epstein D and Plesset MS 1950 On the stability of gas bubbles in liquid–gas solutions, *J Chem Phys* **18**, 1505.
Finch RD 1963 Sonoluminescence, *Ultrasonics* **1**, 87.
Flynn HG 1964 *Physics of Acoustic Cavitation in Liquids*, *Physical Acoustics* Vol 1B Mason WP (ed) Academic Press Chap 9, p 57.
Flynn HG 1975a Cavitation dynamics I A mathematical formulation, *J Acoust Soc Am* **57**, 1379.
Flynn HG 1975b Cavitation dynamics II Free pulsations and models for cavitation bubbles, *J Acoust Soc Am* **58**, 1160.
Frenzel J and Schultes H 1934 Luminescenz im ultraschallbeschickten Wasser. Kurze Mitteilung, *Zeit für Phys Chem* **B27**, 421.
Fyrillas MM and Szeri AJ 1994 Dissolution or growth of soluble spherical oscillating bubbles, *J Fluid Mech* **277**, 381.
Gaitan DF and Crum LA 1990 Observation of Sonoluminescence from a Single Stable

Cavitation Bubble in a Water/Glycerine Mixture Frontiers of Nonlinear Acoustics: Proceedings of 12th ISNA (eds Hamilton MF and Blackstock DT), Elsevier, pp 459–463.
Gaitan DF, Crum LA, Church CC and Roy RA 1992 Sonoluminescence and bubble dynamics for a single, stable, cavitation bubble, *J Acoust Soc Am* **91**, 3166.
Günther P, Heim E and Eichkorn G 1959b Phasenkorrelation von Schallwechseldruck und Sonolumineszenz, *Zeit angew Phys* **11**, 274.
Günther P, Zeil W, Grisar U and Heim 1957 Versuche über die Sonolumineszenz wäßriger Lösungen, *Zeit für Elektrochem* **61**, 188.
Hao Y and Prosperetti A 2002 Rectified heat transfer into translating and pulsating vapor bubbles, *J Acoust Soc Am* **112**, 1787.
Harvey EN, Barnes DK, McElroy WD, Whiteley AH, Pease DC and Cooper KW 1944 Bubble formation in animals, *J Cell Comp Physiol* **24**, 1.
Hatanaka S, Mitome H, Tuziuti T, Kozuka T, Kuwabara M and Asai S 1999 Relationship between a standing-wave field and a sonoluminescing field, *Jpn J Appl Phys* **38**, 3053.
Hatanaka S, Yasui K, Kozuka T, Tuziuti T and Mitome H 2002 Influence of bubble clustering on multibubble sonoluminescence, *Ultrasonics* **40**, 655.
Hatanaka S, Yasui K, Tuziuti T, Kozuka T and Mitome H 2001 Quenching mechanism of multibubble sonoluminescence at excessive sound pressure, *Jpn J Appl Phys* **40**, 3856.
Henglein A 1987 Sonochemistry: historical developments and modern aspects, *Ultrasonics* **25**, 6.
Hsieh D-Y 1965 Some analytical aspects of bubble dynamics, *J Basic Eng ASME* **87D**, 991.
Hsieh D-Y and Plesset MS 1961 Theory of rectified diffusion of mass into gas bubbles, *J Acoust Soc Am* **33**, 206.
King LV 1934 On the acoustic radiation pressure on spheres, *Proc Roy Soc A* **147**, 212.
Lamb Horace 1879 1st edition, 1932 6th edition Cambridge University Press, Art 91a.
Lauterborn W 1976 Numerical investigation of nonlinear oscillations of gas bubbles in liquids, *J Acoust Soc Am* **59**, 283.
Lauterborn W, Kurz T, Mettin R and Ohl CD 1999 Experimental and theoretical bubble dynamics, *Adv Chem Phys* **110**, 295.
Leighton TG 1994 *The Acoustic Bubble*, Academic Press.
Leighton TG, Pickworth MJW, Walton AJ and Dendy PP 1988 Studies of the cavitational effects of clinical ultrasound by sonoluminescence: I Correlation of sonoluminescence with the standing wave pattern in an acoustic field produced by a therapeutic unit, *Phys Med Biol* **33**, 1239.
Leighton TG, Walton AJ arid Pickworth MJW 1990 Primary Bjerknes forces, *Eur J Phys* **11**, 47.
Lepoint T and Lepoint-Mullie F 1998 *Synthetic Organic Sonochemistry*, Luche JL (ed) Plenum, New York, page 1.
Lepoint T and Lepoint-Mullie F 1999 *An Introduction to Sonoluminescence*, Advances in Sonochemistry Vol 5 page 1, JAI Press.
Liger-Belair G and Jeandet P 2002 Effervescence in a glass of champagne: A bubble story, Europhysics News January/February 2002, 10.
Löfstedt R, Weninger K, Putterman S and Barber BP 1995 Sonoluminescing bubbles and mass diffusion, *Phys Rev E* **51**, 4400.
Louisnard O and Gomez 2003 Growth by rectified diffusion of strongly acoustically forced bubbles in nearly saturated liquids, *Phys Rev E* **67**, 036610.
Marinesco N and Trillat JJ 1933 Chimie Physique – Action des ultrasons sur les plaques photographiques, *Comptes Rendus Acad Sci Paris* **196**, 858.
Matula TJ 1999 Inertial cavitation and single-bubble sonoluminescence, *Phil Trans Roy Soc London A* **357**, 225.

Matula TJ 2003 Bubble levitation and translation under single-bubble sonoluminescence conditions, *J Acoust Soc Am* **114**, 775.
Matula TJ, Cordry SM, Roy RA and Crum LA 1997 Bjerknes force and bubble levitation under single-bubble sonoluminescence conditions, *J Acoust Soc Am* **102**, 1522.
Mettin R, Akhatov I, Parlitz U, Ohl CD and Lauterborn W 1997 Bjerknes forces between small cavitation bubbles in a strong acoustic field, *Phys Rev E* **56**, 2924.
Meyer E and Kuttruff H 1959, Zur Phasenbezielung zwischen Sonolumineszenz und Kavitationsvorang bei periodischer Anregung, *Zeit angew Phys* **9**, 325.
Mitome H 2001 Micro bubble and sonoluminescence, *Jpn J Appl Phys* **40**, 3484.
Negishi K 1960 Phase relation between sonoluminescence and cavitation bubbles, *Acustica* **10**, 124.
Negishi K 1961 Experimental studies on sonoluminescence and ultrasonic cavitation, *J Phys Soc Jpn* **16**, 1450.
Neppiras EA 1980 Acoustic cavitation, *Phys Rep* **61**, 160.
Neppiras EA and Noltingk BE 1951 Cavitation produced by ultrasonics: theoretical conditions for the onset of cavitation, *Proc Phys Soc B (London)* **64B**, 1032.
Noltingk BE and Neppiras EA 1950 Cavitation produced by ultrasonics, *Proc Phys Soc B (London)* **63B**, 674.
Oguz HN and Prosperetti A 1990 A generalization of the impulse and virial theorems with an application to bubble oscillations, *J Fluid Mech* **218**, 143.
Pelekasis NA and Tsampoulos JA 1993 Bjerknes forces between two bubbles: Part 2, Response to an oscillatory pressure field, *J Fluid Mech* **254**, 501.
Poritsky H 1952 The Collapse or Growth of a Spherical Bubble or Cavity in a Viscous Fluid, Proc 1st National Congress in Applied Mathematics (ASME), p 813.
Prosperetti A 1984a Bubble phenomena in sound fields: part one, *Ultrasonics* **22**, 69.
Prosperetti A 1984b Bubble phenomena in sound fields: part two, *Ultrasonics* **22**, 115.
Prosperetti A 1999 *Old-Fashioned Bubble Dynamics, Sonochemistry and Sonoluminescence*, LA Crum et al. (eds) Kluwer, p 39.
Putterman SJ and Weninger KR 2000 Sonoluminescence: how bubbles turn sound into light, *Annu Rev Fluid Mech* **32**, 445.
Rayleigh Lord 1917 On the pressure developed in a liquid during the collapse of a spherical cavity, *Phil Mag* **34**, 94.
Roberts PH and Wu CC 1998 On rectified diffusion and sonoluminescence, *Theoret Comput Fluid Dynamics* **10**, 357.
Safar MH 1968 Comment on papers concerning rectified diffusion of cavitation bubbles, *J Acoust Soc Am* **43**, 1188.
Shafer NE and Zare RN 1991 Through a Beer Glass Darkly, *Physics Today* **44**, No. 10, 48.
Skinner LA 1970 Acoustically induced gas bubble growth, *J Acoust Soc Am* **51**, 378.
Temple PR 1970 Sonoluminescence from the gas in a single bubble, M.S. thesis, University of Vermont, USA.
Tian Y, Ketterling JA and Apfel RE 1996 Direct observation of microbubble oscillations, *J Acoust Soc Am* **100**, 3976.
Watanabe T and Kukita Y 1993 Translational and radial motions of a bubble in an acoustic standing wave field, *Phys Fluids A* **5**, 2682.
Wu CC and Roberts PH 1994 A model of sonoluminescence, *Proc Roy Soc London A* **445**, 323.
Yosioka K and Kawasima Y 1955 Acoustic radiation pressure on a compressible sphere, *Acustica* **2**, 167.
Yosioka K, Kawasima Y and Hirano H 1955 Acoustic radiation pressure on bubbles and their logarithmic decrement, *Acustica* **5**, 173.

Yosioka K and Omura A 1962 The light emission from a single bubble driven by ultrasound and the spectra of acoustic oscillations, *Proc Annu Meet Acoust Soc Jpn* (1962) pages 125–126 (in Japanese).
Young FR 1999 *Cavitation*, Imperial College Press.

CHAPTER TWO

# Multibubble Sonoluminescence

And all the people saw the sounds
*Original version of Exodus 20:18*
*Putterman S Physics World 11 No 5, 38 (1998)*

Tiny bubbles, in the wine
Tiny bubbles make you feel fine ...
*"Tiny bubbles", an American lyric*

## 2.1 INTRODUCTION

First, will be discussed in Sec. 2. 2, the various factors upon which sonoluminescence SL depends, as known up to 1989. This SL is often called *multi-bubble sonoluminescence* MBSL.1990 is a very important year in SL, it being the year in which Gaitan (1990), Gaitan and Crum (1990a, 1990b) and Gaitan et al. (1992) discovered *single bubble sonoluminescence*, SBSL, now usually just called SL, as a reproducible phenomenon. SBSL will be discussed in Chapter 3. Lepoint and Lepoint-Mullie (1992), in Fig. 2.1a, give a useful diagram showing the two modes of sonoluminescence.

A typical set-up for studying multibubble sonoluminescence is shown in Fig. 2.1b (Hatanaka et al. (1999)).

## 2.2 DEPENDENCE OF MULTIBUBBLE SONOLUMINESCENCE ON VARIOUS FACTORS

### *2.2.1 Hydrostatic pressure and the sound pressure amplitude*

The dependence on the hydrostatic pressure arises from the dependence on the maximum temperature $T_{max}$ attained by the gas in the bubble during the final phases of transient collapse on $P_A$ and $P_0$. We will first derive an expression for the maximum temperature in a transient bubble according to Walton and Reynolds (1984). Equation (1.18b) was derived

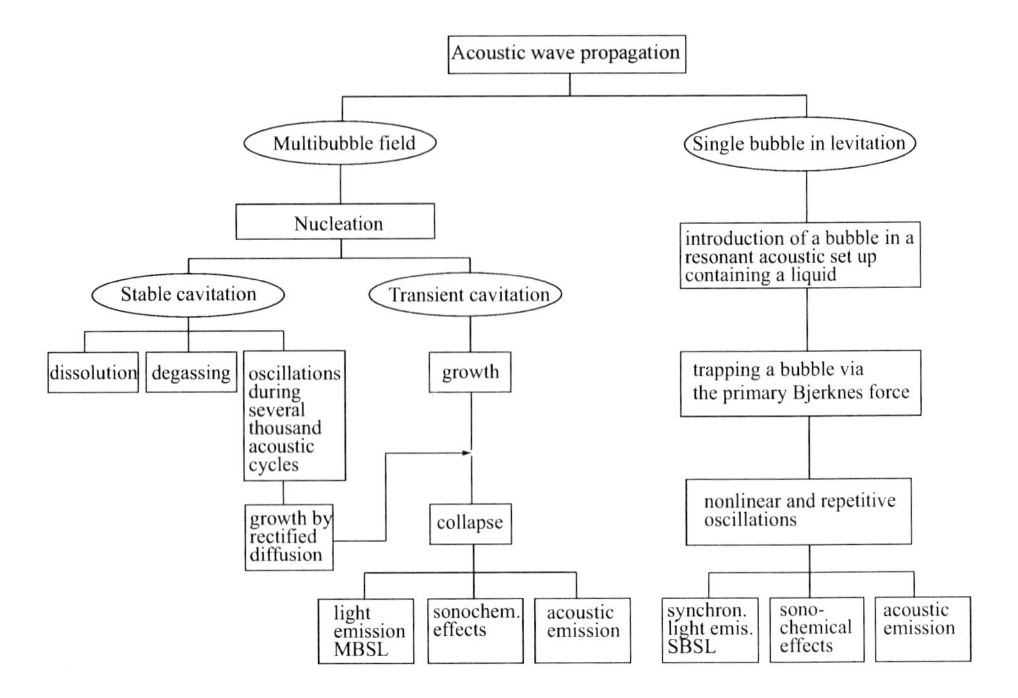


Figure 2.1a Schematic representation of the two modes of sonoluminescence and their relations with bubble dynamics. (Lepoint and Lepoint-Mullie (1999).)

for a gas-filled cavity adiabatically compressed from an initial volume $V_m$ to find the volume $V_{min}$ by a constant hydrostatic pressure $P_m$. Since transient collapse is a rapid process which occurs while $P$ is close to its maximum (see Fig. 3.23 in Young (1999)), we may assume $P_m \simeq P_0 + P_A$. Furthermore, the initial growth from a bubble radius $R_0$ to a maximum bubble radius $R_{max}$ takes place isothermally; the maximum speed of a cavity wall, approximately $(2P_A/3\rho)^{1/2}$, will not exceed Mach 0.01. Applying Boyle's law gives

$$R_0^3\left(P_0 + \frac{2\sigma}{R_0}\right) = R_{max}^3 Q$$

where $Q$ is the gas pressure at the maximum bubble radius $R_{max}$.

Hence Eq. (1.18b) becomes

$$T_{max} = T_0\left(\frac{R_{max}}{R_0}\right)^3 \frac{(P_0 + P_A)(\gamma - 1)}{\left(P_0 + \frac{2\sigma}{R_0}\right)} \tag{2.1}$$

Substituting $R_{max}$ from Eq. (3.50) in Young (1999)) into Eq. (2.1) with the term $2\sigma/R_0$ ignored [it was effectively ignored in deducing Eq. (3.50) in Young (1999)] gives

$$T_{max} = \frac{T_0}{R_0^3}\left(\frac{4}{3\omega}\right)^3\left(\frac{2}{\rho P_A}\right)^{3/2}\frac{(P_A - P_0)^3}{P_0}\left[1 + \frac{2}{3P_0}(P_A - P_0)\right](P_0 + P_A)(\gamma - 1) \tag{2.2}$$

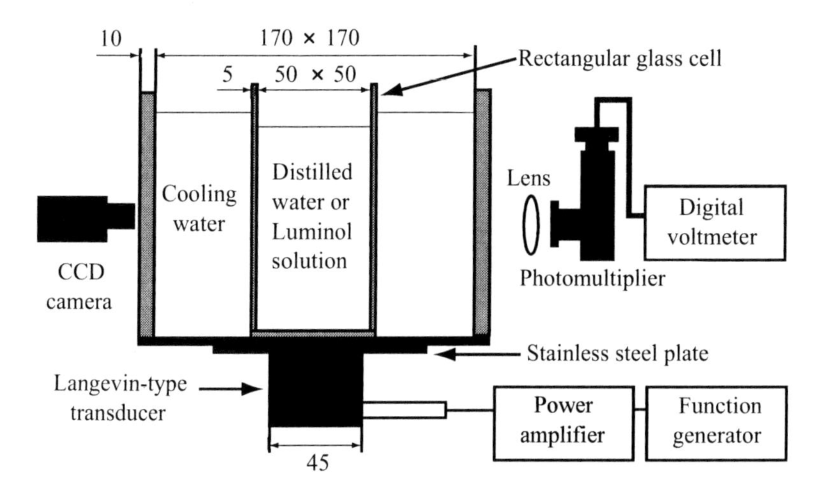


Figure 2.1b Experimental set-up. (Hatanaka et al. (1999).)

It is worth recalling that the $(P_A - P_0)$ terms come from the dependence of the maximum radius $R_{max}$ attained by a transient bubble on $(P_A - P_0)$ (transient cavitation demands $P_A > P_0$) and the $(P_0 + P_A)$ term from the fact that transient collapse occurs while the hydrostatic pressure is close to this value (Fig. 3.23 in Young (1999)).

The dependence of sonoluminescence on the sound pressure amplitude arises because the range of initial bubble radii, which can give rise to transient cavitation, is a function of $P_A$ and $P_0$ (Fig. 3.38 in Young (1999)). The lower limit $R_l$ is determined by the Blake criterion; writing $R_l = R_0$ and $P_A = P_B$ in Eq. (2.2) in Young (1999) gives

$$R_l^3 + \frac{2\sigma R_l^3}{P_0} - \frac{32\sigma^3}{27P_0(P_0 - P_A)^2} = 0 \tag{2.3}$$

Approximate values of $R_l$ valid when $2\sigma/R_0 \gg P_0$ and when $2\sigma/R_0 \ll P_0$ can be obtained from Eq. (2.2) in Young (1999). The upper limit $R_u$ is given by Eq. (3.51) in Young (1999):

$$R_u = \frac{2}{3\omega}(P_A - P_0)\left(\frac{2}{\rho P_A}\right)^{1/2}\left[1 + \frac{2}{3P_0}(P_A - P_0)\right]^{1/3} \tag{2.4}$$

To calculate the total number of bubbles capable of transient growth and collapse one would need to know the number of bubbles per unit range of radii, $n(R)$, at radius $R$. Adopting the semiempirical relation of Gavrilov (1973) (various relations have been proposed),

$$n(R) = \frac{A}{R^3}$$

where A is an empirical constant ($10^{-15}$ m$^3$), the total number of bubbles, $N$, capable of undergoing transient cavitation and is given by

$$N = \int_{R_l}^{R_u} \frac{Q}{R^3} \mathrm{d}R, \quad N = \frac{1}{2} A \left( \frac{1}{R_l^2} - \frac{1}{R_u^2} \right) \tag{2.5}$$

where $R_l$ and $R_u$ are determined by Eqs. (2.3) and (2.4) respectively. The peak intensity of the sonoluminescence produced by each of these $N$ bubbles will, of course, depend on the assumed mechanism for the sonoluminescence emission; if it were to be a black-body emission one would take the peak intensity as proportional to $R_{min}^2 T_{max}^4$, where $R_{min}$ is the minimum radius reached by the bubble during transient collapse.

Walton and Reynolds (1984) give a comprehensive account of the experimental studies of the dependence of sonoluminescence on the hydrostatic pressure and the sound pressure amplitude. Suffice here to describe the work of Finch (1963, 1965) who was the first worker to make a quantitative study of the effect of changing $P_0$ by both hydraulic and gas-applied pressure. In both cases the sonoluminescence is found to pass through a maximum at $P_0 =$ 1.5–1.7 bar (its value depends on the acoustic power, i.e. on $P_A$) and to decrease to near-zero values at still higher values of $P_0$. Finch found that absolute intensities were significantly greater when pressurization was achieved by raising the gas pressure above the liquid. In explaining the effects of the hydraulically applied pressure he appeals to the arguments of Noltingk and Neppiras (1950), noting that on the one hand the velocity of collapse of a cavity wall increases with $P_0$, but on the other hand $R_{max}$ decreases with increasing $P_0$. The maximum, he suggests, occurs when "these two opposite tendencies counterbalance". In essence, Finch is only appealing to Eq. (2.2), although his arguments relate to pressures rather than to bubble temperatures. He accounts for the enhanced sonoluminescence observed via gas pressurization by noting that, as the solubility of a gas in a liquid is proportional to the applied pressure (Henry's law), so the number of cavitation nuclei is likely to be greater with gas than with hydraulic pressurization. Conversely, one might add that during hydraulic pressurization an initially saturated liquid will become non-saturated.

We may add that Chendke and Fogler (1983) and Dezhkunov et al. (1997) measured the sonoluminescence of nitrogen-saturated water at static pressures ranging from 1 to 14.6 atm and found that the SL increased up to a maximum at 6 atm. Further increases in static pressure resulted in a decrease in SL with eventual extinction occurring at 14.6 atm, Fig. 2.1c.

Tuziuti et al. (2002a) found experimentally that multibubble sonoluminescence was increased with external pressures *below* atmospheric pressure due to the effect of the number of bubbles in the supersaturated solution. Bubble dynamics alone has little effect on MBSL intensity. Tuziuti et al. (2002b) found that the sonoluminescence could also be increased by reducing the oxygen content of the solution and that there was an optimum oxygen content for maximum sonoluminescence.

Ciuti et al. (2003) cavitated water with two sound fields simultaneously: a low frequency field of 27.2 kHz and a high frequency field of 880 kHz. The sonoluminescence is enhanced by the generation of new cavitation nuclei upon the collapse of the bubbles driven by the low frequency sound field. Since these fragments are substantially smaller than the sizes of the initial bubbles, these nuclei are suitable for cavitation in the high frequency sound field.

### 2.2.2 *Driving frequency of the sound field*

Equation (2.2) includes $\omega$, the circular frequency of the sound field, so we can conclude that sonoluminescence depends on $\omega$. The sonoluminescence should decrease as the

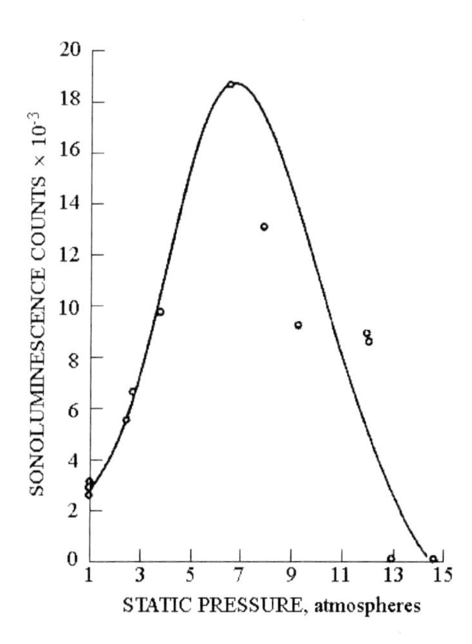


Figure 2.1c Effect of static pressure on the sonoluminescence of water. (Chendke and Fogler (1983).)

frequency increases for a given $P_A$ and $P_0$. The first study of this effect was by Griffing and Sette (1955). They found that the sonoluminescence decreased from $2 \times 10^{-8}$ phot at 660 kHz to $10^{-8}$ phot at 1 MHz and to $3 \times 10^{-9}$ phot at 2 MHz in tap water at a sound intensity of $2 \times 10^4$ W m$^{-2}$. Gabrielli et al. (1967) found that the sonoluminescence decreased from $2 \times 10^{-8}$ phot at 0.7 MHz to $10^{-8}$ phot at 2 MHz in distilled water at an intensity of $2.5 \times 10^4$ W m$^{-2}$.

More recently, Yasui (2002b) has performed computer simulations of bubble oscillations undergoing multibubble sonoluminescence. These reveal that the range of the ambient bubble radius for sonoluminescing bubbles narrows as the ultrasonic frequency increases because the bubble collapse becomes milder due to the shorter period of the bubble expansion. Any sonoluminescing bubble disintegrates into a mass of smaller bubbles in a few or a few tens of acoustic cycles at 1 MHz while at 20 kHz and 140 kHz some sonoluminescing bubbles are shape stable and are widely known as stable SBSL bubbles. Yasui (2002b) gives a useful diagram, Fig. 2.1d, summarizing the life history of bubbles. Also, as the ultrasonic frequency increases, the amount of water vapor trapped inside the bubbles at the collapse decreases.

Yasui (2002b) states that the mechanism of the light emission depends on the ultrasonic frequency; at 1 MHz the light mainly originates in plasma emissions while at 20 kHz it originates both in chemiluminescence of $^{\bullet}$OH radicals and plasma emissions.

A comprehensive study of the influence of ultrasonic frequency on aqueous sonoluminescence and sonochemistry has been undertaken by Beckett and Hua (2001). They measured the multibubble sonoluminescence spectra at 4 discrete driving frequencies in argon saturated water, Fig. 2.1e. With the exception of 1071 kHz, the spectrum in each frequency shows a broad continuum beginning at 200 nm and continuing through 500 nm while peaking at about 300 nm. An extensive discussion of the role of hydroxyl radicals $^{\bullet}$OH in sonobubble chemistry yields the surprising Fig. 2.1f!

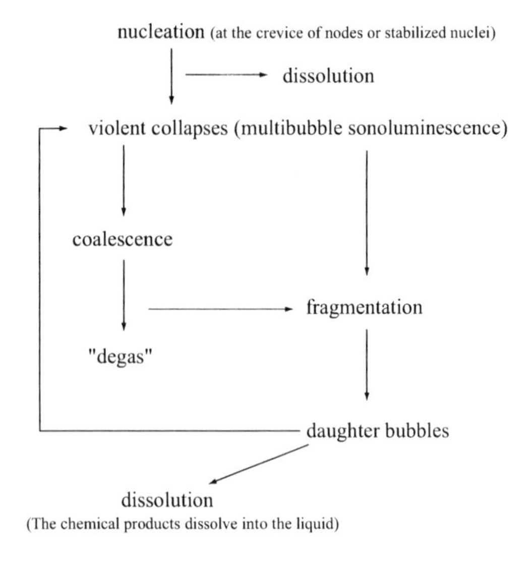


Figure 2.1d Life of bubbles. (Yasui (2002b).)

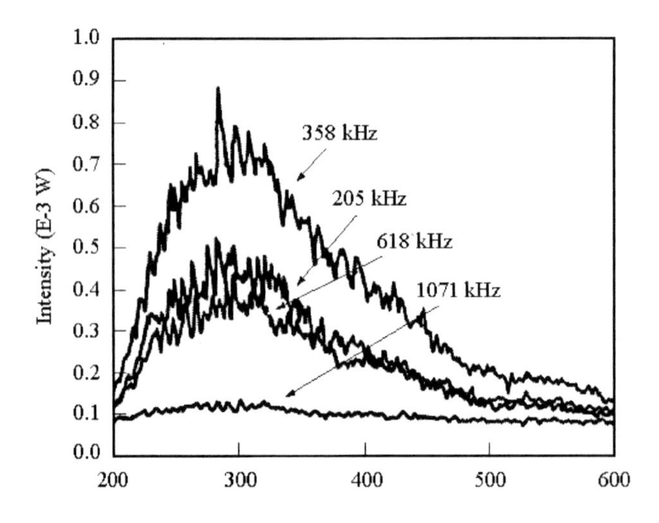


Figure 2.1e SL spectra at four discrete ultrasonic frequencies. Sparge gas = 100% Ar. Significant luminescence species are given by the reactions below the spectra. (Beckett and Hua (2001).)

Johri et al. (2002) cavitated water with a 700 kHz pulsed sound field. An additional 20 kHz pulsed sound field was found to enhance the intensity of the sonoluminescence and subharmonic generation at 350 kHz.

### *2.2.3 Temperature*

Jarman (1959a,b) found an almost linear decrease from 25 to 80°C with several liquids. For water, Taylor and Jarman (1968) found that the sonoluminescence had a maximum of

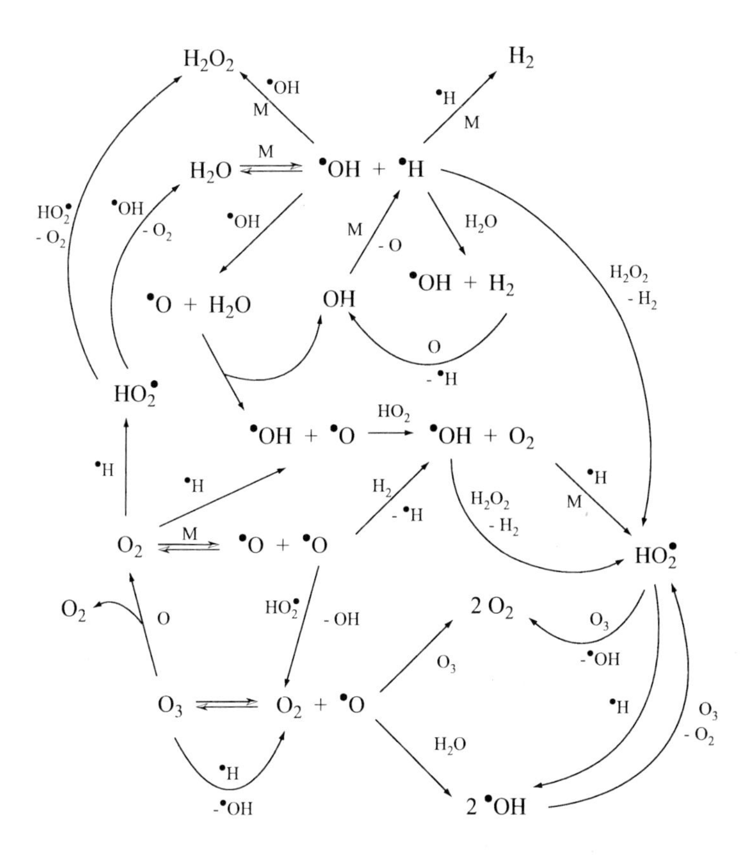


Figure 2.1f Chemical reactions during acoustic cavitation in the presence of oxygen, ozone, or an inert gas (M). M is an inert third molecule such as argon. The ˙OH radical is assumed to play the most significant role in sonochemical reactions although other free radicals ($HO_2^•$, ˙O) may be important oxidizing species as well. (Beckett and Hua (2001).)

13°C and decreased steadily below this temperature, and at 1°C was about one-eighth of the maximum value.

Gabrielli et al. (1967), Iernetti (1972) and Sehgal et al. (1980) reported a linear decrease from 15 to 70°C for air and rare gases in water. Verrall and Sehgal (1988) explained this decrease by relating the SL intensity to the Helmholtz Free energy that bubbles gain during formation and growth and which they release during collapse. Verrall and Sehgal (1988) derived

$$\ln(I/I_r) = -(k\Delta H_v/T_b)[T - T_r] \tag{2.6a}$$

where the SL intensity $I$ is normalized with reference to the SL intensity $I_r$ at 30°C ($T_r$). $k$ is a constant, $\Delta H_v$ is the molar heat of vaporization and $T_b$ is the boiling point of water. Figure 2. 2 is a plot of $\ln(I/I_r)$ vs $T$. The slope should yield the heat of vaporization if $k$ is known. The slope from Fig. 2.2 is $8.6\times10^{-2}\,K^{-1}$. Using the value of $k$ as $2.9\times10^{-3}$ mol cal$^{-1}$ and $T_b$ as 373 K, one obtains $\Delta H_v$ = 46 kJ mol$^{-1}$ which compares with the standard value of 41 kJ mol$^{-1}$ for water.

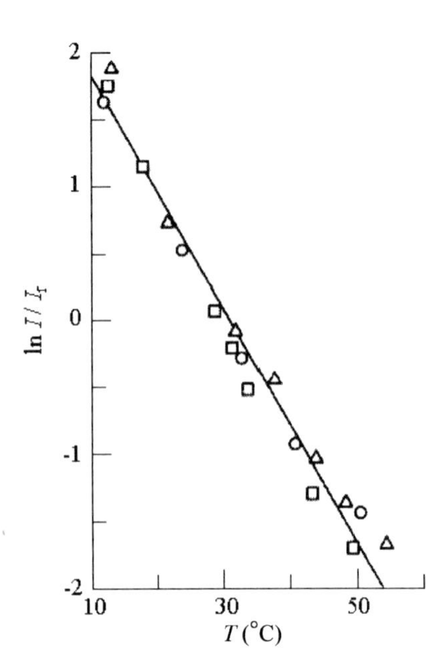


Figure 2.2 Plot of $\ln I/I_r$ vs. $T$ for neon-, argon-, and krypton-saturated aqueous solutions. ○ = Neon; △ = argon; □ = krypton. (Verrall and Sehgal (1988).)

Chendke and Fogler (1985) found that the SL of air saturated water decreased exponentially with increasing water temperatures. SL could be detected at water temperatures up to 90°C. Chendke and Fogler derived a model including decomposition of $H_2O$ with recombination of the H and OH radicals to produce SL.

Barber et al. (1997a) report the liquid temperature variation of the SBSL from a xenon bubble in various organic liquids. They report that the water temperature variation of an air bubble in water where cooling the water from 40°C to −6°C results in an increase in SL of 100 times. They also show that a static ambient pressure of 1 atm provides the most stable and brightest bubble of ethane in water. They mention that tap water can display stable SL at 4–5 atm. Although tap water is saturated as it comes out of the tap it becomes degassed relative to the concentration at $P_0$, as $P_0$ increases.

Hilgenfeldt et al. (1998b) suggest that the variation of SL intensity with water temperature can be accounted for by the temperature dependence of the material constants of water, especially the viscosity, the argon solubility in water, and the vapor pressure. And Erratum: Hilgenfeldt et al. (1998c).

### 2.2.4 *Water vapor*

Finch and Neppiras (1973) made a classical mathematical study of vapor bubble dynamics in Helium I, nitrogen and water.

Yasui (1995) realized the importance of water vapor in bubble dynamics. He developed a simple model in which the evaporation and condensation of water vapor were included. The contribution to the energy flow by evaporating and condensing vapor molecules was found to be very considerable.

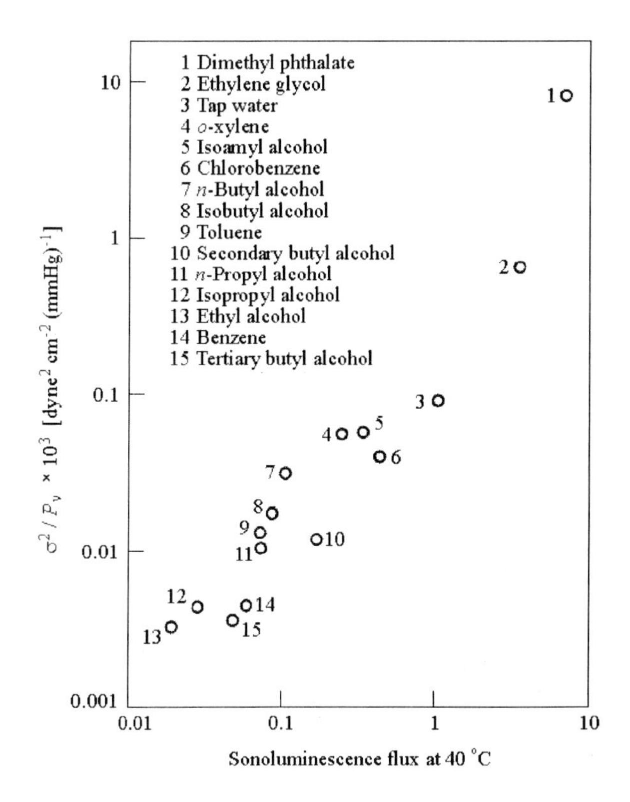


Figure 2.3 Correlation of surface tension/vapor pressure of the liquid with sonoluminescence. (Jarman (1959b).)

Vazquez and Putterman (2000) have recently made a thorough investigation of the dependence of SL on pressure and temperature. They confirm the 100 fold increase as water is cooled. They suggest that water vapor trapped in the collapsing bubble is important. Water vapor is mentioned in Putterman and Weninger's review (2000). And Storey and Szeri (2000) point out that during the intense volume oscillations of a bubble, water is constantly evaporating and condensing, driven by the dis-equilibrium between the partial pressure of water vapor inside the bubble and saturation pressure at the interface. Hence water vapor fills a bubble as it reaches its maximum radius. During collapse, water vapor tends to condense at the wall. They suggest that water is trapped in the interior because the bubble motions become so rapid that the vapor near the center has insufficient time to diffuse to the bubble wall.

See also Sections 3.5.11 and 4.7.15.

### 2.2.5 *Role of the solvent*

Griffing (1952) made a pioneer study of the effects of ultrasonics on the inversion of sucrose.

Jarman (1959a,b) studied 15 liquids with widely different properties. He found that the best correlation was obtained between sonoluminescence and the square of the surface tension $\sigma^2$ divided by the vapor pressure $P_v$ of the liquid, as shown in Fig. 2.3.

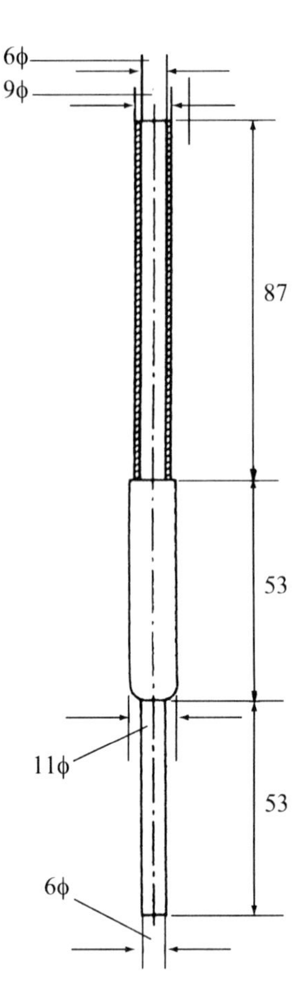


Figure 2.4 Quartz velocity transformer fixed to barium titanate transducer. (Kuttruff (1962).)

The discovery of this empirical relation drew the attention of Kuttruff (1962) to the case of mercury, for which the ratio $\sigma^2/P_v$ is 75,000 times greater than for water. In order to observe the luminescence in this opaque liquid, it is necessary to produce the cavitation adjacent to a transparent surface, and Kuttruff very ingeniously did this by fixing a quartz velocity transformer to a tubular barium titanate transducer working at 25 kHz (Fig. 2.4). The light then traveled up the quartz rod and through the hollow transducer to the photomultiplier. It was found that the sonoluminescence was much greater than that from the ethylene glycol. This experiment led Young (Smith et al. (1967)) to see if he could produce sonoluminescence from the other liquid metals. This required a furnace to melt them, as in Fig. 2.5a. All the liquid metals produced sonoluminescence and Table 2.1 shows that the best correlation with physical parameters is an inverse correlation with thermal diffusivity.

The fact that the sonoluminescence depends on $\sigma$ and $P_v$ should come as no surprise as these terms occur in the Neppiras–Noltingk equation [Eq. (1.13)] describing the growth and collapse of a bubble in liquid. However, the empirical relation does not follow readily from the bubble dynamics. It is quite possible that the correlation would not have been significant if a wider range of liquid types had been examined (a series of alcohols appears

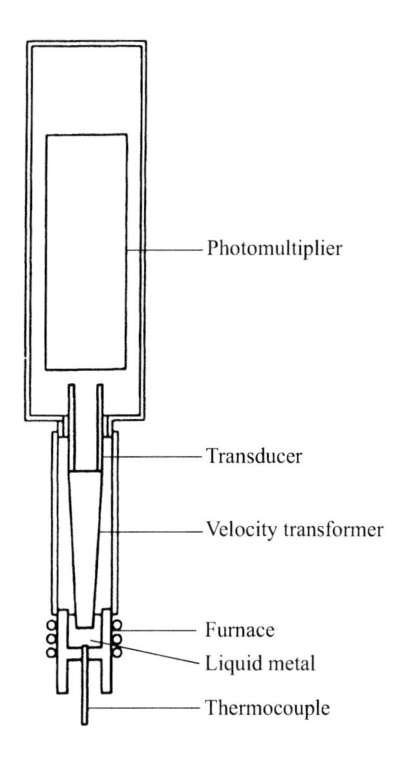


Figure 2.5a Apparatus for generation and detection of sonoluminescence in liquid metals. (Smith et al. (1967).)

in the liquids studied). The possibility that the correlation is fortuitous will be strengthened when we see that the sonoluminescence originates from a series of photochemical reactions occurring within the bubbles. Also, the study employed "off-the-shelf" solvents which presumably contained various unknown amounts of dissolved air. Golubnichii et al. (1971) undertook a similar study of 17 organic liquids and obtained similar results to Jarman, although the free energy of molecular interaction gave a slightly better correlation.

Glycerine was found to give more sonoluminescence than any other liquid and this led Young (1965a) to study glycerine–water mixtures. He found that the sonoluminescence increased with glycerine content, and that this indicated a correlation with viscosity.

**Table 2.1** Sonoluminescence and other parameters for liquid metals. (Smith et al. (1967).)

| Sonoluminescence at 20°C above m.p. | Ng 92 | Ga 51 | In 24 | Sn 24 | Bi 29 |
|---|---|---|---|---|---|
| m.p. (°C) | −39 | 30 | 156 | 232 | 269 |
| Surface tension of pure metal (dyne $cm^{-1}$) | 475 | 735 | 560 | 575 | 390 |
| Vapor pressure at 20°C above m.p. (dyne $cm^{-2}$) | $10^{-14}$ | $10^{-44}$ | $10^{-28}$ | $10^{-30}$ | $10^{-19}$ |
| Thermal conductivity (cal $cm^{-1}$ $s^{-1}$ $deg^{-1}$) | 0.021 | 0.081 | 0.060 | 0.074 | 0.026 |
| Density (g $cm^{-3}$) | 13.6 | 5.9 | 7.3 | 7.3 | 9.8 |
| Specific heat (cal $gm^{-1}$ $deg^{-1}$) | 0.034 | 0.095 | 0.061 | 0.060 | 0.036 |
| Thermal diffusivity ($cm^2$ $s^{-1}$) | 0.045 | 0.144 | 0.133 | 0.169 | 0.074 |

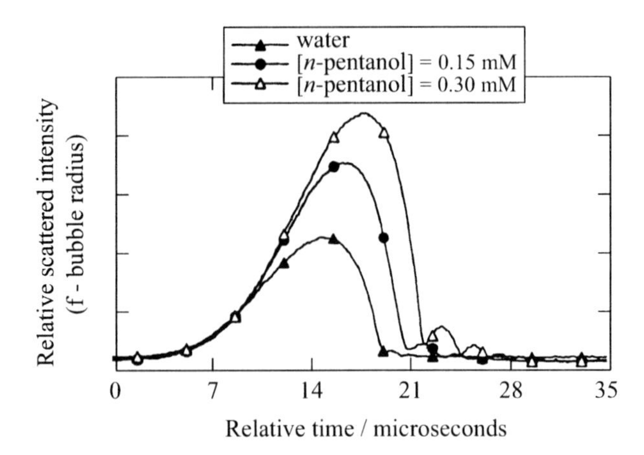


Figure 2.5b Scattered light intensity (proportional to the bubble radius) from a single bubble levitated in water and in *n*-pentanol solutions (driving frequency ~25 kHz) as a function of relative time. (Grieser and Ashokkumar (2001).)

Golubnichii et al. (1970a) confirmed this and obtained a similar result for sucrose–water mixtures. This prompted Young (1965b) to study other very viscous liquids. He measured the sonoluminescence from mixtures of water and glycerine (for comparison and checking), 3-chloropropane 1-2 diol, cyclohexanol and ethylene glycol. Again, the determining parameter seems to be viscosity or possibly the number of OH groups in the molecule.

In 1997, Ashokkumar et al. reported that the MBSL in water decreased with increasing alcohol concentration, and the signal decline was more pronounced with increasing chain length of the alcohol. For all the alcohols studied (methanol, ethanol, propanol, butanol, pentanol), there was a good correlation between the decline in the MBSL signal and the Gibbs surface excess (the equilibrium interfacial concentration of the solute) of the alcohol at the air/water interface.

In 2000, Ashokkumar et al. (2000a) reported that the SBSL in water also decreases with increasing alcohol concentration and the signal decline again increases with chain length. The *quenching* of SBSL by alcohol occurs at a much lower concentration as compared to that for the *quenching* of MBSL. Also, Ashokkumar et al. (2000b) found that MBSL from water was *quenched* by the addition of low levels of unsubstituted or 2-chloro-substituted aliphatic carboxylic acids and the aromatic solutes, phenol and aniline. In all cases, the sonoluminescence intensity could be restored to the level in pure water by adjusting the solution pH to a point where the solutes were completely in their ionized form.

Grieser and Ashokkumar (2001) measured the single bubble sonoluminescence in the presence and absence of low concentration of *n*-pentanol, Fig. 2.5b. It can be seen that the alcohol causes the bubble to expand to a greater size than in pure water before collapse takes place, and that there is an increase in the scattering fluctuations after collapse. This behavior takes place on every acoustic cycle that sweeps past the bubble. This change is consistent with an increase in the pressure in the bubble as would be the case if volatile decomposition products accumulated in the bubble.

Toegel et al. (2000a), in a similar study, added small amounts of ethanol, propanol and butanol to water. In all cases the SBSL was reduced. Toegel et al. (2000a) say that almost all the alcohol molecules at the bubble surface are forced into the bubble *mechanically*. Since the molar heat of the alcohol is much larger than that of argon (which will have displaced the air), this will lead to a decrease in bubble temperature at the collapse.

Yasui (2002a) performed a computer simulation of SBSL oscillations in aqueous methanol solutions under the experimental conditions of Ashokkumar et al. (2002). It was shown that the methanol molecules *evaporate* into the bubble at the bubble expansion, and almost all of them are dissociated inside a heated bubble at the collapse. The bubble temperature at the collapse decreases by the endothermal heat of dissociation of methanol, and by the increase in the molar heat of the gases inside the bubble. The chemistry inside the bubble is very complicated. Inside the bubble we have water vapor and either argon or air. For an *argon* bubble, Yasui (2002a) took into account *25* chemical reactions. For an *air* bubble, Yasui (2002a) took into account *93* chemical reactions.

Ashokkumar et al. (1999) showed that the multibubble sonoluminescence from aqueous solutions containing either alkyl acids or alkylamines was dependent on both the hydrocarbon chain length of the solute and the pH of the solution. Alkylamines quenched the sonoluminescence only at pH values above about 9 whereas alkyl acids quenched only below about pH 7. In the pH ranges where sonoluminescence occurred, the longer the alkyl chain length the greater was the effect. The results were interpreted in terms of the neutral forms of the solutes absorbing at the bubble/solution interface and then evaporating into the bubble core during its growth and compression oscillations, leading to the *quenching* of the sonoluminescence. Possible processes that might be responsible for the *quenching* were discussed.

### *2.2.6 Role of the dissolved gas*

Srinivasan and Holroyd (1961), Parke and Taylor (1956), Günther et al. (1957), Golubnichii et al. (1970b) and Prudhomme and Guilmart (1957a,b) measured the relative intensities of the sonoluminescence from eight different gases dissolved in water. Hickling (1963) explained these results in terms of thermal conduction, observing that, provided the bubbles are sufficiently small, loss of heat from the bubble into the liquid can significantly reduce the temperatures obtained during the collapse, so that there is a consequent reduction in luminous intensity. Hickling (1963) demonstrated this analytically by means of a numerical solution of the equation of motion of the gas in the collapsing bubble. Good agreement was found between the theory and the observed luminous intensities. Hickling confirmed his theoretical analysis, and thus the agreement, to the cases of neon and nitrogen. Young (1968, 1976) made a comprehensive study of 17 different gases dissolved in water, including all the rare gases and the first four saturated hydrocarbons. All these results are shown in Table 2.2 with the method of detection and those obtained by Müller (1965/6) in which the light was measured from glass bulbs of gas imploding in water. A study of the table shows several interesting features. For air, oxygen and nitrogen there is quite a bit of disagreement. This may be due to the fact that the different workers used photomultipliers of different spectral ranges, different transducer frequencies, different operating powers and different acoustic field geometries. Agreement is better for the five rare gases, where a variation from 1 to 540 is recorded for helium to xenon. Why is this variation so large? It must be due to the widely different values of the thermal conductivity of the different rare gases. If sonoluminescence is basically due to an adiabatic compression during the rapid collapse of the cavitation bubbles, then any loss of energy of the gas molecules due to thermal conduction

**Table 2.2** Sonoluminescence from different gases (1 phot = 1 lm $cm^{-2}$).

| | Young (1968, 1976) Phots × $10^{-10}$ at photocathode | Prudhomme and Guilmart (1957) Relative results | Srinivasan and Holroyd (1961) Relative results | Günther et al. (1957) Relative results | Parke and Taylor (1956) Relative results | Golubnichii et al. (1970b) Relative results | Müller (1965/6) Total radiation energy (J) |
|---|---|---|---|---|---|---|---|
| Spectral range<br>Transducer frequency | 165–6500 Å<br>20 kHz | 1900–2800 Å<br>960 kHz | 2800–5500 Å<br>800 kHz | 300 kHz | 3250–600 Å<br>0.5 mHz | 20 kHz | 4000–6000 Å |
| Air | 2.40 | 20 | | | 77 | 1.0 | |
| Nitrogen | 1.22 | 45 | 20 | | 35 | 0.7 | |
| Oxygen | 2.40 | 35 | 35 | 35 | 35 | 0.3 | |
| Helium | 1.16 | 1 | | | | | |
| Neon | 3.20 | 18 | | | | | 0.002 |
| Argon | 30 | 54 | 95 | | 290 | 2.7 | 0.14 |
| Krypton | 50 | 180 | | 2000 | | | |
| Xenon | 125 | 540 | | 6700 | | 7.0 | 1.3 |
| Methane | 0.80 | | | | | | |
| Ethane | 0.85 | | | | | | |
| Propane | 1.0 | | | | | | |
| *n*-Butane | 1.0 | | | | | | |
| Carbon monoxide | 2.40 | | | | | | |
| Carbon dioxide | 0.86 | | | | | | |
| Nitrous oxide | 2.20 | | | | | | |
| Freon ($CC_2Fe_2$) | 3.60 | | | | | | |
| Hydrogen | 0.86 | | | | | | |

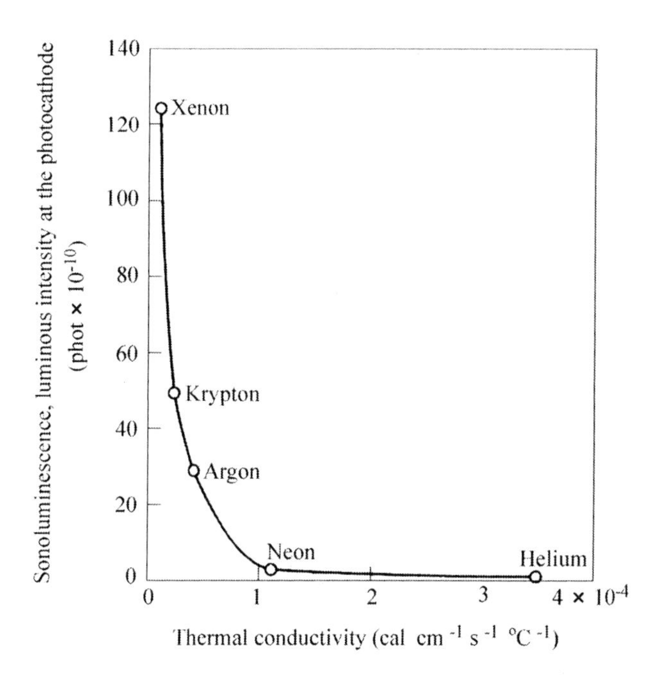


Figure 2.6 Sonoluminescence versus thermal conductivity for the rare gases. (Young (1976).)

will modify the process and cause lower final temperatures. Thus the luminous intensity will be dependent on the thermal conductivity of the gas if there is time for heat conduction to occur. Figure 2.6 is a graph of the luminous intensity against thermal conductivity for the rare gases. A clear inverse relation is seen. In an attempt to evaluate the thermal conduction losses, Finch (1963) assumed that the conduction occurred only in the region of a thin shell of gas near the bubble wall, with a thickness of about three mean free paths $\lambda$. This is justified on the basis of the "temperature jump" (see Present (1958)) which is known to occur near solid boundaries over a distance of the order of $\lambda$. Knudsen (1911) says that the distance involved is $0.95\lambda$ and Langmuir (1915) says $1.2\lambda$. The form of the results would be the same for any chosen value of the order of a few $\lambda$. If at any instant the temperature $T_g$ of the gas within the bubble is uniform but drops linearly over a distance $n\lambda$ (where $n$ is a few $\lambda$) to $T$ and the liquid temperature is assumed to stay constant during the collapse, then the heat loss by conduction in the interval of time $dt$ is

$$k4\pi R^2(T_g - T)\frac{dt}{n\lambda}$$

where $k$ is the thermal conductivity of the gas and $R$ is the radius of the bubble. In the same interval the heat made available through work done on the gas by the liquid is, neglecting work done by surface tension, $p_g 4\pi R^2(-dR)$, where $p_g$ is the gas pressure. Thus the net heat gain is

$$mC_v dT = -p_g 4\pi R^2 dR - \left(\frac{4\pi k}{n\lambda}\right)R^2(T_g - T)dt \tag{2.6b}$$

where $m$ is the mass of gas in the bubble, which is assumed to remain constant during the collapse, and $C_v$ is the heat capacity at constant volume.

Assuming that the perfect gas law holds for the pressures and temperature of interest,

$$p_g v = mR_g T_g \text{ and } R_g = C_p - C_v$$

where $v$ is the volume of the gas, $R_g$ is the gas constant per unit mass and $C_p$ is the heat capacity at constant pressure. Dividing Eq. (2.6b) by $mC_v T_g$ and integrating from the initial radius $R_0$ to the final radius $R_1$ of the bubble, we have

$$\int_T^{T_g} \frac{dT}{T_g} = -\frac{3R_g}{C_v} \int_{R_0}^{R_1} \frac{dR}{R} - \int_0^t \frac{4\pi k}{nmC_v\lambda} R^2 \left(1 - \frac{T}{T_g}\right) dt$$

$$\frac{T_g}{T} = \left(\frac{R_0}{R_1}\right)^{3(\gamma-1)} \exp \int_0^t \left[ -\frac{4\pi k}{nmC_v\lambda} R^2 \left(1 - \frac{T}{T_g}\right) dt \right]$$

where $\gamma = C_p/C_v$, the heat capacity ratio.

For a monoatomic gas, $k = \frac{5}{2}\eta C_v$, where the viscosity $\eta = \frac{1}{3}\rho_g \bar{v}\lambda$, where $\rho_g$ is the density of the gas and $\bar{v}$ is the mean velocity of the molecules. Hence

$$\frac{T_g}{T} = \left(\frac{R_0}{R_1}\right)^{3(\gamma-1)} \exp\left[ -\int_0^t \frac{4\pi \frac{5}{6}\rho_g \bar{v}}{n \quad m} R^2 \left(1 - \frac{T}{T_g}\right) dt \right]$$

However,

$$\rho_g = \frac{m}{\bar{v}}, \quad v = \frac{4}{3}\pi R^3, \quad \bar{v} = \alpha T_g^{1/2}$$

where

$$\alpha = \frac{\bar{v}}{T_g^{1/2}} = \frac{0.92C}{T_g^{1/2}} = \frac{0.92(3p_g/\rho_g)^{1/2} R_g^{1/2}}{(p_g/\rho_g)^{1/2}} = 0.92(3R_g)^{1/2}$$

where $C$ is the r.m.s. velocity of the molecules and $T(R_0/R_1)^{3(g-1)}$ is the adiabatic temperature ($T_{ad}$) that would be reached without thermal conduction or other losses. Thus

$$T_g = T_{ad} \exp\left[ -\int_0^t \frac{5\alpha T_g^{1/2}}{2nR} \left(1 - \frac{T}{T_g}\right) dt \right] \tag{2.7}$$

The dynamic equation of motion for the bubble collapse is

$$R\ddot{R} + \frac{3}{2}\dot{R}^2 = \frac{P_g - P - P_A}{\rho} \tag{2.8}$$

where $P$ is the original ambient pressure in the liquid before the sound field of pressure amplitude $P_A$ is turned on (here we must assume $P_A$ is constant during the time of significant collapse), $P_g$ is the pressure inside the bubble and $\rho$ is the liquid density.

Manipulation of Eq. (2.8) with various approximations gives, with Eq. (2.7)

$$T_g = T_{ad}\exp\left\{-\frac{5\alpha}{4n(2p/3\rho)^{1/2}R_0^{3/2}}\left[\frac{T_g^{1/2}R_1^{1/2}}{[3\ln(R_1/R_0) + 1 + P_A/P]^{1/2}} + \frac{T^{1/2}R_0^{1/2}}{(1 + P_A/P)^{1/2}}\right](R_0 - R_1)\right\} \tag{2.9}$$

where $T_g$ is the final temperature of the gas at $R$.

If we consider the simple case of a bubble of 1 μm radius collapsing to 1/3 mm radius. then

$$R_0 = 10^{-6}\ \text{m}$$

$$R_1 = 3.33 \times 10^{-7}\ \text{m}$$

$$T_{ad} = \left(\frac{R_0}{R_1}\right)^{3(\gamma-1)} = 2637\ \text{K for a monatomic gas.}$$

Assume $P_A = 6 \times 10^5$ N m$^{-2}$ (6 atmospheres). Then

$$\alpha = 0.92(3R_g)^{1/2} = 0.92\left(\frac{3R'}{M}\right)^{1/2}$$

where $R'$ is the gas constant per mole and $M$ is the molecular weight of the gas.

Taking $n = 3$, where $n\lambda$ is the shell thickness, Eq. (2.9) reduces to

$$T_g = 2637\exp\left(-\frac{0.04687}{M^{1/2}}T_g^{1/2} - \frac{1.0153}{M^{1/2}}\right)$$

Taking natural logs yields

$$\ln T_g + \left(\frac{0.04687}{M^{1/2}}\right)T_g^{1/2} = 7.877 - \frac{1.0153}{M^{1/2}}$$

**Table 2.3** Theoretical temperature and experimental sonoluminescence

| Gas | Final temperature of gas $T_g$ (K) | Sonoluminescence $L$ (phots $\times 10^{-10}$) |
|---|---|---|
| Helium | 815 | 1.16 |
| Neon | 1420 | 3.20 |
| Argon | 1650 | 30 |
| Krypton | 1890 | 50 |
| Xenon | 2000 | 125 |

This equation is solved by inspection for $T_g$ for each rare gas. The results are given by Table 2.3 together with Young's data for sonoluminescence from Table 2.2.

If the sonoluminescence can be largely regarded as black-body radiation, then the light intensity $L \propto T_g^4$, and a plot of $\log_{10} L$ against $\log_{10} T_g$, should be a straight line.

Figure 2.7 shows the experimental points with a plot of $\log_{10} L$ against $\log_{10} T_g$ for the rare gases. Most of the points lie on a straight line. Neon is the exception and this may be due to the excitation of the gas, as pointed out by Müller (1965/6). There is also the effect of the solubility of the gas on the sonoluminescence. Hickling (1963) concludes that, provided there is sufficient gas present, it is now certain that the luminous intensity should be strongly dependent on the solubility.

In view of the approximations and assumptions made in the theory, a closer agreement between theory and experiment would not be expected. The correlation between the predicted temperatures and the sonoluminescence for the rare gases affords evidence for the hot-spot theory of sonoluminescence.

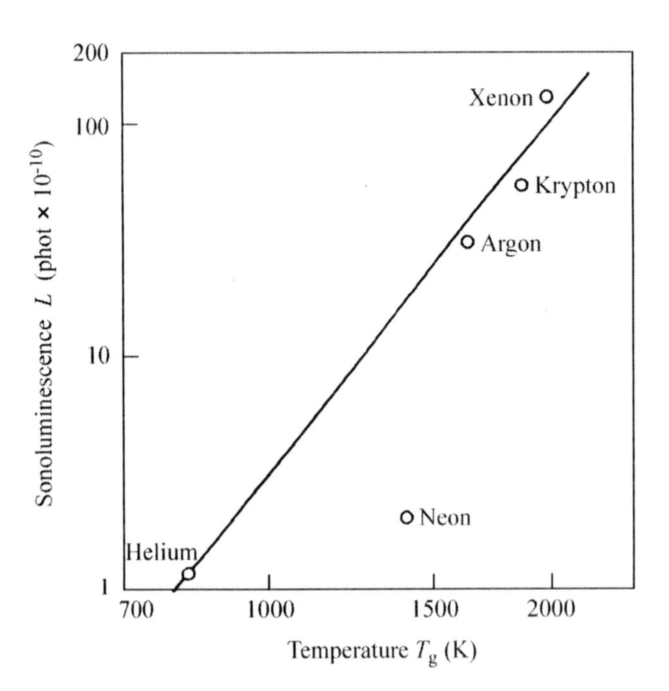


Figure 2.7 Sonoluminescence versus theoretical temperature for the rare gases with experimental points as circles. (Young (1976).)

Basically the above discussion suggests that a bubble gas with increased thermal conductivity will lead to increased heat loss, lower temperatures, less sonoluminescence and less chemical activity. Storey and Szeri (2001) advance an alternative interpretation. They calculate that as the helium content increases, the water vapor has increased mobility and can more easily diffuse out of the bubble into the surrounding water. Therefore, as the helium content increases, the amount of trapped water decreases.

Hammer and Frommhold (2002a) have recently presented an excellent detailed numerical simulation of sonoluminescent rare gases in water, which account for (i) time variations of the water vapor content, (ii) chemical reactions, and (iii) the ionization of the rare gas and the $H_2O$ dissociation products. For water at room temperature with a driving frequency of 33.4 kHz and ambient radius 4.5 μm they calculate 46,000 emitted photons for helium ($P_a$ = 1.40 bar), 138,000 emitted photons for argon ($P_a$ = 1.39 bar) and 297,000 emitted photons for xenon ($P_a$ = 1.36 bar). Hammer and Frommhold's (2002a) model accounts for the light emission by electron–atom, electron–ion, and ion–atom bremsstrahlung, recombination radiation, radiative attachment of electrons to hydrogen and oxygen atoms, which are all more or less important for single bubble sonoluminescence. Spectral shapes, spectral intensities and durations of the light pulses are computed for helium, argon, and xenon bubbles. They generally obtain good agreement with the observations for photon numbers and pulse durations. Hammer and Frommhold (2002a) provide a useful list of 79 references.

## 2.3 SPECTRA FROM MULTIBUBBLE SONOLUMINESCENCE

Paounoff (1939) was the first to record the spectra of SL. He used a conventional spectrograph and with exposures of up to 2 days (!) he obtained the SL spectra of water saturated with a variety of gases. Günther et al. (1959a) and Heim (1960) adopted a similar approach. Heim (1960) obtained a well-resolved spectrum of the sodium D lines in the SL of a NaCl solution but this required an exposure time of 3 weeks (!!). The system used by Günther

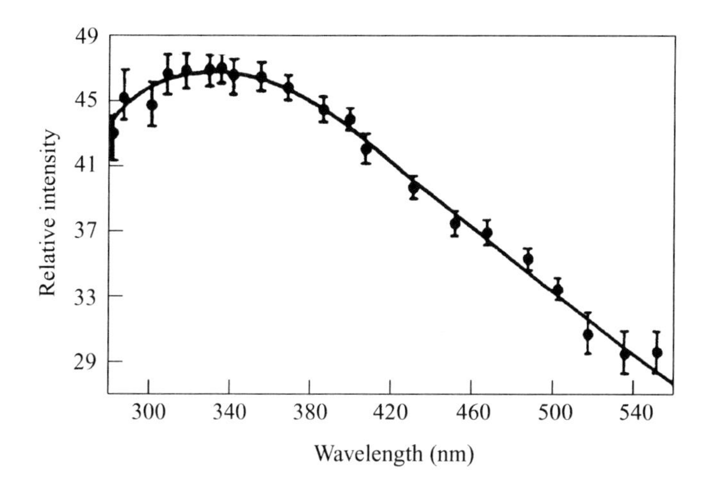


Figure 2.8 Spectral distribution for oxygen-saturated water. Solid line is the theoretical curve for black-body temperature of 8800 K. (Srinivasan and Holroyd (1961).)

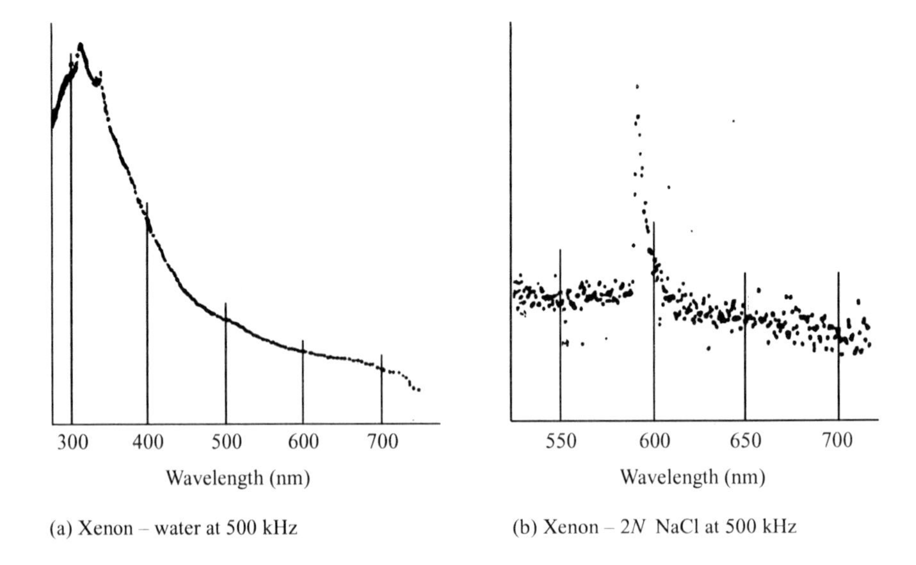


Figure 2.9 Spectrum distribution of xenon-saturated solution of sodium chloride at 500 kHz. (Taylor and Jarman (1970).)

(1957) for observations in the ultraviolet and infrared consisted of 8 color glass filters, each a $2\pi/8$ segment of a circle, laid over a high speed plate.

The spectrum of sonoluminescence is generally a continuum. Figure 2.8 shows that the spectral intensity distribution fits closely to the curve of black-body radiation at 8800 K for oxygen-saturated water irradiated at 800 kHz (Srinivasan and Holroyd (1961)).

Taylor and Jarman (1970) confirm this for air-saturated water at 500 kHz. The continuum can be explained in two ways. First, that ionization recombination of the contents causes thermal excitation and then radiation. Second, that chemical reactions involving photons cause chemiluminescence.

The bubbles are always surrounded by a cold wall of liquid at approximately the temperature of the liquid. Thus we have a hot central core of gas surrounded by a relatively cool shell which would be expected to produce absorption of lines and bands. However Fig. 2.9 for xenon in 2N NaCl, from Taylor and Jarman (1970), shows no very pronounced flattening of the tops of the spectral lines. Thus any absorption of the lines must have been very weak, indicating that the cool shell was optically quite thin.

The sonoluminescence from distilled water containing the rare gases shows a peak at 310 nm superimposed on a broad continuum extending from 240 nm to the near infra-red (Taylor and Jarman (1970), Sehgal et al. (1977, 1980)), Verrall and Sehgal (1987)).

We can conclude that the spectrum is mainly due to black-body radiation, but with some chemiluminescence occurring.

Suslick and Flint (1987) measured the multibubble SL spectra from hydrocarbon and halocarbon liquids. The spectra indicate the formation of excited state molecules by acoustic cavitation. These high energy species probably result from the recombination of radical and atomic species generated during the high temperatures and pressures of cavitation.

Flint and Suslick (1989) did a pioneering experiment in measuring the multibubble SL spectrum at 20 kHz from dodecane during a continuous Ar purge, Fig. 2.10. The peaks are

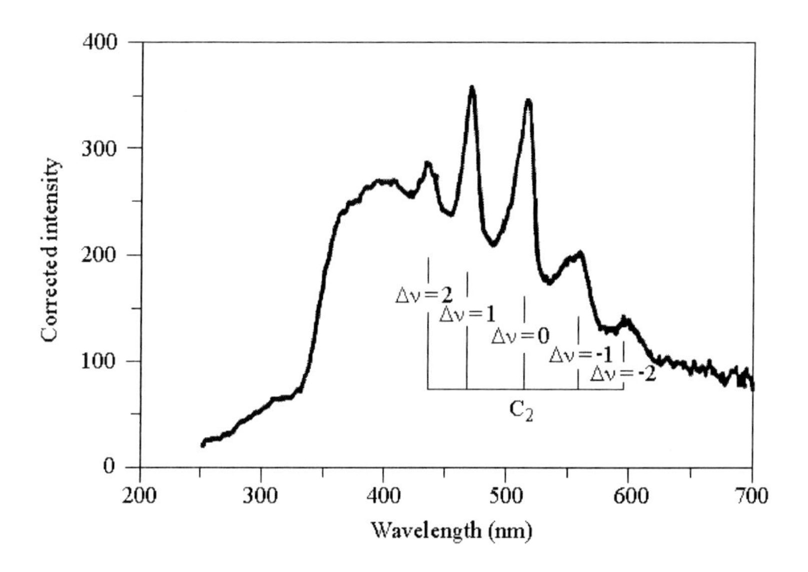


Figure 2.10 Sonoluminescence spectrum of dodecane under Ar at 4°C, vapor pressure = 0.012 torr. Spectrum is the concatenation of four spectra, each of the average of 10 100-s spectra. (Flint and Suslick (1989).)

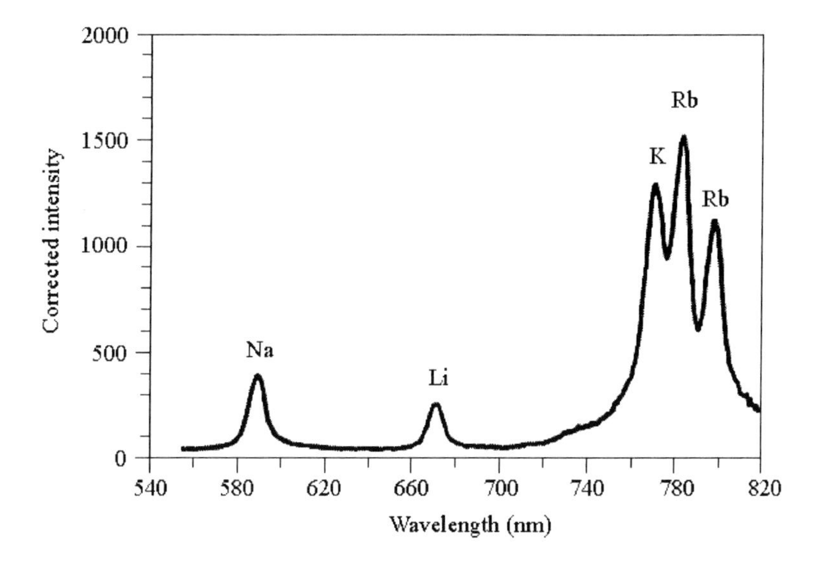


Figure 2.11 Low-resolution sonoluminescence spectrum of the alkali metals. Emission shown is from a 0.1 M $Li^+$, 0.05 M $Na^+$, 0.05 M $K^+$, 0.05 M $Rb^+$ (as the 1-$OC_8H_{17}^-$ salts) in 1-octanol under Ar. Average of 10 100-s spectra. (Flint and Suslick (1991a).)

the Swan rotovibrational bands of $C_2$. These Swan bands are seen in flames, plasmas, and even in the implosion of' Xe-filled glass spheres in paraffin oil (Fig. 2.28).

Flint and Suslick (1991a) then recorded the multibubble SL spectrum at 20 kHz from a 1-octanol solution of $Li^+$, $Na^+$, $K^+$, and $Rb^+$ (as the 1-octyl alkoxide salts) (Fig. 2.11).

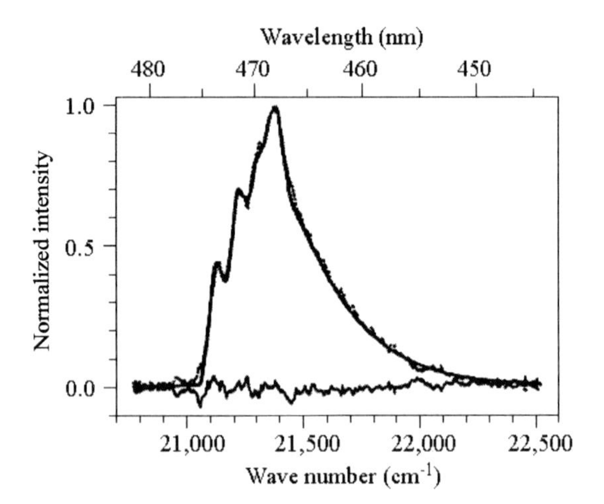


Figure 2.12 Emission from the $\Delta v = +1$ manifold of the $d^3\Pi_g - a^3\Pi_u$ (Swan band) of $C_2$. Dotted line shows observed sonoluminescence from silicone oil (polydimethylsiloxane, Dow 200 series, 50 centistokes viscosity) under a continuous Ar sparge at 0°C, vapor pressure ~0.01 torr. Boldface line shows the best-fit synthetic spectrum, with $T_v = T_r = 4900$ K. Thin line shows the difference spectrum. Sonoluminescence spectra were collected with equipment described in Flint and Suslick (1989). Light emitted during ultrasonic irradiation (Heat Systems Ultrasonics 375 W sonicator equipped with a half-inch horn operating at ~100 W/cm$^2$) was collimated by a lens onto the slit of a 0.25-m Thermo-Jarrell Ash MonoSpec-18 spectrograph equipped with a 50-μm entrance slit and an 1800-groove-per-millimeter grating; the resolution of the spectrometer was 5.3 cm$^{-1}$. Dispersed light was detected with a Princeton Instruments IRY 512N diode array detector. Each spectrum is the average of four 100-s spectra and is corrected for variable detector response as a function of wavelength and for light absorption of the solution. An underlying continuum, assigned to $C_2H$ emission (*12*), has been subtracted from each corrected spectrum. (Flint and Suslick (1991b).)

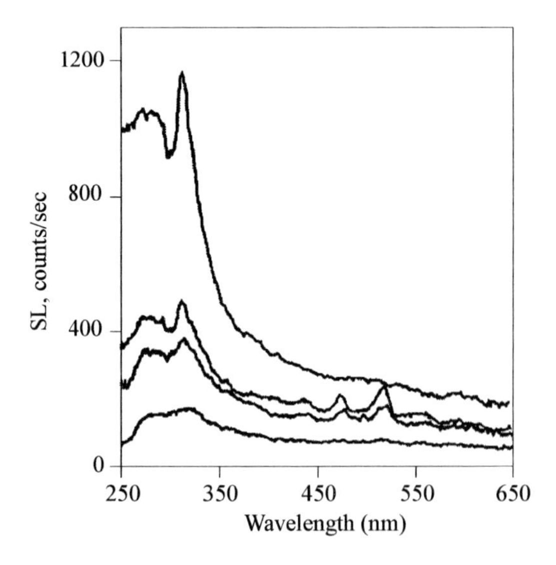


Figure 2.13 Sonoluminescence spectra of water–benzene mixtures at 278 K under Ar at 20 kHz and 50 W/cm$^2$. The spectra from top to bottom are for MBSL from pure water, 0.01% v/v benzene in water, 0.04% v/v benzene in water, and 0.08% v/v benzene in water, respectively. (Didenko et al. (1999).)

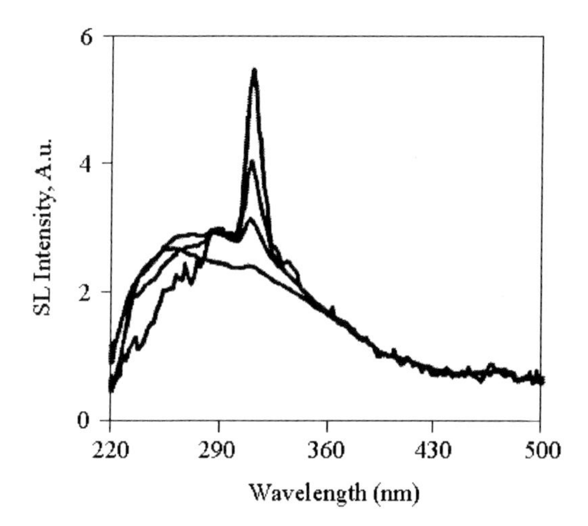


Figure 2.14 MBSL spectra of water saturated with different noble gases: xenon, krypton, and argon, from the near field and with argon from the far field. From the bottom to the top of the figure. (Didenko and Gordeychuk (2000).)

Flint and Suslick (1991b) measured the multibubble SL spectra from silicone oil (polydimethylsiloxane) under a continuous sparge of Ar at 0°C. Figure 2.12 shows the theoretical emission from the $\Delta\nu = +1$ manifold of the $d^3\Pi_g - a^3\Pi_u$ Swan band transition of $C_2$. The dotted line shows the observed sonoluminescence. The thin line at the bottom shows an underlying continuum, assigned to $C_2H$ emission, which has been subtracted from each corrected spectrum. From a comparison of the synthetic to the observed spectra, Flint and Suslick (1991b) found the effective cavitation temperature to be $5075 \pm 156$ K!

Didenko and Pugach (1993) measured the multibubble SL spectra at 20 kHz from water saturated with He, Ar, Cr, and Xe.

Didenko et al. (1999) measured the multibubble SL spectra at 20 kHz from water doped with several organic liquids at low concentrations. As an example, Fig. 2.13 is the multibubble SL spectrum at 20 kHz for a water–benzene mixtures at 278 K under Ar.

Didenko and Gordeychuk (2000) measured the multibubble SL from water in the presence of different rare gases in the far field of an ultrasound horn at about 40 W. Figure 2.14 shows the strong $OH^+$ radical emission at 310 nm. Surprisingly the emission is weakest for xenon, and highest for argon. Didenko and Gordeychuk (2000) state that this MBSL spectra of water in the presence of xenon resembles the SBSL as reported by Hiller et al. (1998). See also Fig. 3.52.

Weninger et al. (2000b) measured the Swan bands from multibubble SL at 27 kHz from argon and xenon cavitation clouds in 3C dodecane, Fig. 2.15.

See also Sec. 4.6.2 The hot spot theory – Bremsstrahlung.

Miyoshi et al. (2001) measured the pH dependence of the multibubble sonoluminescence excitation of *luminol* at 141 kHz.

Lepoint-Mullie (2001) recorded the emission spectra in the vicinity of the resonance lines of alkali metals from cavitating solutions of argon dissolved in NaCl, RbCl or rubidium 1-octanolate. The acoustic frequency was 20 kHz. It was shown that (i) the emission from alkali metals arose from the gas phase of the bubbles, (ii) the blue satellite and line

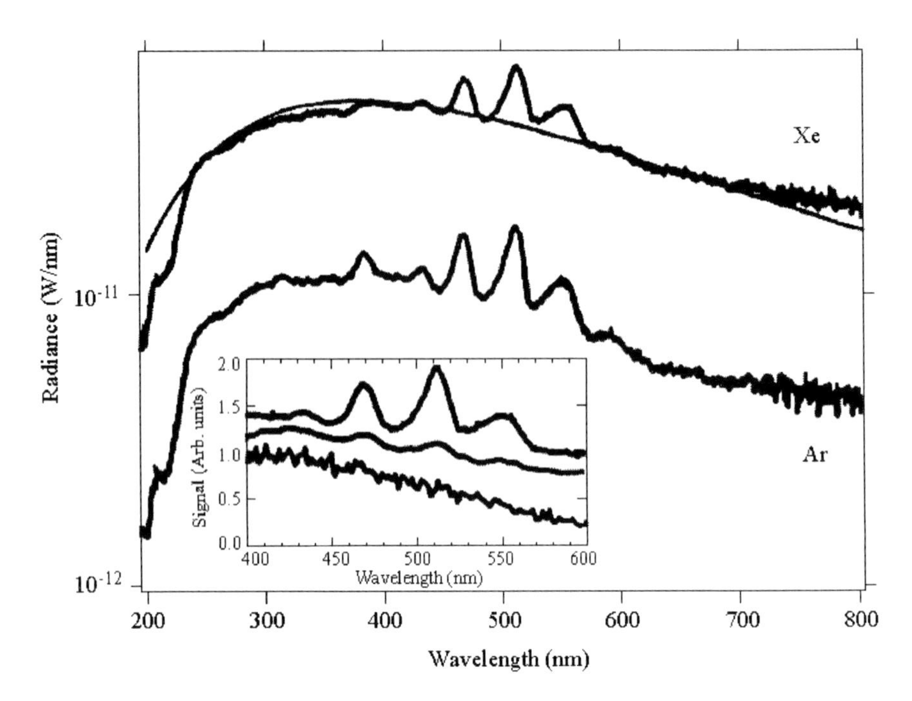


Figure 2.15 SL spectra from argon and xenon cavitation clouds in 3C dodecane driven at 27 kHz with 30-μm displacement amplitude (12 nm FWHM). The falloff in the UV (below 260 nm) is due to absorption in the liquid which has not been corrected. The inset shows the $C_2$ Swan line transition for xenon dissolved in dodecane. (Weninger et al. (2000b).)

distortions were induced, respectively, by $B^2\Sigma^+–X^2\Sigma^+$ and $A^2\Pi–X^2\Sigma^+$ transitions of "alkali–metal/rare gas" van der Waals molecules and (iii) excitation/de-excitation mechanisms were actually chemiluminescence.

Lepoint et al. (2003) measured multibubble sonoluminescence spectra in the 300–350 nm wavelength range in the case of $H_2O$/Ar, $D_2O$/Ar and $H_2O$/Kr. The acoustic frequency was 20 kHz and the spectral resolution was up to 0.34 nm. Three groups of rotational components were identified in the OH/D $A^2\Sigma^+–X^2\Pi_i(0,0)$ transitions via the substitution of $H_2O$ for $D_2O$. The congestion of the bands and the origin of a red shading extending up to 350 nm were discussed.

Using 515 kHz ultrasound, Ashokkumar and Grieser (1999) generated multibubble sonoluminescence in air-saturated solutions and non-aqueous solutions, to vibronically excite the fluorescent solutes: fluorescein, eosin, pyranine and pyrene. In these cases, the sonoluminescence is absorbed by the solutes in solution which in turn emit fluorescence characteristic of their excited state emission spectra. Ashokkumar and Grieser (1999) name this new emission *sonophotoluminescence*. Absorption and emission spectra are measured for the above cases.

Hatanaka et al. (2002) observed *sonophotoluminescence* (SCL) from luminol solutions. No SCL was observed from a stable single bubble that emitted high intensity SI. Instead, SCL was observed from an unstable dancing single bubble, which grows and ejects single bubbles. The parent bubble is known as a *dancing bubble*. Dancing bubbles were first observed by Willard (1953) and mentioned by Saksena and Nyborg (1970) and Barber et al. (1995). A dancing bubble grows in size with every acoustic cycle and emits tiny

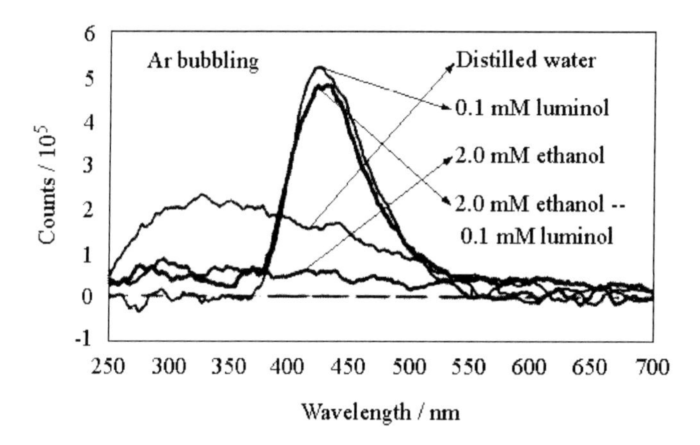


Figure 2.16 Effect of 2.0 mM ethanol additive on the spectra from dancing bubbles in distilled water and 0.1 mM aqueous luminol solution with Ar-bubbling pretreatment. (Hatanaka et al. (2002).)

daughter bubbles repeatedly, making the parent bubble dance round the pressure antinode by reaction. Hatanaka et al. (2002) obtained dancing bubbles by bubbling argon through the liquid. Figure 2.16 shows the spectra from dancing bubbles in distilled water and luminol solutions in the presence and absence of ethanol. Note that the spectral intensity in the luminol solutions greatly exceeds that in distilled water at 420 nm.

## 2. 4 LIGHT FROM HYDRODYNAMIC CAVITATION

Konstantinov in 1947 was the first to observe flashes of light some 0.2–0.3 mm long behind the trailing edge of a cavitating cylindrical obstacle. These flashes were bright with a bluish tinge. Around the junction of the cavitation region, flashes with a diffusive yellowish tinge occasionally appeared. The most definitive observation of hydrodynamic cavitation was made by Jarman and Taylor (1964, 1965) who placed a photomultiplier tube opposite a polystyrene Venturi tube (Fig. 2.17). Tap water at mains pressure was sent through the Venturi tube and vaporous cavitation occurred by the fall in pressure as it passed through the constriction. The light appeared to come from region C in the figure. No appreciable

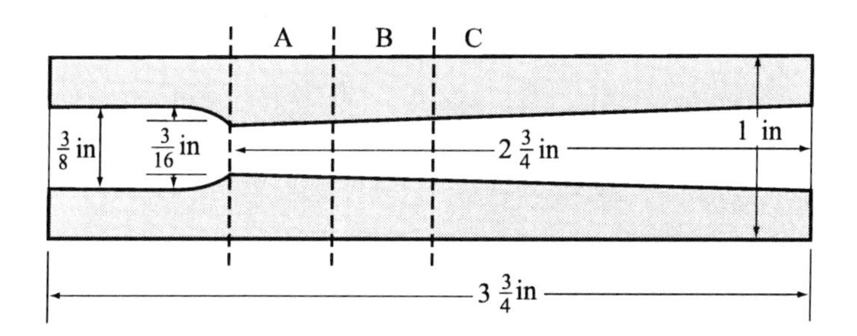


Figure 2.17 Profile of Venturi tube. (Jarman and Taylor (1964).)

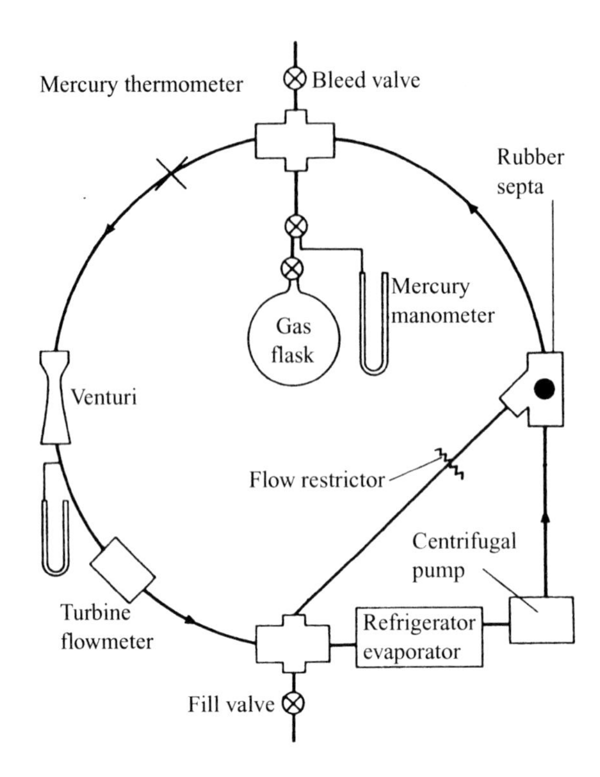


Figure 2.18 Flow system. (Peterson and Anderson (1967).)

light came from regions A and B. The light was very weak. The side pipe to a Bunsen-type pump attached to the water pump was used to increase the proportion of air in the tap water. If the side pipe was partially clamped so as to restrict this inflow of the air, the cavitation and the light emission were increased up to 15 times. However, if the side pipe was unclamped, the resulting air inflow caused the cavitation and light emission to cease. The method of cessation of cavitation and light emission confirms the fact that cavitation in hydraulic machinery can be suppressed in this manner, thus preventing cavitating erosion (Streeter (1961)).

Peterson and Anderson (1967) produced cavitation in the free stream of water using a Venturi tube in a closed-loop system (Fig. 2.18). The system was filled with vacuum deaerated water and then either oxygen or xenon was introduced to raise the gas partial pressure to the desired level. In this study there was no difficulty in observing the light emission in a darkened room. When xenon was introduced upstream from the Venturi tube discrete bluish-white flashes were observed. When gas admission was terminated and the cavitation became self-sustaining, the collapse region took on a bluish-white glow with occasional discrete flashes still discernible. The light emission occurred at the end of the cavity collapse region in a volume approximately 1 cm long and 1.2 cm in diameter. Peterson and Anderson determined the spectral distribution of the light using a spectrophotometer. Their results (Fig. 2.19) show that the spectral distribution cannot be associated with a black-body radiator. Based on this fact alone, it cannot be stated that the individual light pulse itself has no relation to a black-body radiation source. The pulse height spectra show that the intensity of each light pulse is not the same. Theoretical calculations

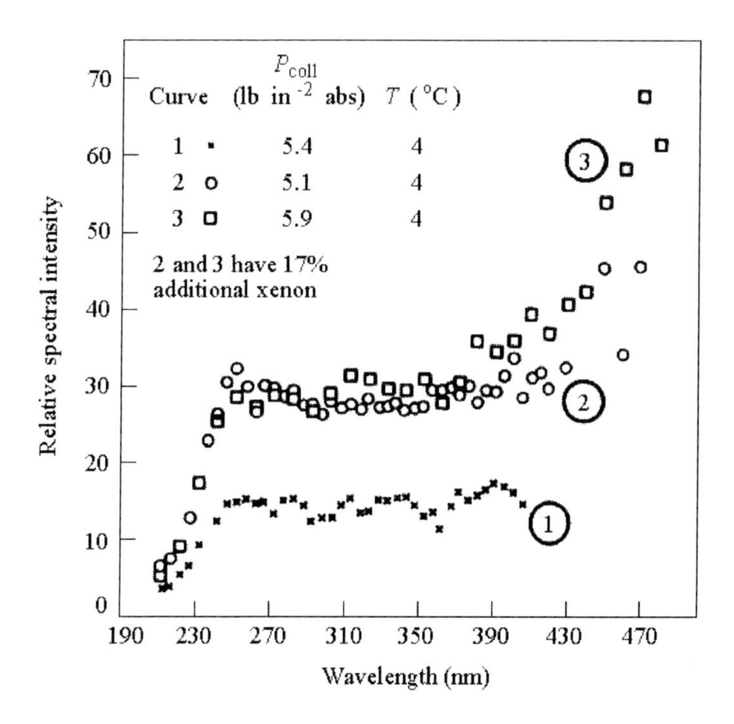


Figure 2.19 Relative integrated spectral intensity of xenon light pulses. (Peterson and Anderson (1967).)

have shown that the size of the cavity before collapse significantly affects the temperature within the cavity during a non-isothermal collapse. If this is the case, then one cannot expect a spectrum determined by many pulses to have a black-body distribution.

Light has also been reported in flow cavitation, involving large energy distributions in field installations, by Knapp, Daily and Hammitt ((1970) on p. 353). An example is that visible light has been observed at night in the tailrace at Boulder Dam, USA, when sudden changes of load necessitated passing large quantities of high pressure water into an energy-dissipating structure, which cavitates severely under these conditions. Another example is described by Jarman (1959a). Light was observed at night from the tailrace of a hydro-electric station at Errochty, Scotland. The luminescence appeared for up to 10 s shortly after the relief valve was fully open. The head of water was 200 m. The luminescence appeared as a shimmering light stretching for some square metres over the surface of the water; it appeared to originate from the gas phase above the water. It could just be seen in daylight, but at night it was very clearly visible; its color appears definitely to be blue.

Weninger et al. (1997b) measured the SL from isolated cavities that form spontaneously on solid objects in a liquid-filled acoustic resonator. The sound period was 33.6 μs. The subnanosecond light flashes have a smooth ultraviolet spectrum similar to that from single bubbles, although with organic liquids low intensity Swan lines ride on the continuum. The hemispherical cavities reach maximum radii about five times larger than are realized from single bubbles. The SL continues to increase as the liquids are cooled to temperatures as low as 160 K.

Leighton et al. (2000) describe a conical bubble collapse which can be controlled. No sound is used, and SL is produced.

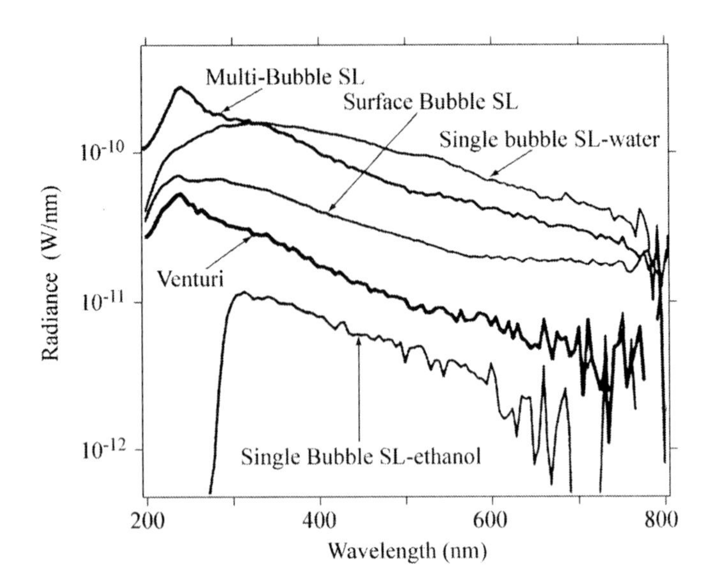


Figure 2.20 Spectrum of light emitted from xenon bubbles in a Venturi tube at 28°C and 50 Torr partial pressure. Also shown for comparison are xenon spectra from surface bubble SL (300 Torr, 12°C), single bubble SL in ethanol (325 Torr, −12°C), single bubble SL in water (3 Torr, 20°C), and multibubble SL from saturated water (760 Torr, 16°C). (Weninger et al. (1999).)

Weninger et al. (1999) circulated water round a closed loop of $\frac{1}{4}$ in wall tygon tubing and a polyvinylchloride pipe of inner diameter 1 $\frac{3}{8}$ in. The loop contained a quartz Venturi tube with entrance/exit of inner diameter 34.5 mm tapering in at 40° full angle to a 5 mm long constriction of inner diameter 9 mm and then tapering out at a full angle of 10°. With pressure drops in excess of one atmosphere, they found that cavitation induced by the converging flow is dramatically strongest for xenon gas. Figure 2.20 shows that the spectrum of flow-induced cavitation matches the properties of resonant SL. Figure 2.21 shows the intensity of light emission as a function of flow rate for several concentrations of xenon in water. Inset shows the maximum intensity for various gases. Note that argon is down by a factor over 100 from xenon. Also, He, $D_2$, $O_2$ and air all yield signals that are down from Xe by a factor of at least 5,000, which places a possible signal (at 50 torr) below the noise flow.

Leighton et al. (2003) described a photon-counting study of the cavitation luminescence produced by flow over a hydrofoil. Significant photon counts were recorded when leading-edge cavitation took place and U-shaped vortices (cavities) were shed from the main cavity.

Su et al. (2003) detected luminescence from collapsing bubbles in a *water hammer*. A *water hammer* occurs when the flow of water through a pipe is suddenly stopped. Accounts of the *water hammer* are given in Batchelor (1967) and Young (1999). Chesterman (1952) described experiments to produce and photograph bubbles produced in a *water hammer*. Su et al. (2003) dissolved a small amount of xenon gas in the water to enhance the luminescence. They detected a stream of predominantly ultraviolet subnanosecond flashes of light. Each flash could exceed $10^8$ photons and have a peak power of more than 0.4 W.

Lohse (2003), Lohse et al. (2001) and Versluis et al. (2000) have described various species of *snapping shrimp* (*Alpheus heterochaelis* and others) that can stun and even kill

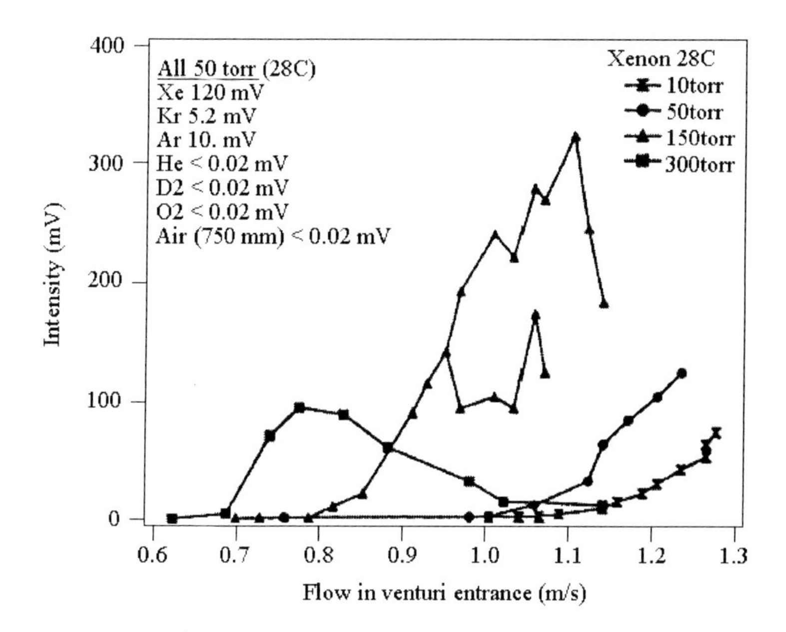


Figure 2. 21 Intensity as a function of flow rate for Venturi cavitation with xenon in water at 28°C and various levels of saturation. At low concentration we cannot rule out the possibility that the intensity continues to increase with flow rate. Although the 150 Torr system is the brightest, it was undesirable because on the scale of seconds it would alternate between two states of different intensities at a fixed flow. Shown as a table are the maximum intensity for various gases at 50 Torr and maximum flow. For helium, deuterium, oxygen and undegassed water any possible signal was not detectable over our noise floor (0.02 mV). (Weninger et al. (1999).)

prey. This shrimp has giant claw that can be closed rapidly, producing a thin water jet. This jet is so fast that, according to Bernoulli's law, a cavitation bubble develops as shown on the third frame of Fig. 2.22. When the bubble collapses, sound is emitted in the form of a shock wave – with fatal effect on the prey. The victim is then picked up by a second, normal-sized claw and eaten. The sound generated at bubble collapse has a broadband spectrum corresponding to the very narrow peak in time over which the bubble collapses. A short intense flash of light is produced as shown in Fig. 2.23. This light is very weak (one or two orders of magnitude less that the sonoluminescence typical of a sonoluminescing bubble) and so cannot be detected by the naked eye. Because the snapping shrimp live in large colonies (in California's San Diego Bay or around Florida, for example) they can generate noise so loud that it can disturb submarine communication.

## 2.5 LIGHT FROM AGITATED MERCURY

In 1675 Jean Bicard noticed a glow above the mercury in a barometer when carried about in a dark room. In 1700 Johan Bernoulli (Bernoulli (1700), Harvey (1957)) observed that mercury in a phial well exhausted of air gave a brilliant light whenever the tube was shaken. Francis Hausksbee in 1709 noticed that when mercury was shaken violently in a globe containing air at atmospheric pressure, "particles of light appeared plentifully, about the

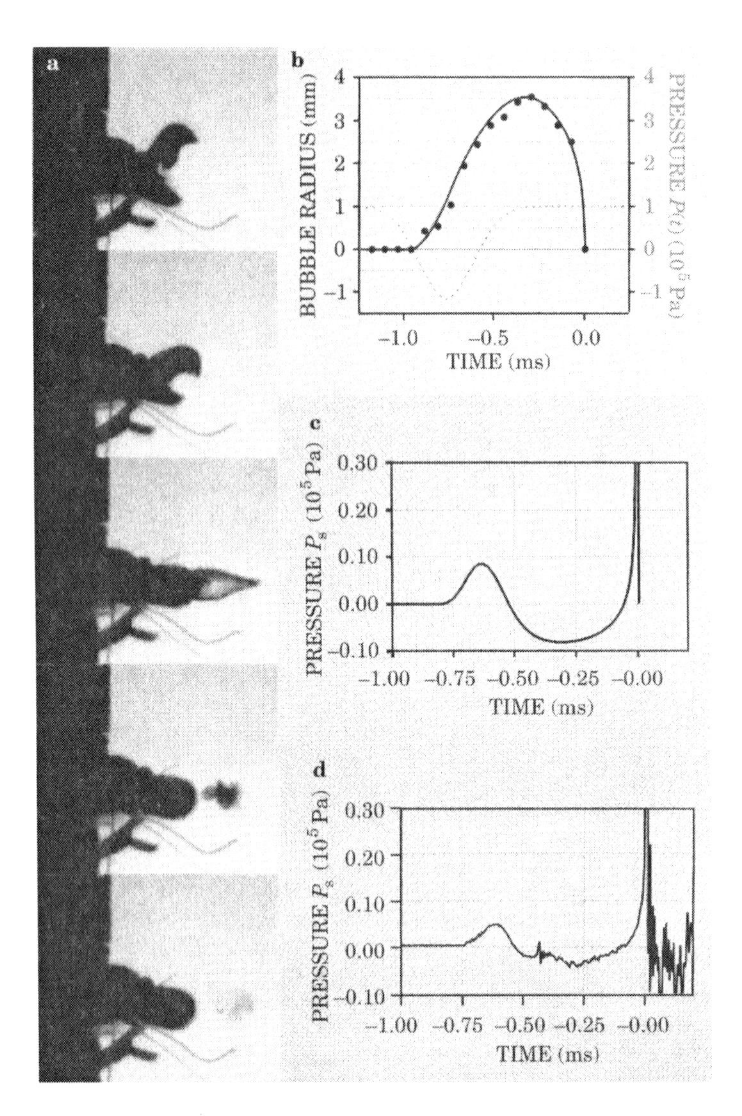


Figure 2.22 *Snapping shrimp* use bubble cavitation to stun and even kill prey. (a) Frames taken every 0.5 ms from a high-speed video recording capture the claw closure of a snapping shrimp. A fast water jet is emitted and generates cavitation. In the last frame, the bubble has collapsed; only microbubbles remain. (b) The measured and calculated bubble radius. The points are the measured data; the solid line shows the results from our Rayleigh–Plesset based theory (equation (1.14)). The dashed line is the assumed pressure reduction $P(t)$ due to the jet. (c) The corresponding sound emission $P_s$ from the bubble. (d) The measured sound emission. From Versluis et al. (2000). (Lohse (2003).) (SEE ALSO COLOR PLATES)

bigness of small pinheads, very vivid, resembling bright twinkling stars". When the same vessel was exhausted, "the mercury did then appear luminous all round, not as before, like little bright sparks, but as a Continued Circle of light during that motion".

Kuttruff (1962) has more recently examined the effect. He evacuated a glass tube, partially filled with mercury, and shook it backwards and forwards in the dark, when an extensive weak bluish light was seen. The light formed a border to the glass adjoining the

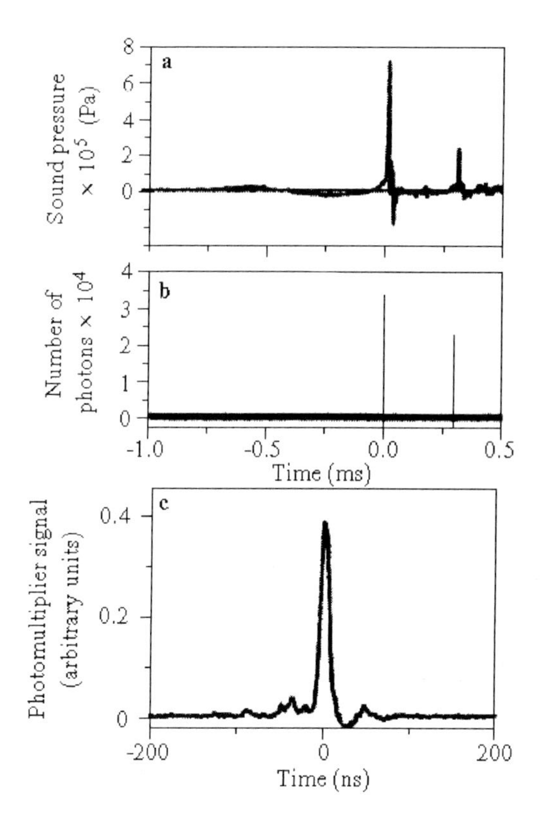


Figure 2.23 Sound and light from a snapping shrimp (*Alpheus heterochaelis*). a, Hydrophone signal and b, light emission from the snap of a shrimp. The principal light-emission event coincides with the bubble collapse at $t = 0$. A second light flash coincides with the subsequent collapse of a cloud of bubbles, which is formed upon collapse of the principal bubble. c, Expanded view of the photomultiplier signal, showing the short pulse duration of "shrimpoluminescence". Within a range of 100 nanoseconds around the main peak, dimmer flashes of light are also evident; these may be due to emission of light from small bubble fragments formed during the violent collapse of the asymmetric cavitation bubble. Data analyses, R. M. Nelissen. (Reprinted with permission from *Nature*, Lohse et al. **413**, 427. Copyright 2001 Macmillan Magazines Limited.)

mercury and followed this edge during the rolling movement of the mercury. The light appeared as soon as the mercury was moving with sufficient speed. With stronger shaking, the mercury formed cavities between the liquid and the glass wall which rebounded the mercury with a sharp crack. This cavitation was accompanied by a second light effect: occasional "pin-point" flashes seen in the cavitation zone. These were irregular and were a bright reddish-yellow. Spectral observations with a direct vision prism and a 2-h exposure of the light from a tube continuously rotated by a motor showed that the bluish light had lines with little background continuum, the 436-nm line being particularly bright. The reddish light, however, showed a strong continuum with a maximum in the red. From this, Kuttruff concluded that the bluish light was electroluminescence, whilst the reddish light was sonoluminescence.

A very interesting paper by Dybwad and Mandeville (1967) describes experiments in which glass balls containing helium and mercury were spun round by a motor. Light was emitted as a discharge occurred between the mercury and electrons trapped in surface states

Figure 2.24 Photograph of the "barometer light", which is the orange line of light generated at the intersection of the mercury meniscus and the wall of the rotating glass cylinder. The rotation speed is $30° s^{-1}$; the cylinder is made from Supracil with a wall thickness of 1 mm, a length of 5 cm, and an outer diameter of 2 cm. The cell is filled with 15 ml of mercury and above the mercury is neon gas at 340 torr. The light is emitted on the side where the wall leaves the mercury. With dimmed ambient lighting this effect is easily seen with the unaided eye, provided that the mercury is properly cleaned. Each black Delrin endcap is joined to the glass with an O-ring seal. We note that the Delrin–mercury interface also emits light. (Reprinted with permission from *Nature*, **391**, 266, Budakian et al. (1998) Copyright Macmillan Magazines Limited.) (SEE ALSO COLOR PLATES)

on the glass. Spectrograms of the light show lines due to Hg, He I, He II, C I, C II, CO, $CO^+$, $CO_2$, $CO_2^+$, H, Si and B.

Ikenoue and Sasada (1962) studied the luminescence produced when a pool of mercury in the bottom of an evacuated U-shaped quartz tube is slightly moved, from the outside. The light was analyzed by a spectrograph and found to contain line and continuous spectra, and quenchings in the visible and ultraviolet regions. It was suspected that electrification occurred and probes showed that a maximum potential difference of about 6 V was formed and that the fundamental mechanism of the electrification was entrainment changing produced by the formation of droplet and double-layer separation.

Budakian et al. (1998) have extended the study quantitatively, using the apparatus shown in Fig. 2.24 in which a cylindrical glass cell was sealed and mounted so that its axis was in the horizontal plane perpendicular to the direction of gravity. It was filled to ~10% of its volume with mercury and sealed in a neon atmosphere at a pressure of 340 torr. The cell was rotated about its axis at constant speed. As seen in Fig. 2.24 light was emitted along the line of contact of the mercury and glass on the side where the glass leaves the mercury. The orange emission was easily seen with the unaided eye. Budakian et al. (1998), in a thorough investigation, found that the repetitive emission of light from mercury moving over glass was accompanied by the collective picosecond transfer of large numbers of electrons. When brought into contact with glass, electrons hop on to sites on the glass where they remain as the glass separates from the mercury. This trapped static charge exerts a force on its image in the mercury which is strong enough to drag the mercury against gravity until a sudden picosecond electrical discharge releases the force, allowing the mercury to fall. This discharge is accompanied by a flash of light. This repetitive build-up and discharge of static electricity thus gives rises to stick-slip motion.

## 2.6 LIGHT FROM COLLAPSING GLASS SPHERES

Beccaria (1716–1781) observed that glass spheres containing air at reduced pressure emitted luminescence when broken in air. Priestly (in Harvey (1957)) gave the following description of Beccaria's experiments in 1769:

> Signor Beccaria observed that hollow glass vessels, of a certain thickness, exhausted of air, gave a light when they were broken in the dark. By a beautiful train of experiments, he found, at length, that the luminous appearance was not occasioned by the breaking of the glass, but by the dashing of the external air against the inside, when it was broke.

A model bubble experiment on sonoluminescence at Göttingen was designed to study the collapse of a glass sphere of gas under controllable conditions of purity, of initial gas pressure, of known gas content and of radius of sphere. Schmid (1962) used thin-walled glass spheres of 7-cm diameter which were evacuated, filled with various gases at low pressures and then immersed in a liquid. On breaking the glass wall by a striking plate, the sphere imploded and a flash of light was emitted. A high speed cinematograph record of the implosion, which lasted for 4 mins, showed that the walls accelerated until the volume of gas was effectively divided into a number of parts, each giving rise to strong shock waves in the liquid and a flash of light near the end of each implosion. Using glycerine as the liquid, a number of experiments were made with various filling gases at different pressures and the results are shown in Fig. 2.25.

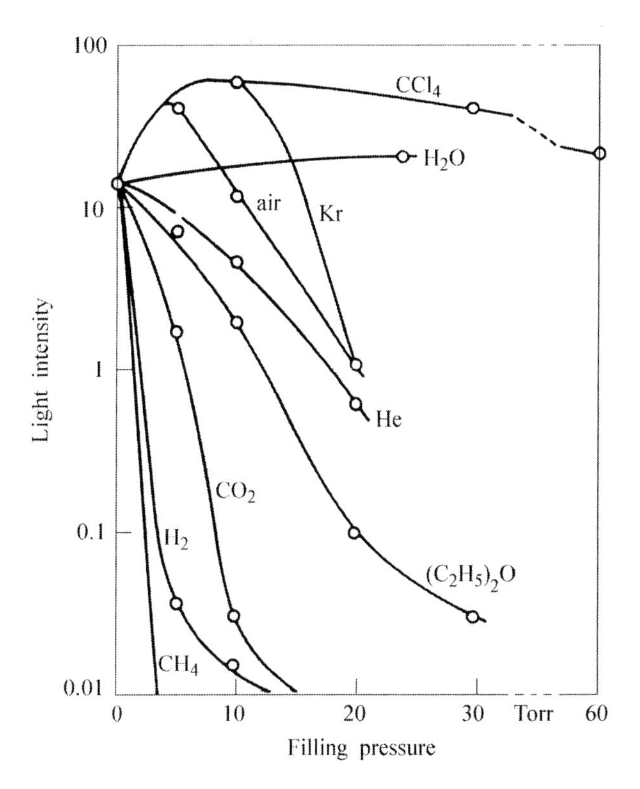


Figure 2.25 Light intensity against filling pressure for implosion in 88 percent glycerine. (Schmid (1962).)

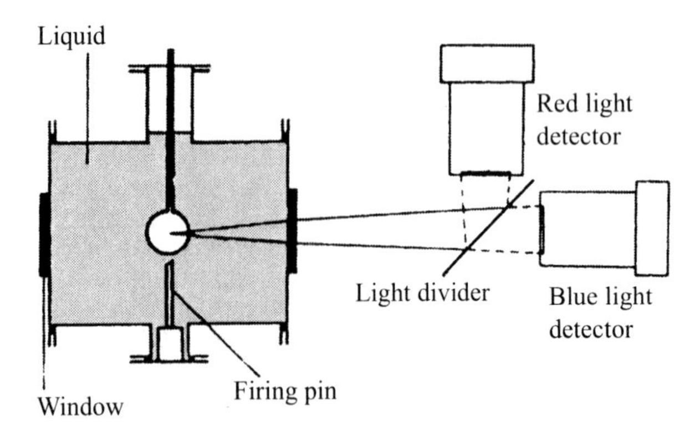


Figure 2.26 Apparatus for observing light from the model bubbles. (Müller (1965/6).)

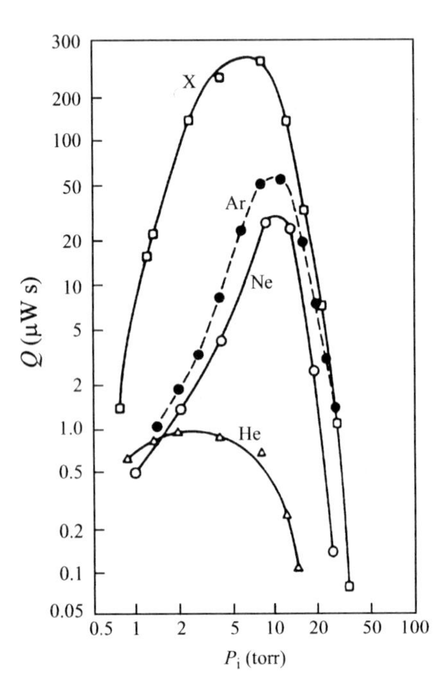


Figure 2.27 Total light output against filling pressure for implosion in paraffin. (Müller (1965/6).)

Müller (1965/6) continued this work using smaller spheres of 5-cm diameter as these showed less instability in the collapse. Figure 2.26 shows his apparatus. In *paraffin* he imploded spheres containing helium, neon, argon, and xenon, and Fig. 2.27 shows the variation with filling pressure of the emitted radiation in the wavelength band 400–600 nm. Müller next imploded spheres of helium, neon, argon and xenon in *water* and obtained curves showing a similar maximum but of much smaller values of total radiated energy (Fig. 2.28). The difference in the light output in *paraffin* and *water* is due to the influence of the liquid molecules and their excitation, despite the fact that no allowance was made for light absorption by the liquid. Müller split the emitted light into blue and red components

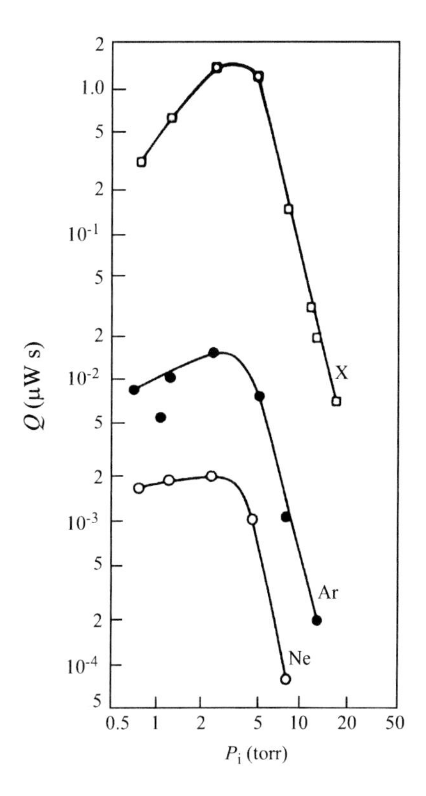


Figure 2.28 Total light output filling pressure for implosion in water. (Müller (1965/6).)

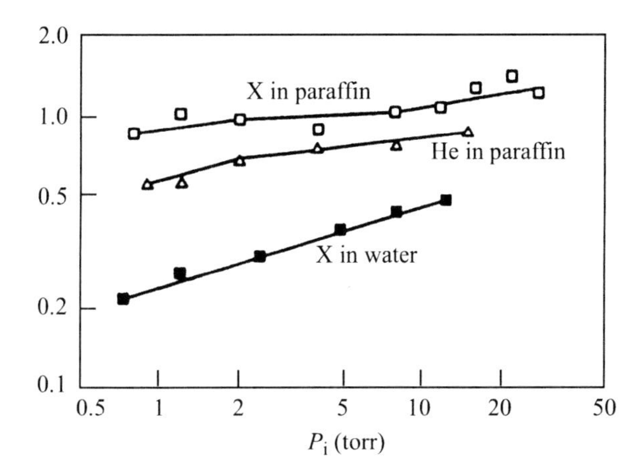


Figure 2.29 Ratio $V$ of red and blue light ($V = Q_{red}/Q_{blue}$) for xenon and helium in paraffin, and for xenon in water. (Müller (1965/6).)

by a special filter and measured the components with blue- and red-sensitive photomultipliers (Fig. 2.29). Figure 2.29 shows $V = Q_{red}/Q_{blue}$, where $Q$ is light energy against filling pressure. It is surprising how little $V$ changes with filling pressure. If the implosion light were from a black-body radiator, then the fall in the xenon curve from 24 to 8 torr should

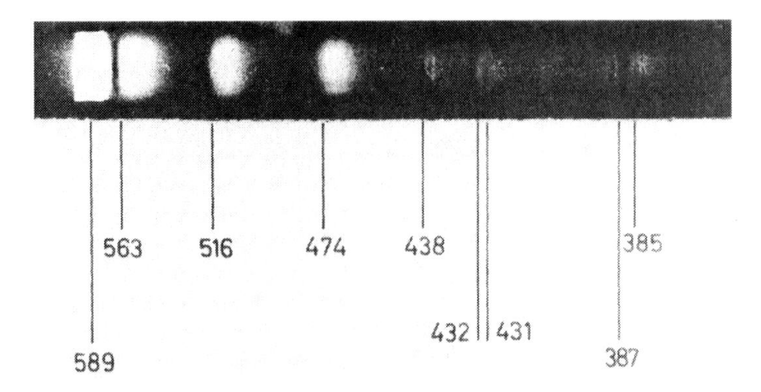


Figure 2.30 Spectrum of implosion light for xenon-filled spheres in paraffin (nm). (Müller (1965/6).)

correspond to a fivefold increase in the relative red fraction of the radiation. Actually, $V$ as a function of filling pressure hardly changes, so we would conclude that the implosion light is not emitted from a black-body source. Müller (1965/6) obtained further information about the light from a spectrogram of the xenon-filled glass sphere (filling pressure 8 torr) in paraffin. Even though this gas gives the greatest light output, it is still necessary to use the total light of 15 implosions to obtain a photograph. Figure 2.30 shows a continuum with very distinct sodium D-lines and the strong Swan band of $C_2$ radicals (563, 516, 474 and 438 nm) as well as the weaker CH bands (432, 431 and 387 nm). The xenon lines were not seen. All the involved substances (gas, liquid and glass) and especially the sodium in the glass and the paraffin molecules in the liquid influence the nature of the light emitted. Müller (1965/6) concludes that the light from the imploding model bubble is dependent on the heating of the filling gas during the bubble collapse and also on the excitation of the involved substances produced by thermal processes or chemiluminescence. Comparison of the results of the model bubble and the sonoluminescence shows a difference in the spectral distribution. Günther, Heim and Borgstedt (1959a) and Srinivasan and Holroyd (1961) obtained a continuous sonoluminescence spectrum, but Müller points out that this was because of the large slit widths of their photoelectric arrangements (up to 13 and 25 nm respectively).

Moreover, Heim (1960) produced spectrum photographs of the resonance lines of the alkali and alkaline-earth metals from their salts dissolved in water.

## REFERENCES

Ashokkumar M, Hall R, Mulvaney P and Grieser F 1997 Sonoluminescence from aqueous alcohol and surfactant solutions, *J Phys Chem B* **101**, 10845.

Ashokkumar M and Grieser F 1999 Sonophotoluminescence from aqueous and non-aqueous solutions, *Ultrasonics Sonochemistry* **6**, 1.

Ashokkumar M, Mulvaney P and Grieser F 1999 The effect of pH on multibubble sonoluminescence from aqueous solutions containing simple organic weak acids and bases, *J Am Chem Soc* **121**, 7355.

Ashokkumar M, Crum LA, Frensley CA, Grieser F, Matula TJ, McNamara III WB and Suslick KS 2000a Effect of solutes on single-bubble sonoluminescence in water, *J Phys Chem A* **104**, 8462.

Ashokkumar M, Vinodgopal K and Grieser F 2000b Sonoluminescence quenching in aqucous solutions containing weak organic acids and bases and its relevance to sonochemistry, *J Phys Chem B* **104**, 6447.
Barber BP, Weninger K, Löfstedt R and Putterman SJ 1995 Observation of a new phase of sonoluminescence at low partial pressures, *Phys Rev Lett* **74**, 5276.
Barber BP, Hiller RA, Löfstedt R, Putterman SJ and Weninger KR 1997a Defining the unknowns of sonoluminescence, *Phys Rep* **281**, 65.
Batchelor GK 1967 *An Introduction to Fluid Mechanics*, Cambridge University Press, page 183.
Beckett MA and Hua I 2001 Impact of ultrasonic frequency on aqueous sonoluminescence and sonochemistry. *J Phys Chem A* **105**, 3796.
Bernoulli JI 1700(5–8), 1701(1–8) *Sur le phosphore du baromètre*, Histoire Acad Roy, Paris.
Budakian R, Weninger K, Hiller RA and Putterman SJ 1998 Picosecond discharges and stick-slip friction at a moving meniscus of mercury on glass, *Nature* **391**, 266.
Chendke PK and Fogler HS 1983 Effect of static pressure on the intensity and spectral distribution of the sonoluminescence of water, *J Phys Chem* **87**, 1644.
Chendke PK and Fogler HS 1985 Variation of sonoluminescence intensity of water with liquid temperature, *J Phys Chem* **89**, 1673.
Chesterman WD 1952 The dynamics of small transient cavities, *Proc Phys Soc* **B65**, 846.
Ciuti P, Dezhkunov NV, Francescutto A, Calligaris F and Sturman F 2003 Study into mechanisms of the enhancement of multibubble sonoluminescence emission in interacting fields of different frequencies, *Ultrasonics Sonochemistry* **10**, 337.
Dezhkunov N, Iernetti G, Francescutto A, Reali M and Ciuti P 1997 Cavitation erosion and sonoluminescence at high hydrostatic pressures, *Acustica* **83**, 19.
Didenko YT, McNamara III WB and Suslick KS 1999 Temperature of multibubble sonoluminescence in water, *J Phys Chem A* **103**,10783.
Didenko YT and Gordeychuk TV 2000 Multibubble sonoluminescence spectra in water which resemble single-bubble sonoluminescence, *Phys Rev Lett* **84**, 5640.
Didenko YT and Pugach 1993 Sonoluminescence from noble gas saturated water, Proc 13th Int Symposium on Non-linear Acoustics, Advances in Non-linear Acoustics, ed Hobaek, World Scientific.
Dybwad GL and Mandeville CE 1967 Generation of light by the relative motion of contiguous surfaces of mercury and glass, *Phys Rev* **161**, 527.
Finch RD 1963 PhD thesis, Imperial College, London.
Finch RD 1965 The dependence of sonoluminescence on static pressure, *Brit J Appl Phys* **16**, 1543.
Finch RD and Neppiras EA 1973 Vapor bubble dynamics, *J Acoust Soc Am* **53**, 1402.
Flint EB and Suslick KS 1989 Sonoluminescence from nonaqueous liquids: emission from small molecules, *J Am Chem Soc* **111**, 6987.
Flint EB and Suslick KS 1991a Sonoluminescence from alkali-metal salt solutions, *J Phys Chem* **95**, 1484.
Flint EB and Suslick KS 1991b The temperature of cavitation, *Science* **253**, 1397.
Gabrielli I, Iernetti G and Lavenia A 1967 Sonoluminescence and cavitation in some liquids, *Acustica* **18**, 173.
Gaitan DF 1990 An experimental investigation of acoustic cavitation in gaseous liquids, PhD thesis, University of Mississippi.
Gaitan DF and Crum LA 1990a Observation of Sonoluminescence from a Single, Stable Cavitation Bubble in a Water/Glycerine Mixture, Frontiers of Nonlinear Acoustics: Proceedings of 12th ISNA, eds. Hamilton MF and Blackstock DT, Elsevier, pp 459–463.
Gaitan DF and Crum LA 1990b Sonoluminescence from single bubbles, *J Acoust Soc Am* Supplement 1 Vol 87, S141 (Abstract).

Gaitan DE, Crum LA, Church CC and Roy RA 1992 Sonoluminescence and bubble dynamics for a single, stable, cavitation bubble, *J Acoust Soc Am* **91**, 3166.
Gavrilov LR 1973 *Physical Principles of Ultrasonics Technology*, Rosenberg (ed.) Vol 2 Part IV, Plenum Press, New York.
Golubnichii PI, Goncharov VD and Protopopov KhV 1970a Sonoluminescence of aqueous solutions of sucrose and glycerine, *Sov Phys-Acoust* **16**, 115.
Golubnichii PI, Goncharov VD and Protopopov KhV 1970b Sonoluminescence in liquids: influence of dissolved gases, departures from thermodynamic theory, *Sov Phys-Acoust* **15**, 464.
Golubnichii PI, Goncharov VD and Protopopov KhV 1971 Sonoluminescence in various liquids, *Sov Phys-Acoust* **16**, 323.
Grieser F and Ashokkumar 2001 The effect of surface solutes on bubbles exposed to ultrasound, *Adv Colloid Interface* **89–90**, 423.
Griffing V 1952 The chemical effects of ultrasonics, *J Chem Phys* **20**, 939.
Griffing V and Sette D 1955 Luminescence produced as a result of intense ultrasonic waves, *J Chem Phys* **23**, 503.
Günther P, Heim E and Borgstedt HU 1959a Über die kontinvierlichen Sonolumineszenzspektren wäßriger Lösungen, *Zeit für Elekrochem* **63**, 43.
Günther P, Zeil W, Grisar U and Heim E 1957 Versuche über Sonolumineszenz wäßriger Lösungen, *Zeit für Elektrochem* **61**, 188.
Hammer D and Frommhold L 2002a Light emission of sonoluminescent bubbles containing a rare gas and water vapour, *Phys Rev E* **65**, 046309.
Harvey EN 1957 *A History of Luminescence*, American Philosophical Society, Philadelphia, USA.
Hatanaka S, Mitome H, Tuziuti T, Kozuka T, Kuwabara M and Asai S 1999 Relationship between a standing-wave field and a sonoluminescing field, *Jpn J Appl Phys* **38**, 3053.
Hatanaka S, Mitome H, Yasui K and Hayashi S 2002 Single-bubble sonoluminescence in aqueous luminol solutions, *J Am Chem Soc* **124**, 10250.
Heim E 1960 Asymmetrisch verbristerte Emissionslinien in den Sonolumineszenz-spektren wäßriger Salzlösungen, *Zeit für Phys* **12**, 423.
Hickling R 1963 Effects of thermal conduction in sonoluminescence, *J Acoust Soc Am* **35**, 967.
Hilgenfeldt S, Lohse D and Moss WC 1998b Water temperature dependence of single bubble sonoluminescence, *Phys Rev Lett* **80**, 1332.
Hilgenfeldt S, Lohse D and Moss WC 1998c Erratum: water temperature of single bubble. sonoluminescence, *Phys Rev Lett* **80**, 3164.
Iernetti G 1972 Temperature dependence of sonoluminescence and cavitation erosion in water, *Acustica* **26**, 168.
Ikenoue K and Sasada Y 1962 Memoirs of Defense Academy, Japan, Vol II, No. 2, 49.
Jarman PD 1959a Sonoluminescence, PhD thesis, Imperial College, London, page 22.
Jarman PD 1959b Measurements of sonoluminescence from pure liquids and some aqueous solutions, *Proc Phys Soc (London)* **73**, 628.
Jarman PD and Taylor KJ 1964 Light emission from cavitating water, *Brit J Appl Phys* **15**, 321.
Jarman PD and Taylor KJ 1965 Light flashes and shocks from a cavitating flow, *Brit J Appl Phys* **16**, 675.
Johri GK, Singh D, Johri M, Saxena S, Iernetti G, Dezhkunov N, and Yoshino K 2002 Measurement of the Intensity of Sonoluminescence, Subharmonic Generation and Sound Emission Using Pulsed Ultrasonic Technique, *Jpn J Appl Phys* **41**, Part 1, 5329.
Knapp RT, Daily JW and Hammitt FG 1970 *Cavitation*, McGraw-Hill.
Knudsen M 1911 Molekulare Warmeleitung der Gase usw, *Ann Physik* **34**, 593.
Konstantinov VA 1947 *Dokl Akad Nauk* **56**, 259.

Kuttruff H 1962 Über den Zusammenhang zwischen der Sonolumineszenz und der Schwingungskavitation in Flussigkeiten, *Acustica* **12**, 230 (Akustische Beihefte).

Langmuir I 1915 Dissociation of hydrogen into atoms Part II Calculation of the degree of dissociation and the heat of formation, *J Am Chem Soc* **37**, 417.

Leighton TG, Cox BT and Phelps AD 2000 The Rayleigh-like collapse of a conical bubble, *J Acoust Soc Am* **107**, 130.

Leighton TG, Farhat M, Field JE and Avellan F 2003 Cavitation luminescence from flow over a hydrofoil in a cavitation tunnel, *J Fluid Mech* **480**, 43.

Lepoint T and Lepoint-Mullie F 1999 *Advances in Sonochemistry*, Vol 5 pages 1–108, JAI Press.

Lepoint T, Lepoint-Mullie F, Voglet N, Labouret S, Pétrier C, Avni R and Luque J 2003 OH/D $A^2\Sigma^+-X^2\Pi_i$ rovibronic transitions in multibubble sonoluminescence, *Ultrasonics Sonochemistry* **10**, 167.

Lepoint-Mullie F, Voglet N, Lepoint T and Avni R 2001 Evidence for the emission of "alkali–metal–gas" van der Waals molecules from cavitation bubbles, *Ultrasonics Sonochemistry* **8**, 151.

Lohse D, Schmitz B and Versluis M 2001 Snapping shrimp make flashing bubbles, *Nature* **413**, 477.

Lohse D 2003 Bubble puzzles, *Physics Today*, February 2003, page 36.

Miyoshi N, Hatanaka S, Yasui K, Mitome H and Fukuda M 2001 Effects of pH and surfactant on the ultrasound-induced chemiluminescence of luminol, *Jpn J Appl Phys* **40**, 4097.

Müller HM 1965/6 Investigation of the light from imploding glass spheres in liquids, *Acustica* **16**, 22.

Noltingk B and Neppiras EA 1950 Cavitation produced by ultrasonics: theoretical conditions for the onset of cavitation, *Proc Phys Soc B (London)* **64B**, 1032.

Paounoff P 1939 La luminescence de l'eau sous l'action des ultrasons, *CR Hebd Séance Acad Sci* **209**, 33.

Parke AVM and Taylor D 1956 The chemical action of ultrasonic waves, *J Chem Soc* **4**, 4422.

Peterson FB and Anderson TP 1967 Light emission from hydrodynamic cavitation, *Phys Fluids* **10**, 874.

Present RD 1958 *Kinetic Theory of Gases*, McGraw-Hill, page 190.

Prudhomme RO 1957a Actions oxydantes des ultrasons sur l'eau en presence des gaz rares, J *Chim Phys* **54**, 332.

Prudhomme RO and Guilmart T 1957b Photogenèse ultraviolette par irradiation ultrasonore de l'eau en presence des gaz rares, *J Chim Phys* **54**, 336.

Putterman SJ and Weninger KR 2000 Sonoluminescence: how bubbles turn sound into light, *Ann Rev Fluid Mech* **32**, 445.

Saksena TK and Nyborg WL 1970 Sonoluminescence from stable cavitation, *J Chem Phys* **53**, 1722.

Schmid J 1962 Gasgehalt und Lumineszenz einer Kavitationsblase (Modellversuche an Glaskugeln), *Acustica* **12**, 70.

Sehgal C, Steer RP, Sutherland RG and Verrall 1977 Sonoluminescence of aqueous solutions, *J Phys Chem* **81**, 2618.

Sehgal C, Sutherland RG and Verrall RE 1980 Sonoluminescence intensity as a function of bulk solution temperature, *J Phys Chem* **84**, 525.

Smith RT, Webber GMB, Young FR and Stephens RWB 1967 Sound propagation in liquid metals, *Adv Phys* **16**, 515.

Srinivasan D and Holroyd LV 1961 Optical spectrum of the sonoluminescence emitted by cavitated water, *J Appl Phys* **32**, 446.

Storey BD and Szeri AJ 2000 Water vapour, sonoluminescence and sonochemistry, *Proc Roy Soc London A* **456**, 1685.
Storey BD and Szeri AJ 2001 A reduced model of cavitation physics for use in sonochemistry, *Proc Roy Soc London A* **457**, 1685.
Streeter VL (ed) 1961 *Handbook of Fluid Dynamics*, McGraw-Hill, page 12.
Su C-K, Camara C, Kappus B and Putterman SJ 2003 Cavitation luminescence in a water hammer: Upscaling sonoluminescence, *Physics of Fluids* **15**, 1457.
Suslick KS and Flint EB 1987 Sonoluminescence from non-aqueous liquids, *Nature* **330**, 553.
Taylor KJ and Jarman PD 1968 The temperature dependence of sonoluminescence from water, *Brit J Appl Phys* (J Phys D) Ser 2, 1.
Taylor KJ and Jarman PD 1970 The spectra of sonoluminescence, *Australian J Phys* **23**, 319.
Toegel RB, Hilgenfeldt S and Lohse D 2000a Squeezing alcohols into sonoluminescing bubbles: the universal role of surfactants, *Phys Rev Lett* **84**, 2509.
Tuziuti T, Hatanaka S, Yasui K, Kozuka T and Mitome H 2002a Effect of ambient-pressure reduction on multibubble sonochemiluminescence, *J Chem Phys* **116**, 6221.
Tuziuti T, Hatanaka S, Yasui K, Kozuka T and Mitome H 2002b Influence of dissolved oxygen content on multibubble sonoluminescence with ambient-pressure reduction, *Ultrasonics* **40**, 651.
Vazquez GE and Putterman SJ 2000 Temperature and pressure dependence of sono-luminescence, *Phys Rev Lett* **85**, 3037.
Verrall RE and Sehgal CM 1987 Sonoluminescence, *Ultrasonics* **25**, 29.
Verrall RE and Sehgal CM 1988 *Ultrasound: Its Chemical, Physical and Biological Effects*, Suslick KS (ed), VCH, New York, page 227.
Verslius M, Schmitz B, von der Heydt A and Lohse D 2000 How snapping shrimp snap: through cavitating bubbles, *Science* **289**, 2114.
Walton AJ and Reynolds GT 1984 Sonoluminescence, *Adv Phys* **33**, 595.
Weninger KR, Camara CG and Putterman SJ 1999 Energy focusing in a converging fluid flow: implications for sonoluminescence, *Phys Rev Lett* **83**, 2081.
Weninger KR, Camara CG and Putterman SJ 2000b Observation of bubble dynamics within luminescent clouds: sonoluminescence at the nano-scale, *Phys Rev E* **63**, 016310.
Weninger KR, Cho H, Hiller RA, Putterman SJ and Williams GA 1997b Sonoluminescence from an isolated bubble on a solid surface, *Phys Rev E* **56**, 6745.
Willard GW 1953 Ultrasonically induced cavitation in water: a step-by-step process, *J Acoust Soc Am* **25**, 669.
Yasui K 1995 Effects of thermal conduction on bubble dynamics near the sonoluminescence threshold, *J Acoust Soc Am* **98**, 2772.
Yasui K 2002a Effect of volatile solutes on sonoluminescence, *Phys Rev E* **56**, 6750.
Yasui K 2002b Influence of ultrasonic frequency on multibubble sonoluminescence, *J Acoust Soc Am* **112**, 1405.
Young FR 1965a Sonoluminescence from glycerine–water mixtures, *Nature* **206**, 706.
Young FR 1965b Sonoluminescence from Viscous Liquid–Water Mixtures, Proc 5th Int Congress on Acoustics, Liege, Paper C 37.
Young FR 1968 Sonoluminescence from Water containing Dissolved Gases, Proceedings of the 6th Int Congress on Acoustics, Tokyo, Paper H-5-3.
Young FR 1976 Sonoluminescence from water containing dissolved gases, *J Acoust Soc Am* **60**, 100.
Young FR 1999 *Cavitation*, Imperial College Press, London.

CHAPTER THREE

# Single Bubble Sonoluminescence

Experiments are the true teachers which one must follow in physics
*Blaise Pascal 1663*

I hope I shall not shock the experimental physicists too much if I add that it is also a good rule not to put over-much confidence in the observational results that are put forward until they have been confirmed by theory
*Arthur Eddington*
*New Pathways in Science*

## 3.1 HISTORY

Mitome (2001) states that credit should be given to Yosioka and Omura (1962), researchers of acoustic radiation pressure (Yosioka and Kawasima (1955)), at Osaka University, Japan, for being the first to produce *single bubble sonoluminescence* (SBSL). Yosioka and Omura's 1962 paper is in Japanese. I am grateful to Dr. Yasui for supplying me with a copy of the paper and its translation. Figure 3.1 shows the apparatus used by Yosioka and Omura (1962). G is a spherical flask about 36 cm in diameter filled with water, which was driven by the magnetostrictive oscillator M in radial mode of the second harmonic. A spherically symmetric acoustic field was formed by adjusting the volume of the water and the sound frequency. The resonance frequency ranged from 16.0 kHz to 16.2 kHz depending on the water temperature. P is a barium titanate hydrophone connected to an amplifier and frequency analyzer. Using degassed water, Yosioka and Omura (1962) often observed light with the naked eye from a bubble a few tens of percents of 1 mm diameter at the center of the flask.

Not long after this discovery, and unaware of it, in 1970, Temple made a remarkable set of experiments for his master's thesis at the University of Vermont. He trapped a *single gas bubble* at a velocity node of a resonant sound field and then observed the emission of light as the acoustic drive was increased. Temple found that one flash was emitted with each cycle of sound and that in water the flash widths were shorter than could be determined by his instruments (20 ns). He also noted that the bubble disappeared when the pressure was

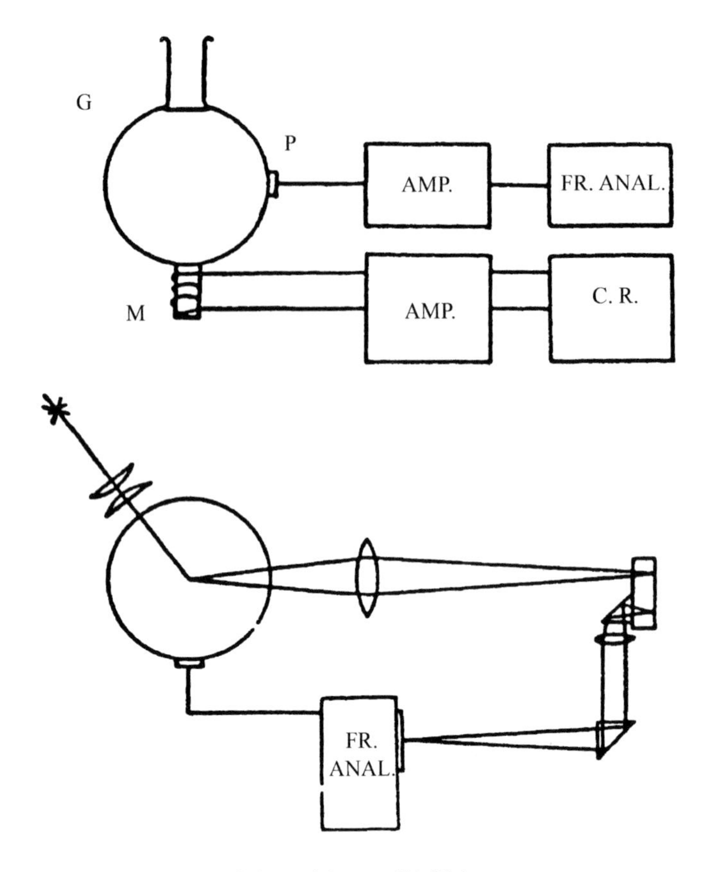


Figure 3.1 Experimental setup. (Yosioka and Omura (1962).)

higher than 1.4 atmospheres, and that it required 5–10 s to "turn on." All this is recorded by Putterman (1999). Curiously enough, the very next year, in 1971, Rozenberg in Fig. 3.2 shows the familiar waveform for single bubble sonoluminescence. Rozenberg's Figure is reproduced as Fig. 12 in Neppiras's famous report (1980) and before all these reports, as Fig. 7 in Hund (1969). Why were these discoveries not pursued? I suggest that it was because multibubble sonoluminescence had been known for 37 years, so there was nothing special about getting SL from a single bubble.

Gaitan (1990) rediscovered SBSL as part of his PhD research at the University of Mississippi. As Moss et al. (1999) says:

> "In the beginning, Gaitan said:
> 'I need a small spot of light to get my PhD.'
>
> Then ...
> Gaitan made the light and saw that it was good.
> Crum saw the light and quickly signed the thesis.
> Putterman saw the light ...
> – it was hot
> – it was brief
> ... and for a while we all believed in magic."

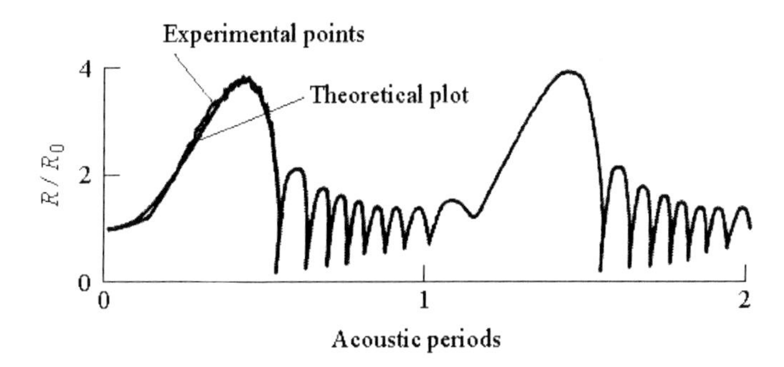


Figure 3.2 Radius–time curve for a bubble showing oscillations at the bubble resonance frequency. $R_0 = 10^{-3}$ cm; $f = 28$ kHz; $P_A$ is near the transient threshold. (Rozenberg (1971).)

Gaitan was surprised to discover that a single bubble could emit a steady glow of bluish light – sometimes for days.

The discovery of single bubble sonoluminescence as a *reproducible well understood phenomenon* must be attributed to Larry Crum and his PhD student Felipe Gaitan [Gaitan et al. (1992)]. As with many physical phenomena, the conditions (here, acoustic pressure, gas content of the water etc) must be right. The story is that Gaitan varied the conditions many times and, one day, he left the system running while he went out for lunch and found when he came back the beautiful sight of SL from a single bubble (Nyborg (2002)).

Gaitan's work (Gaitan (1990), Gaitan and Crum (1990a,b), Gaitan et al. (1992)) led to the appreciation that the pressure, gas content and other properties of a single bubble could be controlled leading to a systematic study of SBSL. Great efforts have been made to explain how the light and high temperatures at the moment of extreme collapse occur. SL has generated tremendous interest and excitement (Cheeke (1997)) and in the period 1990–2003 over 500 papers on SBSL have appeared.

## 3.2 INTRODUCTION

I can do no better than quote the abstract from Barber et al. (1997a).

As the intensity of a standing sound wave is increased the pulsations of a bubble of gas trapped at a velocity node attain sufficient amplitude so as to emit picosecond flashes of light with a broadband spectrum that increases into the ultraviolet. The acoustic resonator can be tuned so that the flashes of light occur with a clocklike regularity: one flash for each cycle of sound with a jitter in the time between flashes that is also measured in picoseconds. This phenomenon is remarkable because it is the only means of generating picosecond flashes of light that does not use a laser and the input acoustic energy density must be concentrated by twelve orders of magnitude in order to produce light. Experiments indicate that the collapse is remarkably spherical, water is the best fluid for SL, some rare gas is essential for stable SL, and that the light intensity increases as the ambient temperature is lowered. In the extremely stable experimental configuration consisting of an air bubble in water, measurements indicate that the bubble chooses an ambient radius that is not explained by mass diffusion. Experiments have not yet been able to map out the complete spectrum because above 6 eV it is obscured by the cutoff imposed by water, and

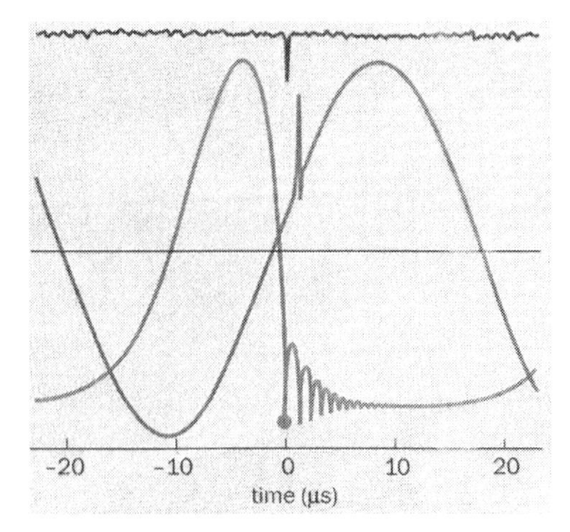


Figure 3.3 During a single cycle of the sound field, the pressure exerted on the bubble (green) follows a sinusoidal pattern. The bubble radius (red) expands during the rarefaction part of the sound field and collapses during the ensuing compression. At the minimum radius, a photomultiplier trained on the bubble records a flash of light (blue). The implosion also generates an outgoing pulse of sound detected by a microphone about 1 mm from the bubble, as shown by the *spike* on the sound wave. The time delay between the collapse and this *spike* is due to the finite speed at which sound propagates in water. (Putterman (1998).) (SEE ALSO COLOR PLATES)

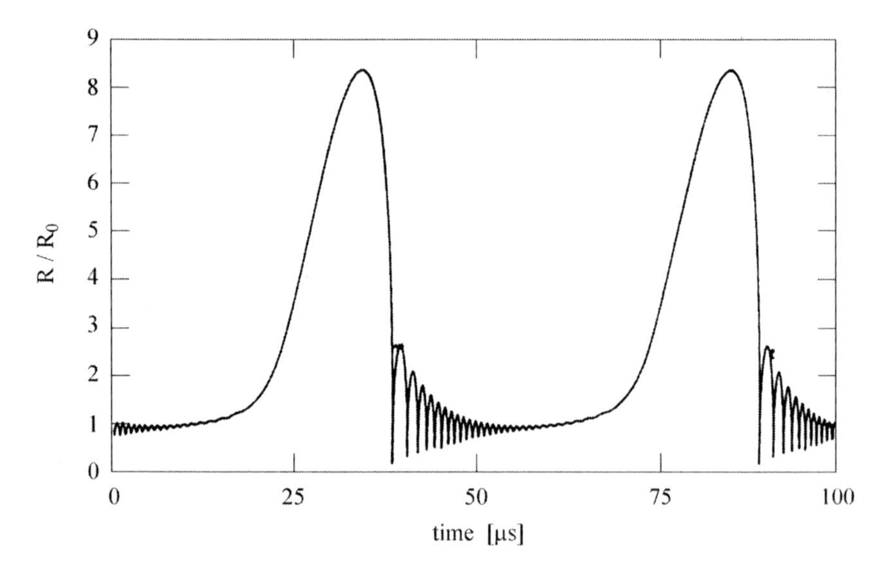


Figure 3.4 Steady-state solution in the giant-response region. The bubble rest radius is 5 μm, the driving frequency is 20 kHz, and the driving amplitude is 130 kPa. (Lauterborn et al. (1999).)

furthermore experiments have only determined an upper bound on the flash widths. In addition to the above puzzles, the theory for the light emitting mechanism is still open. The scenario of a supersonic bubble collapse launching an imploding shock wave which ionizes the bubble contents so as to cause it to emit Bremsstrahlung radiation is the best candidate theory but it has not been shown how to extract from it the richness of this phenomenon.

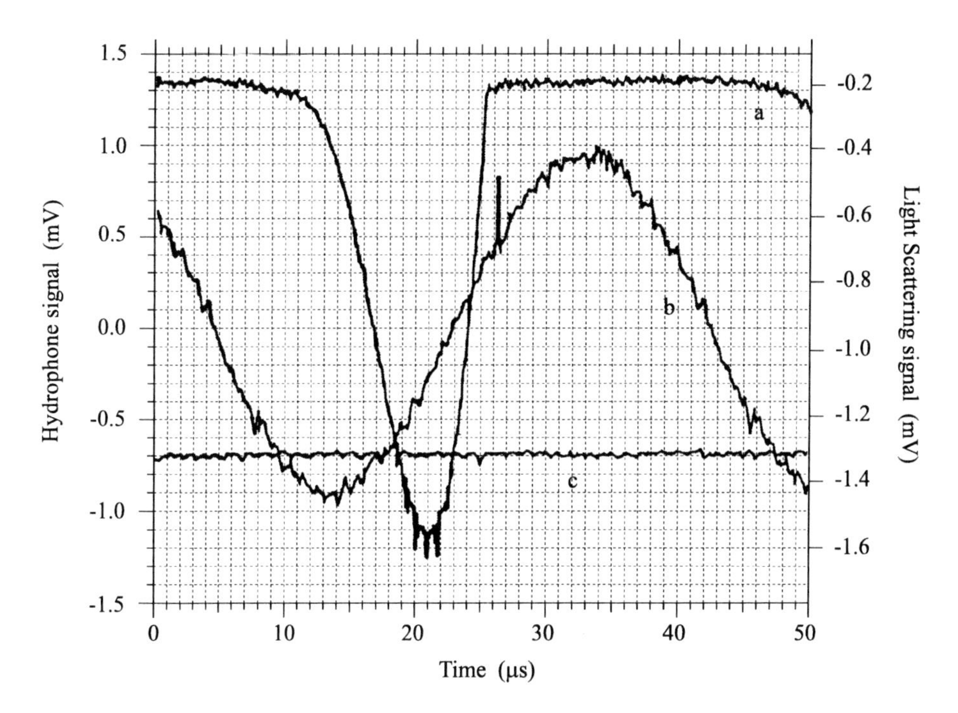


Figure 3.5 Relative timing of sonoluminescence (c) from the stressed interior of a collapsing air bubble whose radius squared is proportional to the magnitude of the intensity of scattered laser light (a). The high pressures reached during the collapse launch an outgoing *spike* recorded by a microphone that measures the driving sound field (b) inside the acoustic resonator. The scale for SL has been offset and the phase of the 26 kHz sound wave has been shifted by 3 μs to correct for the phase delay introduced by the AC-coupled preamplifier. The needle microphone is located about one mm from the bubble, and this accounts for the 1 μs delay between the flash of light and the spike. The flash of SL is less than 50 ps long, and the spike is less than 20 ns wide. The maximum radius of the bubble is about 45 μm and the amplitude of the sound field is about 1.2 atm. This process repeats with each cycle of sound. For an air bubble in water each flash of SL yields about $2 \times 10^5$ photons. (Reprinted from *Physics Reports* **281**, 65, Barber et al. 1997a, with permission from Elsevier Science.)

Most exciting is the issue of whether SL is a classical effect or whether Planck's constant should be invoked to explain how energy which enters a medium at the macroscopic scale holds together and focuses so as to be emitted at the microscopic scale.

Putterman (1998), in his graphic article in *Physics World* entitled "Sonoluminescence: the star in a jar", shows Fig. 3.3 in which the bubble is trapped by the sound field. The successive rarefactions and compressions cause it to pulsate rapidly. During a rarefaction, when the pressure goes negative, the radius of the bubble expands from its equilibrium value of about 4 μm to a maximum of about 40 μm.

The ensuing compression causes the bubble to collapse suddenly. The collapse is only stopped when the gas inside the bubble is compressed to its van der Waals hard-core radius of 0.5 μm. At this moment, the flash of light due to sonoluminescence is emitted. The characteristic form of the steady-state bubble radius vs. time is also shown in the beautifully simple Fig. 3.4 by Lauterborn et al. (1999). The implosion also launches a pulse of sound when the bubble reaches its minimum radius. This is shown by Barber et al. (1997a) in Fig. 3.5 trace (b). The frontispiece is a photograph of a sonoluminescence flash by Crum (1994) – it is a blue dot and can be seen in an undarkened room with the unaided eye as a steady starlike light.

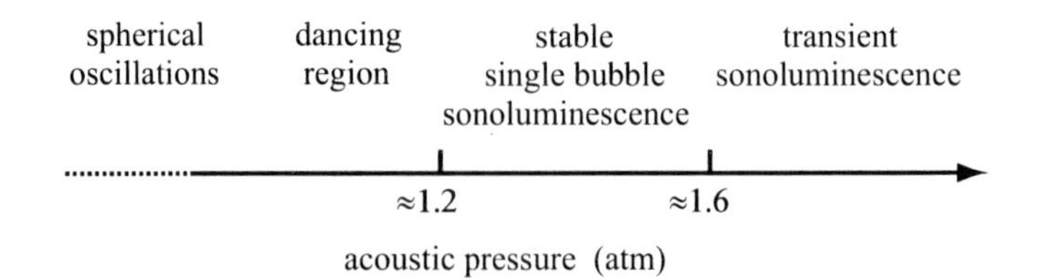


Figure 3.6a With increasing driving pressure the bubble goes through various unstable and one stable regimes. Stable sonoluminescence occurs with driving pressures between roughly 1.2 and 1.6 atm, depending on the exact experimental conditions. (Hammer and Frommhold (2001).)

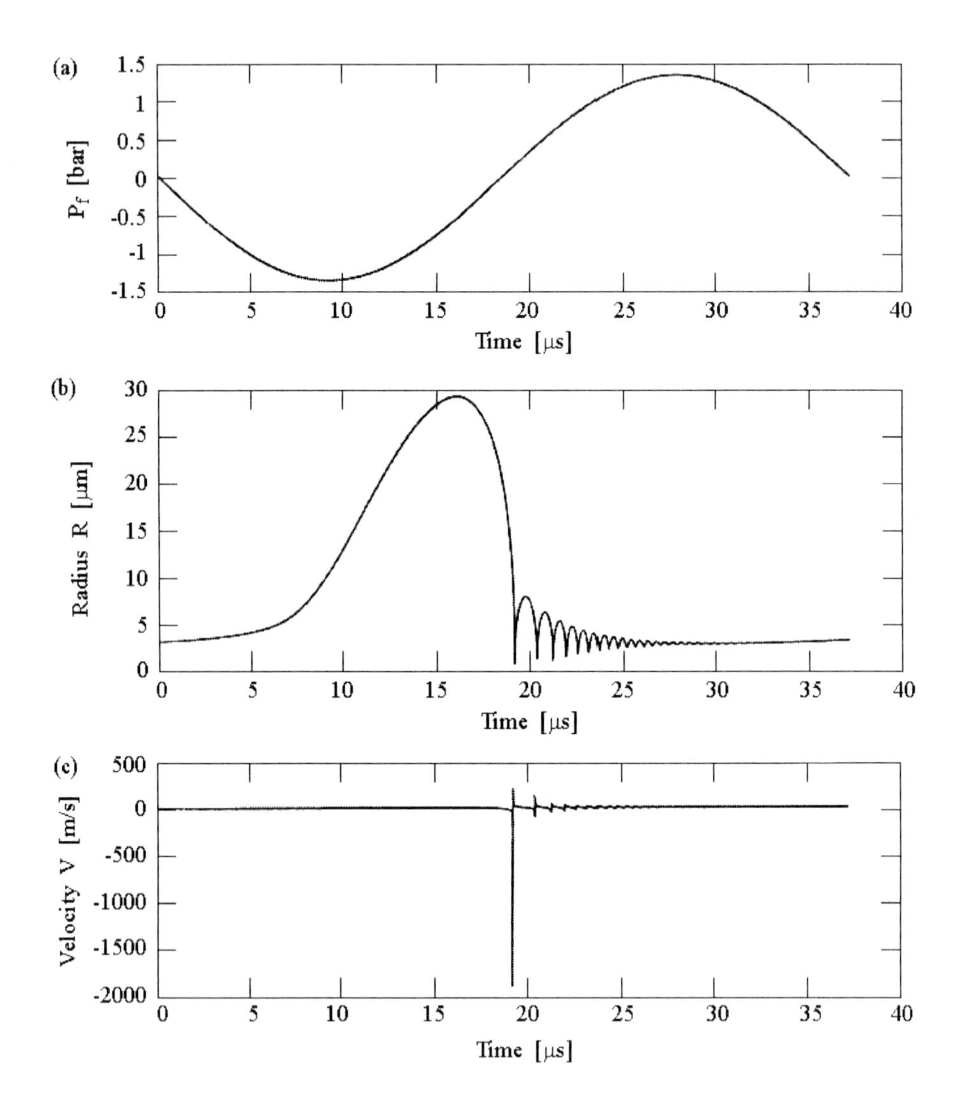


Figure 3.6b The driving pressure ($a$), the radius of the bubble ($b$), and the velocity of its interface ($c$) as a function of time for one acoustic cycle. The parameters are $P_a = 1.36$ bar, $R_0 = 3$ μm, $\gamma = 5/3$, and $\omega/2\pi = 27$ kHz. (Simon et al. (2002.))

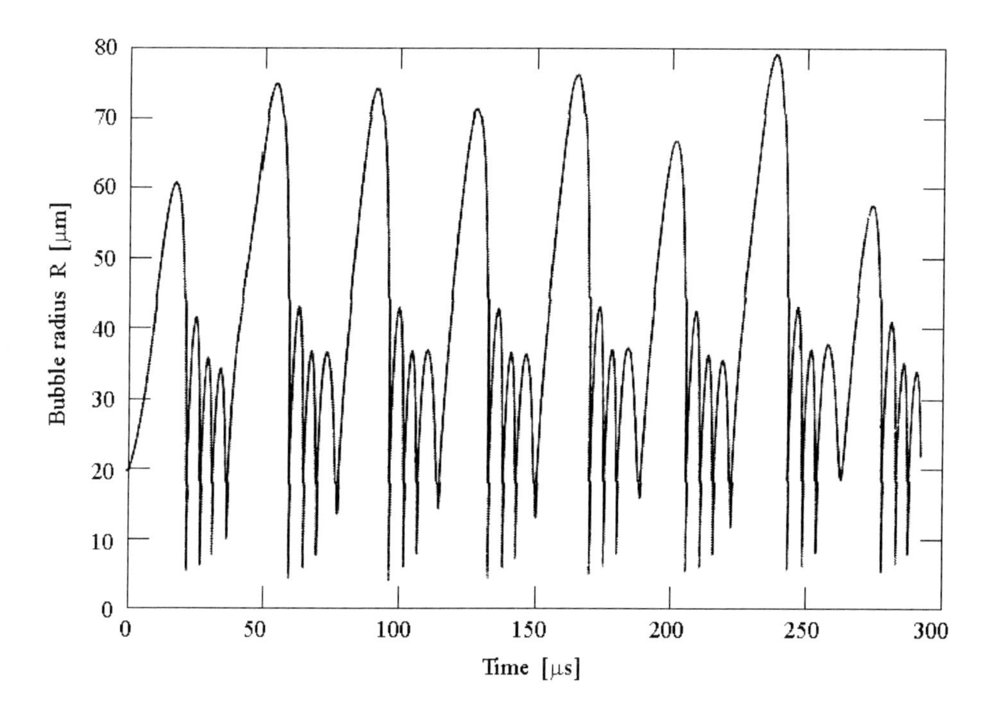


Figure 3.6c Radius of the bubble over eight acoustic cycles. The parameters are $P_a = 1.26$ bar, $R_0 = 20$ μm, $\gamma = 5/3$, and $\omega/2\pi = 27$ kHz. (Simon et al. (2002.))

Figure 3.6a, from Hammer and Frommhold (2001), shows the various unstable and one stable regime that the bubble goes through with increasing driving pressure.

Cheeke (1997) gives an excellent introductory review of single bubble sonoluminescence.

Simon et al. (2002), in Fig. 3.6b, clearly show the forcing pressure, bubble radius and bubble wall velocity for similar values of the parameters, for a stable solution to the Rayleigh–Plesset equation. The main collapse is followed by a series of afterbounces with decreasing amplitude which shows how remarkably elastic the bubble is. It is rather like a rubber ball dropped on the ground and then having decreasing smaller bounces. The whole process then repeats itself in the next acoustic cycle. The angular frequency of the afterbounces is then given by Simon et al. (2002) as approximately

$$\omega_b \approx \sqrt{\frac{1}{\rho_w R_0^2}\left(3\gamma\left(P_0 + \frac{2\sigma}{R_0}\right) - \frac{2\sigma}{R_0} - \frac{4\eta^2}{\rho_w R_0^2}\right)}$$

An important feature of Fig. 3.6c, which is also the reason for the stability of the cycle, is that the amplitude of the bounces dies away completely before the beginning of the next cycle and thus the radius and the interface velocity at the end of the cycle equal the initial values (see Fig. 3.6b).

Simon et al. (2002) also show that if the ambient radius is increased to $R_0 = 20$ μm, the solution to the Rayleigh–Plesset equation shifts to the chaotic regime. Figure 3.6c shows

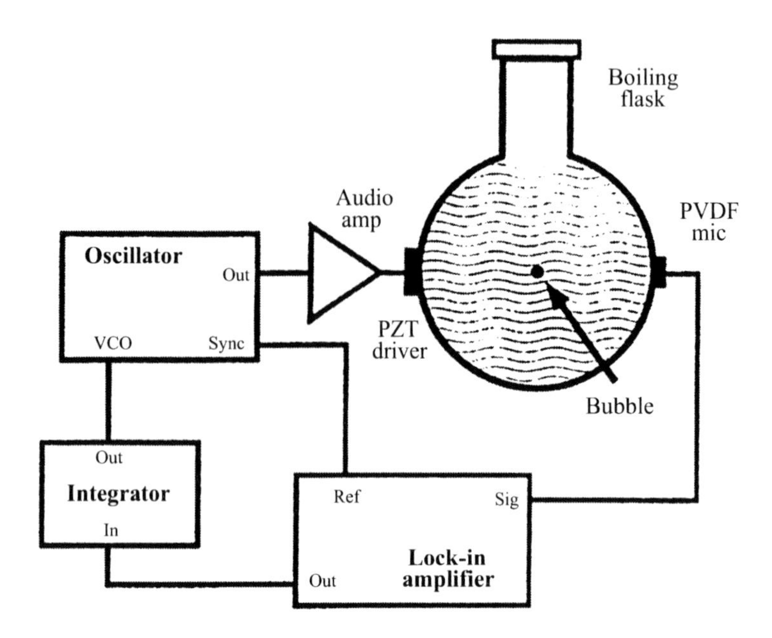


Figure 3.7 Basic apparatus for generating and modelocking sonoluminescence. The amplified output of a sine-wave generator drives a water-filled flask at resonance so that a trapped bubble of gas pulsates at sufficient amplitude to generate light. The modelocking circuit tracks the resonance. (Reprinted from *Physics Reports* **281**, 65, Barber et al. 1997a, with permission from Elsevier Science.)

the oscillation of such a bubble over 8 acoustic cycles. The regime differs from the stable solution in several ways. Probably the most important is that the angular frequency of the afterbounces is decreased. As a result there is not enough time for the afterbounces to be damped out completely before the next expansion begins. Thus each new acoustic cycle will have the memory of the undamped amplitude from the previous cycle, resulting in large variations in the maximum radii. In particular, the $\omega_b = 7\omega$ in Fig. 3.6c is in good agreement with the equation above.

## 3.3 PRODUCTION OF A SINGLE SONOLUMINESCING BUBBLE

This section is from the excellent review from Professor Putterman's group at UCLA, Barber et al. (1997a).

An overview of the basic SL apparatus is shown in Fig. 3.7. An oscillating voltage across a piezoelectric ceramic (PZT) causes it to vibrate and acts as a transducer to drive the water-filled flask at an acoustic resonance. The lowest breathing mode of a spherical resonator is described by $\varphi = j_0(kr)$, where $\varphi$ is the velocity potential such that $v = \nabla\varphi$ and and $j_0$ is the spherical Bessel function. The wavenumber $k$ is chosen to satisfy the pressure-release boundary condition presented by the air–glass interface, viz. $\varphi(R_f) = 0$ where $R_f$ is the radius of the flask. This solution is perturbed slightly by the difference in acoustic impedance between the fluid and the glass. In this geometry, a gas bubble in the water will be forced to the pressure antinode at the center of the flask by the acoustic radiation pressure of a sufficiently intense sound field (Löfstedt et al. (1995)). The second spherically symmetric resonance of the flask creates a spherical velocity-nodal shell which can trap many bubbles at once. This shell is located at about 0.7 $R_f$ and appears when $kR_f = 2\pi$. The corresponding resonance frequency $\omega_a$ is determined by $\omega_a/k = c$.

At the lowest acoustic drive level at which a bubble is trapped, the bubble will slowly dissolve away. At higher drive levels, the trapped bubble is stable against dissolution but emits no light. Still higher drive levels are the parameter space for stable SL: the bubble suddenly shrinks in size and emits light. The author first saw the SBSL spot of light demonstrated by Dagmar Krefting (2002) at Professor Lauterborn's laboratory at Göttingen. It was very exciting and unbelievable and produced with the simplest apparatus of Fig. 3.7. Increasing the drive level leads to more intense light emission. Eventually an upper drive threshold is reached, at which pressure the sonoluminescing bubble will abruptly disappear (Gaitan (1990), Barber and Putterman (1991)). Sometimes, the abrupt upper threshold is replaced by the tendency for the bubble to wander away from the center of the flask with increasing drive levels; whether this is a property of the specific resonator design is unknown.

Holt et al. (1994) discussed the very slight parameter variation that can eliminate the sonoluminescence altogether. Thus, very sensitive experimental control was needed for the external driving parameters $P_A$, $f$. Controlling $R_0$ was even more difficult due to mass diffusion. Further, departures from strictly singly periodic dynamical behavior occurred via small variations in $P_A$, $f$, $R_0$ within the already minuscule window for single bubble sonoluminescence. The resonator itself must have a very high $Q$ in order to reach the pressures necessary for single bubble sonoluminescence. Thus as the driving frequency is varied, the pressure also varies. In the experiments of Holt et al. (1994), fine control was achieved only via frequency control.

The resonant frequency of the flask is determined by searching for peaks in the amplitude of the sound field in the flask by means of a microphone, or by tracking the phase difference between voltage and current of the PZT (Barber (1992)). The resonance itself is surprisingly narrow, typically about 30 Hz wide (Greenland (1999)). Various spurious resonances, such as flexing modes of the glass flask, do not trap a bubble. Variations in the ambient temperature of a few degrees Celsius correspond to shifts in the resonant frequency by 200 Hz which is larger than the width of the resonance of a typical flask. The drive frequency may be continuously adjusted to accommodate such changes in the resonance by using a mode-locking scheme, as shown in Fig. 3.7 (Hiller (1995)). Underlying the operation of such a method is the fact that the phase of the response of an oscillator relative to a sinusoidal drive shifts by $\pi$ as the frequency is tuned through resonance. At low frequencies of drive the response is dominated by the spring constant so that it is "in" phase, whereas at high frequency the response is dominated by inertia and is therefore "out" of phase with the drive. The phase shift occurs over a frequency range determined by the damping coefficient of the oscillator. To maintain SL as the drive frequency drifts off resonance, the phase difference between the output of the signal generator, which drives the PZT, and the standing wave in the resonator, as measured by a PVDF microphone, is determined by means of a lock-in amplifier. This phase difference is integrated and used as the voltage-controlled oscillator input to the signal generator. The quality factor of a typical SL apparatus is between 300 and 1500 (Barber and Putterman (1991)). Typical voltages across the PZT required to drive the acoustic resonance and sustain a sonoluminescing

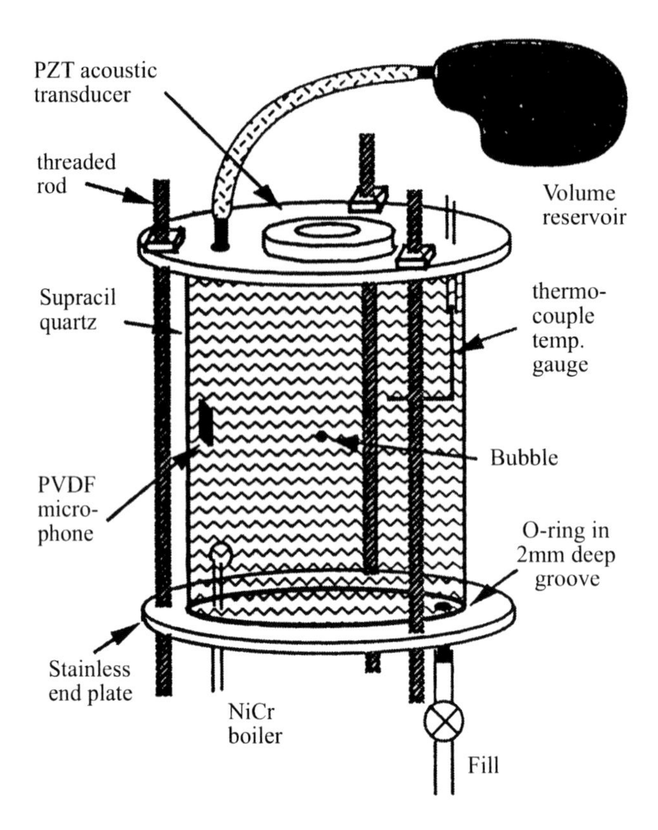


Figure 3.8 Cylindrical resonator for obtaining sonoluminescence from a sealed system. This is important for controlling the composition and the partial pressure of the gas content in the resonator. The NiCr (toaster) wire is used to seed a bubble by boiling the liquid locally. The vaporous cavity fills with whatever gas is dissolved in the liquid and is at the same time yanked to the velocity node of the sound field where it emits light at a sufficiently high acoustic drive. (Reprinted from *Physics Reports* **281**, 65, Barber et al. 1997a, with permission from Elsevier Science.)

bubble are 50 to 150 V (which can be generated easily, but expensively, with high voltage linear amplifiers). Alternatively, impedance-matching the capacitance-dominated PZT with inductors allows the SBSL cell to be driven with less than 10 V (Barber (1992)).

A spherical flask such as the one in Fig. 3.7 can be seeded with bubbles by simply poking the open surface. However, the quality of the resonance is very sensitive to the level of fluid in the neck. Achieving some degree of experimental reproducibility requires a sealed system, which allows the liquid and the gas contents to be controlled accurately. Such a system is shown in Fig. 3.8 where a sealed cylindrical resonator is driven by piezo-electrics mounted on the steel endcaps (Hiller et al. (1994)). In this apparatus the free surface is eliminated by filling the system entirely; changes in density are accommodated by a variable volume which is provided by a polyethylene bag. The bubble is seeded by passing a brief current through a loop of NiCr wire, which boils the surrounding liquid, leaving vaporous bubbles, into which flows the gas dissolved in the liquid. These gas bubbles are then carried to the pressure antinode by acoustic radiation pressure, where they coalesce into a single bubble. The ambient liquid temperature is measured by means of thermocouple gauges.

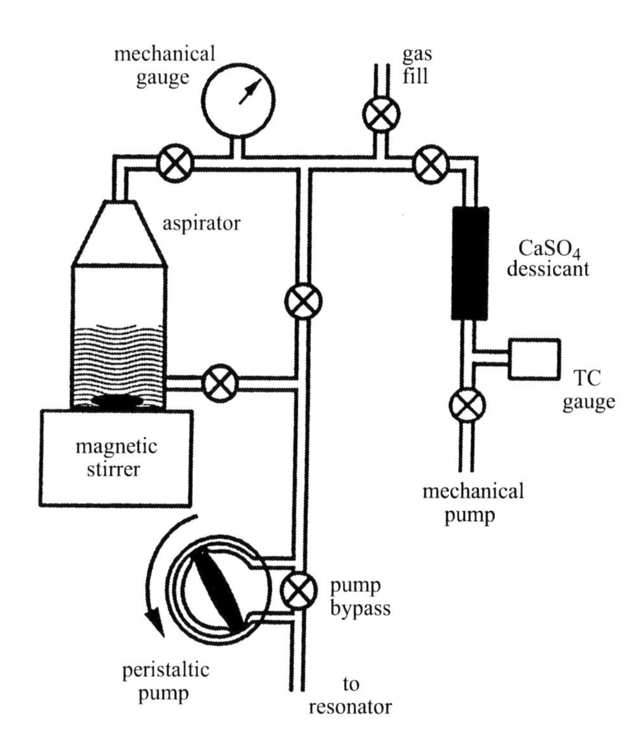


Figure 3.9 Gas manifold apparatus used for the purpose of degassing water or another liquid and then preparing it with a desired partial pressure of a particular gas mixture. (Reprinted from *Physics Reports* **281**, 65, Barber et al. 1997a, with permission from Elsevier Science.)

The fill lines of the resonator are connected to the gas manifold shown in Fig. 3.9 (Hiller (1995)). The designated liquid is degassed to tens of millitorr of partial pressure of the gas (plus the vapor pressure) by means of a mechanical pump. A magnetic stirrer accelerates the degassing turbulent voids which increase the area of the fluid/gas interface. The chosen gas mixture is prepared in a large reservoir, and admitted to the degassed liquid through the gas fill line at the chosen pressure. Stirring the liquid accelerates the establishment of equilibrium between the gas dissolved in the liquid and the gas pressure above it. Although equilibrium appears to be established in less than 5 min, stirring is usually carried on for about 20 min. The liquid finally pumped into the SL resonator has a partial pressure of gas dissolved in it which is accurate to about 1/2 torr.

The modal structure which can trap bubbles in the cylindrical cell in Fig. 3.8 is approximated by the pressure amplitudes

$$P_a = P'_a J_0(k_n r)\cos(2q\pi z/H)\cos(\omega_a t)$$

where $H$ is the height of the cylinder and

$$\omega_a^2 = c^2\left(k_n^2 + \frac{4q^2\pi^2}{H^2}\right)$$

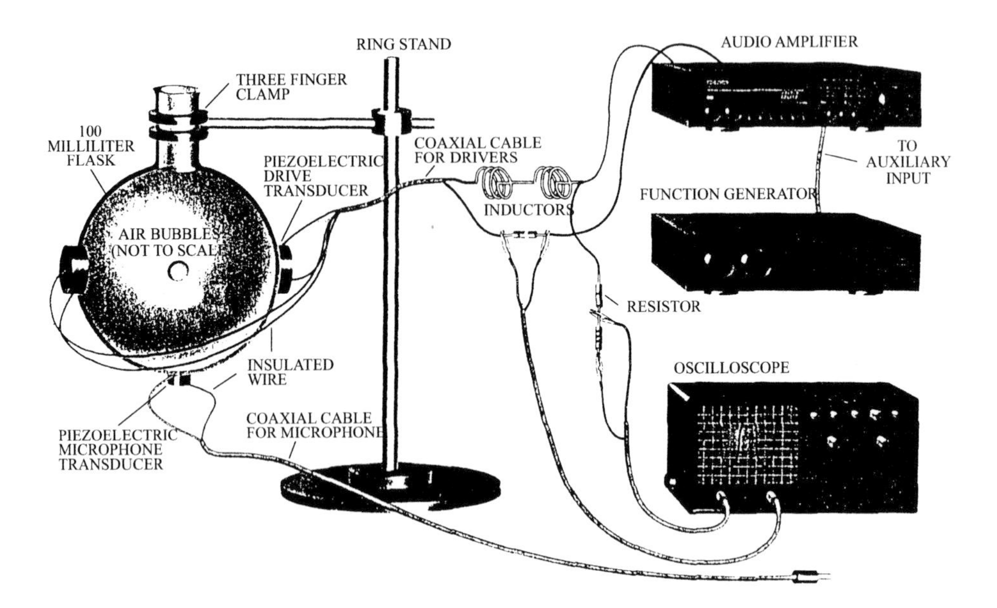


Figure 3.10 Sonoluminescence can be demonstrated with a few electronics parts and equipment typically found in a school laboratory. (Hiller and Barber (1995).)

and the surface of the cylinder is taken as a pressure node, so that $J_0(k_n R_f) = 0$, and the $z$ dependence has been chosen so that $z = 0$, $+H/2$, and $-H/2$, are all velocity nodes. This fixes $q$ to be an integer, as can be seen from the velocity potential, $\varphi$, which is related to the pressure by

$$P_a = -\rho \partial \varphi / \partial t$$

Also the tendency of an SL bubble to walk off-center as the drive level is increased varies from resonator to resonator. They say that the appearance and disappearance of the dipole component may be connected with the input acoustics. And slight variations in the acoustic properties of the resonator can be critical. Matula (1999) discusses the balance between the Bjerknes and buoyancy forces on a small bubble.

Barber et al. (1997a) describe their experiences with the acoustic resonator. Their best resonator has been custom made by GM from GE quartz. They have worked with four such spheres each of which has a sphericity accurate to 1%. Three of these spheres have produced the most stable reproducible SL that they have measured. The fourth looked identical in every way to the other three yet it produced a poor acoustic resonance and even for air in water the SL signal was highly unstable.

Chen et al. (2002) have produced single bubble sonoluminescence with rectangular and triangular sound waveforms as well as sinusoidal waveforms. The triangular wave was the most effective, while the rectangular was the least effective with the sinusoidal wave in the middle.

Hiller and Barber (1995) described a homemade set-up for SBSL. With an oscilloscope, a moderately precise sound generator, a home stereo amplifier and about $100, sound can be turned into light (Fig. 3.10).

Metcalf (1998) gives a similar but less detailed description of this set-up.

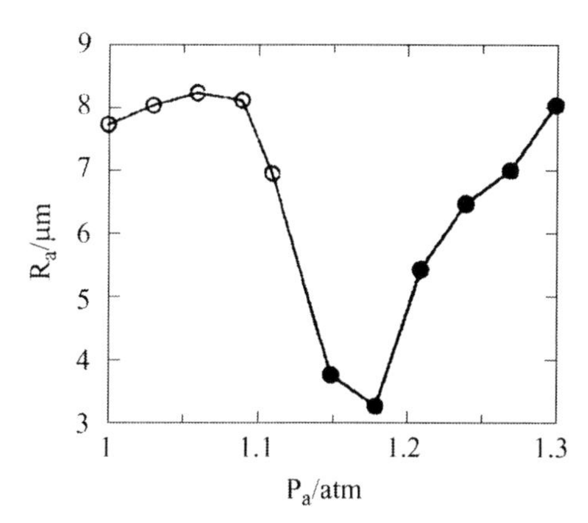


Figure 3.11 The ambient bubble radius as a function of forcing pressure $P_a$ for a gas mixture of 5% argon and 95% nitrogen at a pressure overhead of 150 mm. For sonoluminescing bubbles the symbols are filled; for nonglowing bubbles they are open. Note the abrupt decrease in bubble size right before the sonoluminescence threshold. The figure is a sketch from Fig. 38 of Barber et al. (1997). In that paper the ambient radius is obtained from a fit of the Rayleigh–Plesset equation to the $R(t)$ curve. In that fit heat losses are not considered explicitly, but material constants are considered as free parameters. Therefore the values for $R_0$ are only approximate. (Brenner et al. (2002).)

## 3.4 MEASUREMENT OF THE BUBBLE MOTION – MIE SCATTERING

Barber et al. (1994) showed that both the light intensity and the amplitude of the oscillations of the bubble depended sensitively not only on the forcing pressure amplitude, but also on the concentration of the gas dissolved in the liquid, the temperature of the liquid, or small amounts of surface active impurities (Weninger et al. (1995), Ashokkumar et al. (2000) and Toegel (2000a)). As an example Fig. 3.11 is Fig. 7 from Brenner et al. (2002) and shows the dependence of $R(t)$ and the total light intensity on the increasing drive level for an air bubble in water. As the forcing is increased, the bubble size abruptly decreases, and then the light turns on. For some years, the precise reasons for this sensitivity (observed repeatedly in experiments) were difficult to understand, mostly because varying one of the experimental parameters, such as the water temperature, would tend to change others as well.

Early experiments on measuring the bubble motion are reviewed by Crum and Cordry (1994).

Mie (1908) scattering has been extensively used for measuring the bubble radius $R(t)$, (van de Hulst (1957), Dave (1969), Kerker (1969), Marston (1979, 1991), Wiscombe (1980), Hansen (1985) (a very good introduction), Gaitan (1990), Barber and Putterman (1992), Holt and Crum (1992), Lentz et al. (1995), Weninger et al. (1997, 2000), Barber et al. (1997a,b), Jeon et al. (2000), Pecha and Gompf (2000) with Comment by Weninger et al. (2001).

A laser beam falls on the bubble. The light scattered is proportional to $R^2(t)$. Figure 3.12 shows the apparatus from Matula (1999). Light from a 10 or 30 mW HeNe laser is scattered by the bubble through a large angle. A large lens focuses the scattered light on to a photomultiplier tube (PMT). The dynamic range of the PMT is not sufficient to observe

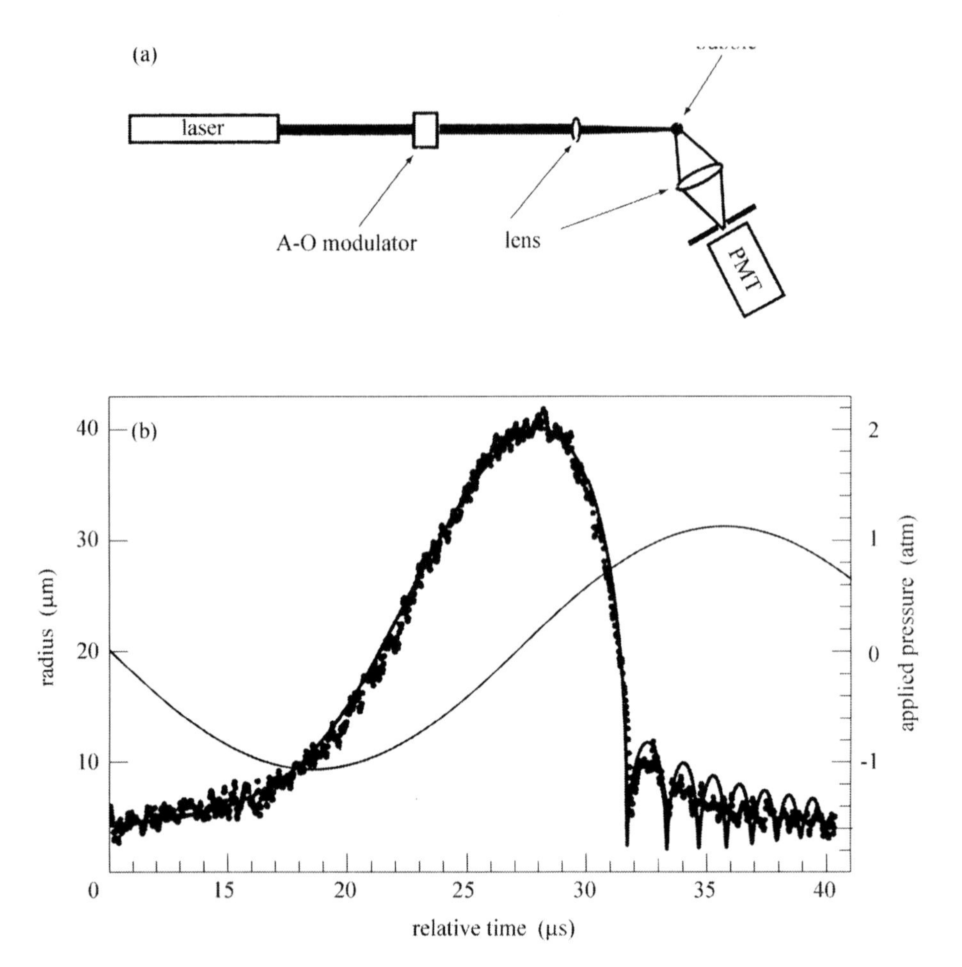


Figure 3.12 (a) Light-scattering technique for measuring the radial oscillations of a sonoluminescence bubble. (b) Instantaneous scattered intensity collected from a pulsating bubble. In the geometrical optics limit, the scattered intensity is proportional to the square of the bubble radius. The normalized drive pressure is shown as well. The data (non-averaged) fit nicely with the Keller–Miksis nonlinear bubble-dynamics equation (Keller and Miksis (1980)). Reproduced with permission from Matula (1999) *Phil Trans Roy Soc Lond A* **357**, 225.

in high quality both the maximum size and rebounds, or afterbounces. If both regions are to be studied, an acoustic-optic modulator (AOM) can be used. Also, the finite band width of the PMT will limit the resolution of the collapse so that probing the bubble near its minimum radius requires a fast photodetector, such as a streak camera (see for example Moran et al. (1995) and Pecha et al. (1998)). Figure 3.13 is from the latter. (a) shows the result of 762 streak images integrated by jitter correction. (b) shows the pulse shape is asymmetric. Gompf et al. (1997), Pecha et al. (1998) and Lohse and Hilgenfeldt (1999) record the remarkable result that the light pulse in the UV spectral regime is as long as in the red spectral regime. The Signal PMT from Fig. 3.12(a) records a signal $V(t)$ proportional to $R^2(t)$ plus the background level $\tilde{V}(t)$. The noise level is measured by sending the laser light through the flask without the bubble. The radius is then $\sqrt{V(t)-\tilde{V}(t)}$, as plotted in Fig. 3.12(b).

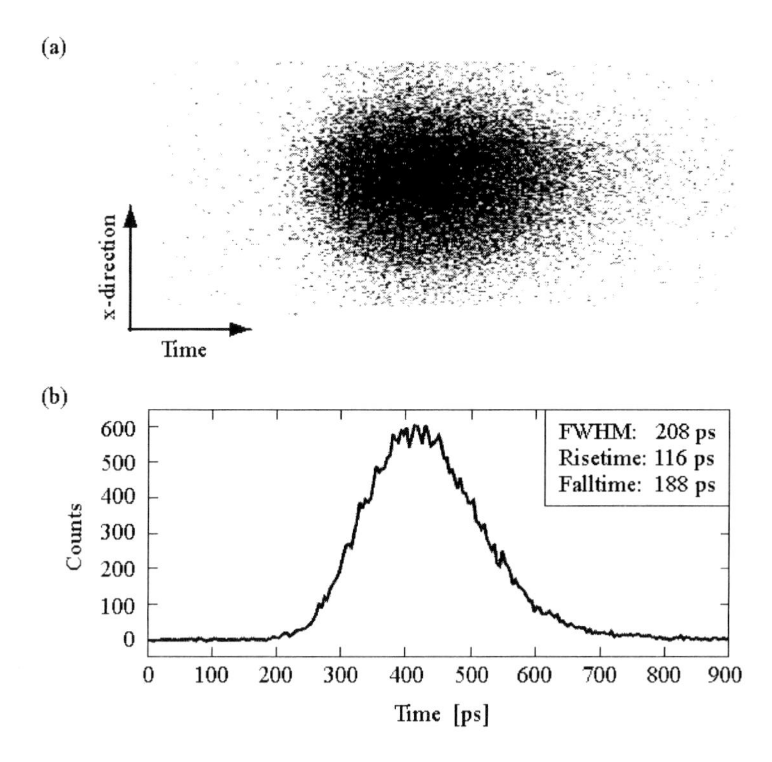


Figure 3.13 (a) 762 single SBSL-pulse streak images integrated by jitter correction (water temperature, 8°C; $O_2$ concentration, 2.1 mg/l (190 torr air); amplitude of the driving pressure $P_a$ (1.3 bar). (b) Horizontal profile of the SBSL pulse. The pulse shape is asymmetric with a rise time (10%–90%) of 116 ps and a fall time (90%–10%) of 188 ps. (Pecha et al. (1998).)

This clearly shows the slow expansion of the bubble to the rarefied sound field, the runaway collapse of the bubble and the well defined rebounds. The bubble then sits dead in the water waiting for the next sound refraction when it expands again as shown in Fig. 3.14 from Gaitan et al. (1992). Figure 3.15 shows the radius of a bubble at the collapse. Barber et al. (1997a) point out that the bubble is collapsing inward at Mach 4, relative to the ambient speed of sound in the gas, and that the acceleration which brings the bubble to a halt at its minimum radius exceeds $10^{11}$ g! At this moment of great stress and energy focusing the flash of light is emitted; and we go beyond the range of Navier–Stokes hydrodynamics (Barber et al. (1997b)).

Brenner et al. (2002) state that to fit the theoretical Rayleigh–Plesset or Keller and Miksis (1980) equation to the experimental curves is difficult. Neither the ambient bubble radius $R_0$ nor the driving pressure $P_A$ are known *a priori*. $R_0$ can change through diffusion of gas as well as evaporation or condensation of water vapor, and the driving pressure $P_A$ is very sensitive to variations of the flask geometry such as a small hydrophone attempting to measure $P_A$. Also, the precision of such a hydrophone is limited to about 0.05 bar.

If we consider Matula's results in Fig. 3.12(b), the dots are the measured $R(t)$ values. The solid curve is a solution to the Keller–Miksis equation, under the assumption of isothermal heating $\gamma = 1$. Superimposed as a thin line is the forcing pressure.

A difficulty is that if one adjusts $R_0$ and $P_A$ such that the bubble's maximum is well fitted, the afterbounces are always overestimated as in Fig. 3.12(b). Better fits can be

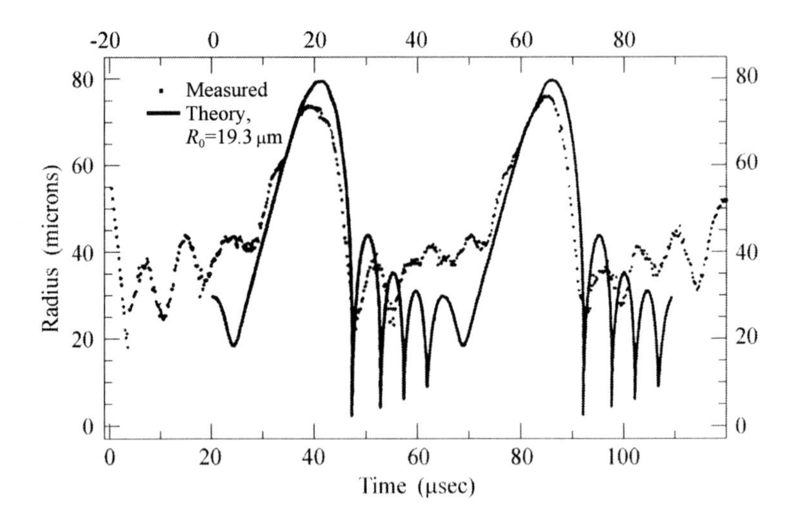


Figure 3.14 Theoretical (solid) and experimental (dotted) radius–time curve obtained with the light-scattering apparatus in 21% glycerine at $P_A = 1.22$ atm and $f = 22.3$ kHz. (Gaitan et al. (1992).)

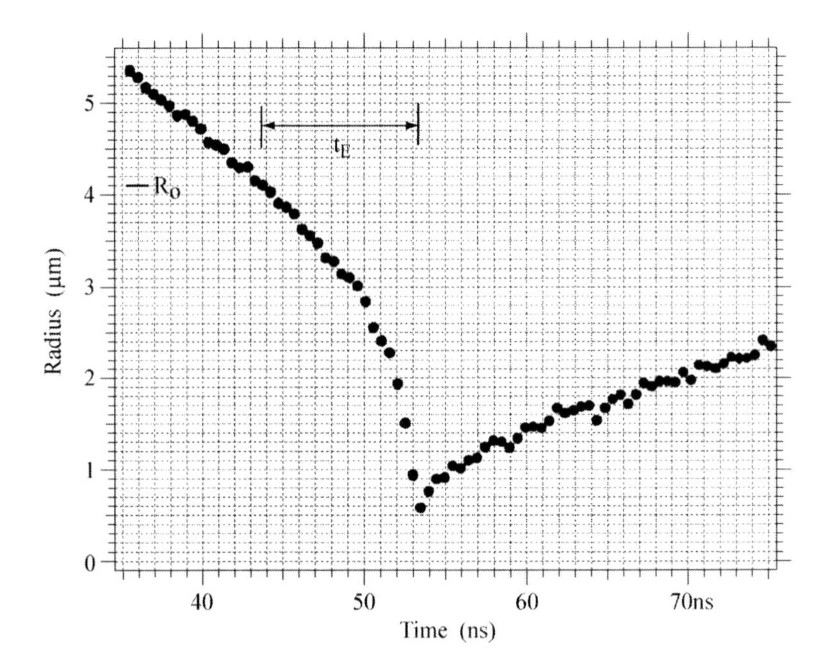


Figure 3.15 Radius of a sonoluminescing bubble (1% xenon in oxygen at 150 mm) as the moment of collapse is approached. The bubble is collapsing over 4 times the ambient speed of sound in the gas. For this bubble $R_0 \approx 4.1$ μm; $P'_a \approx 1.45$ atm. (Reprinted from *Physics Reports* **281**, 65, Barber et al. 1997a, with permission from Elsevier Science.)

achieved by adjusting the material constants. For example, Barber and Putterman (1992) used 0.07 gm s cm$^{-1}$ for the kinematic viscosity of water instead of the usual value of 0.01 gm s cm$^{-1}$ to achieve a fit to the afterbounces.

Prosperetti and Hao (1999) explain this larger viscosity as an effective parametrization of other damping mechanisms not accounted for in simple Rayleigh–Plesset models, for example, thermal losses as in the model of Prosperetti (1991).

Another parameter which must be considered when fitting the experimental $R(t)$ curves to RP models is the invasion of water vapor at bubble maximum. This leads to an ambient size of the bubble which varies over the bubble cycle, being larger at maximum than at collapse and during the afterbounces. As many early fits of $R(t)$ curves (summarized by Barber et al. (1997a)) did not consider these effects, the resulting values for $R_0$ and $P_A$ can only be considered as approximate.

Mie scattering data near the collapse are difficult to interpret because the bubble radius $R$ becomes of the order of the wavelength of light. The simple proportionality of Mie intensity and $R^2$, valid for larger $R$, gets lost and the relation even becomes non-monotonic (Gompf and Pecha (2000)). Moreover, at collapse the light is not only reflected at the bubble wall, but also at the shock wave emitted from the bubble at collapse.

Arakeri and Giri (2001) measured SL from a nonaqueous alkali-metal salt solution at low acoustic drive levels with the ratio of the acoustic pressure amplitude to the ambient pressure being about 1. For krypton-saturated sodium salt ethylene glycol solution they observed an optical pulse width varying from 10 ns to 165 ns with the most probable value being 82 ns. With argon, the variation is similar to that of krypton but the most probable value is reduced to 62 ns. The range is much less for helium, being from 22 ns to 65 ns, with the most probable value reduced to 42 ns.

Hayashi et al. (2001) examined a possible cause of the background level of the Mie scattering by a sonoluminescing single bubble. They considered *double scattering* where the light initially scattered by the bubble is subsequently scattered by dust particles or the wall. They calculate the amount of double scattering to be ~1 to 10%. They suggest ways of reducing double scattering.

Verraes et al. (2000) used sulphur particles produced *in situ* by a bubble itself, fuchsin spots and dust to observe the water flow near a single sonoluminescence bubble.

An alternative method of measuring $R$ and $\dot{R}$ at certain points in time is by interfering scattered and unscattered light (Delgadino and Bonetto (1997)).

Tian et al. (1996) devised a direct imaging system to study the oscillations of single bubbles acoustically levitated in water. The bubble was illuminated with a light emitting diode which was strobed at a frequency different from the driving sound field frequency, Fig. 3.16. The technique slowed the moving image of the bubble, allowing one to observe the shape of a bubble oscillating between 5 to 100 μm in diameter. The bubble shapes were recorded on video tape and then analyzed by an automatic image analysis system which gave the variations of bubble diameter with time. Experiments were performed with and without the bubbles *sonoluminescing*. Asymmetric shapes of the bubbles were observed in some cases.

Kozuka et al. (2000) used the single flash of a stroboscope, with a flash time much shorter than the acoustic cycle, and a charge coupled device (CCD) camera, to capture the instantaneous image of a bubble including its dancing condition (see §2.3). Changing the flash timing of the stroboscope slowly made it possible to observe the periodic expansion and collapse of the bubble. Kozuka et al. (2002) improved the method by adding a photomultiplier tube using a beam splitter to keep everything on the same optical axis, and using a laser, as in Fig. 3.17. This method enabled detailed observation of the bubble shape and precise measurement of the bubble radius. Figure 3.18 is an example of their excellent results.

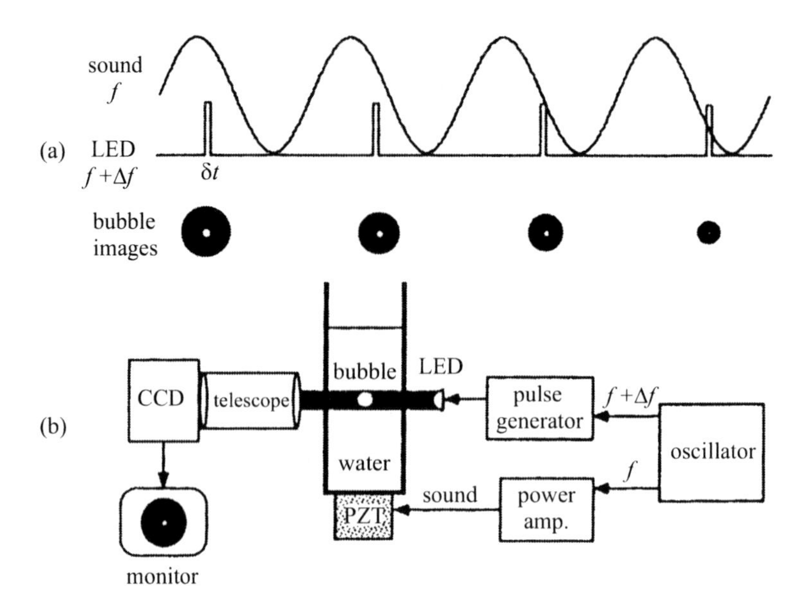


Figure 3.16 Diagram of (a) the experimental principle and (b) the experimental apparatus. (Tian et al. (1996).)

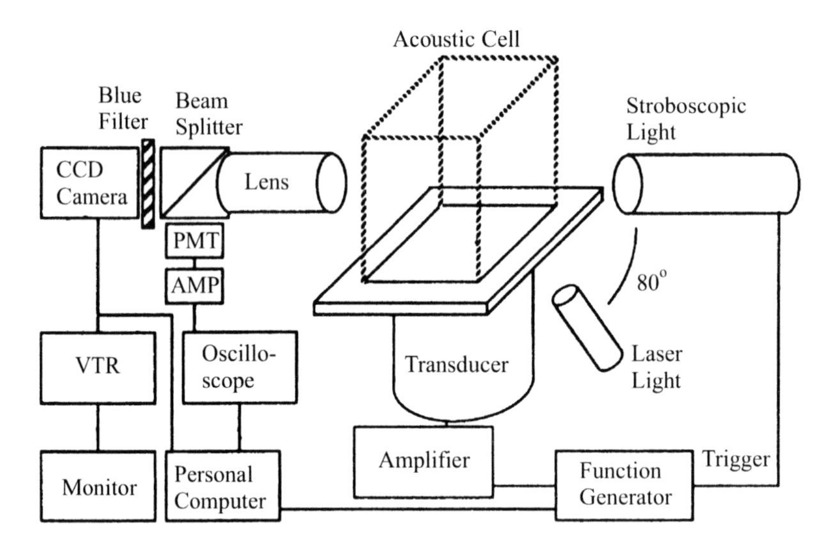


Figure 3.17 Experimental apparatus. (Kozuka et al. (2002).)

Kaji et al. (2002) point out that the method of Kozuka et al. (2000) takes a few seconds to capture enough images to reproduce an entire cycle of bubble oscillations due to the slow recording rate of the CCD camera. Kaji et al. (2002) improved the method by means of a high-speed video camera and a continuous light source, Fig. 3.19. Because a high speed video camera can capture enough images in the millisecond range, enough detailed observation to detect minute variations in the rebounding phase is realized. Radius–time curves obtained from captured images are fitted to the theoretical results calculated from the

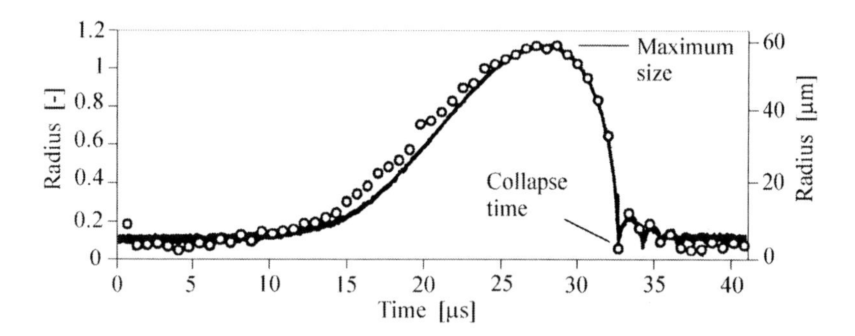


Figure 3.18 The full curve shows the bubble radius determined from the captured images on the CCD camera. The circles are the radii from laser light scattering. (Kozuka et al. (2002).)

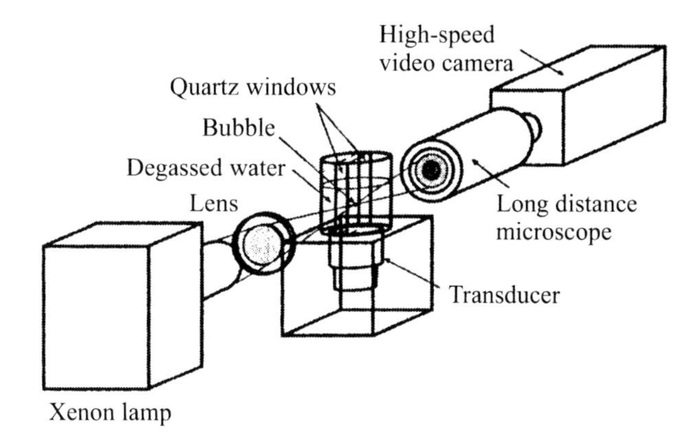


Figure 3.19 Diagram of the imaging system. (Kaji et al. (2002).)

equation of Keller and Miksis (1980). Figure 3.20 shows these results for degassed water for a driving frequency of 27 kHz, sound pressure 1.4 atm and ambient radius 7.0 μm. They considered that the major cause of the disagreement in the rebounding phase was because although Keller and Miksis's equation takes into account the compressibility of the liquid, it does not allow for thermal conduction and gas diffusion. Also, the vapor pressure of the water in the bubble was regarded as constant.

Hayashi (1999) suggests, from theoretical considerations, that for Mie-scattering measurements, detectors with a wide acceptance area are advantageous, provided that the scattering angle is greater than 90° where the square root of the intensity is proportional to the bubble radius.

Recently, Simon and Levinsen (2003a) have shown how the acoustic pressure amplitude, ambient radius and radius–time curve of a single sonoluminescing bubble can be easily measured by a digital oscilloscope, avoiding the cost of the expensive lasers or ultrafast cameras of previous methods.

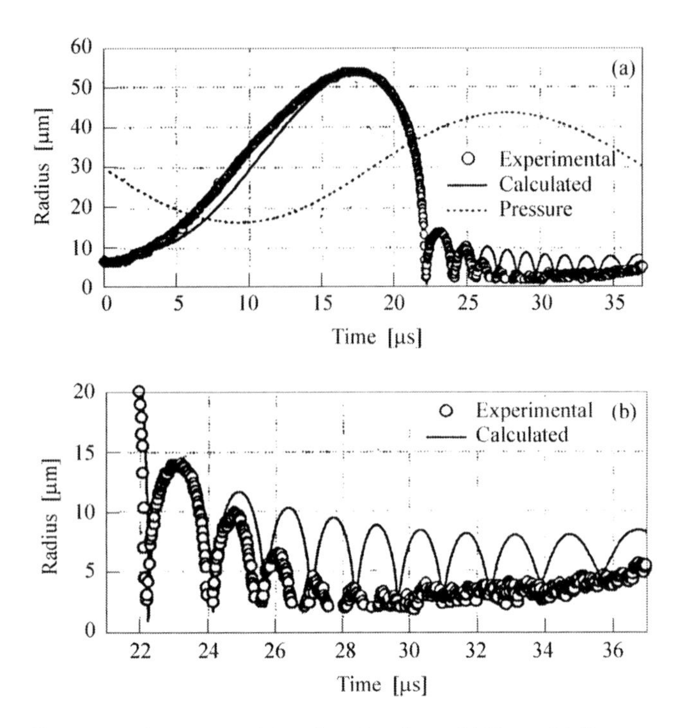


Figure 3.20 Radius versus time curve with an imaging time of 133 ms; (*a*) for an acoustic cycle (37 ms) (*b*) at the rebounding phase. (Kaji et al. (2002.))

## 3.5 BUBBLE DYNAMICS OF SINGLE BUBBLE SONOLUMINESCENCE

A lot of this section is taken from the recent excellent review by Brenner et al. (2002).

### *3.5.1 Introduction*

They start off by emphasizing that the very existence of a single sonoluminescence (SL) bubble depends critically on a subtle balance of hydrodynamic and acoustic forces. During SL, a diverse array of physical effects influence this balance such as the pressure that oscillates from low enough that the liquid air-interface vaporizes, to extremely high values producing very high temperatures so that the gas inside the bubble emits light. Gas is continually exchanged between the bubble and the surrounding liquid, causing the number of molecules in the bubble to vary. A scholarly background to this section is entitled "Old Fashioned Bubble Dynamics" by Prosperetti (1999).

### *3.5.2 Rayleigh–Plesset equation*

The Rayleigh–Plesset (RP) equation was derived in Chapter 1 as Eq. 1.14 and is stated here again for reference

$$R\ddot{R} + \frac{3}{2}\dot{R}^2 = \frac{1}{\rho}\left(P_g - P_0 - P(t) - 4\eta\frac{\dot{R}}{R} - \frac{2\sigma}{R}\right)$$

where $P_g$ = pressure of gas
$P_0$ = background static pressure = 1 bar
$P(t)$ = sinusoidal driving pressure = $-P_A \sin \omega t$
$\eta$ = shear viscosity of liquid
$\sigma$ = surface tension of liquid

A historical review of the development of this equation has been given by Plesset and Prosperetti (1977).

To close the equation, it is necessary to calculate the pressure in the gas. Roughly speaking, when the bubble wall moves slowly with respect to the sound velocity in the gas, the pressure in the gas is uniform throughout the bubble. In this regime, how strongly the pressure depends on the bubble volume is governed by the heat transfer across the bubble wall (Prosperetti et al. (1988)). The pressure–volume relation is given by

$$R_g(t) = \left(P_0 + \frac{2\sigma}{R_0}\right)\frac{(R_0^3 - h^3)^\gamma}{(R(t)^3 - h^3)^\gamma} \tag{3.1}$$

where $R_0$ is the ambient radius of the bubble (i.e. the radius at which an unforced bubble would be in equilibrium, and $h = R_0/8.86$ is the van der Waals hard core radius for argon given by the collective excluded volume of all gas particles inside the bubble. $\gamma$ is the polytropic index (see Prosperetti and Hao (1999)).

If the heat transfer is fast (relative to the time scale of the bubble motion), then the gas in the bubble is maintained at the temperature of the liquid, and the pressure is determined by an isothermal equation of state with $\gamma = 1$. On the other hand, if the bubble wall moves very quickly relative to the time scale of heat transfer, then the heat will not be able to escape from the bubble, and the bubble will heat (cool) adiabatically on collapse (expansion). For a monatomic rare gas, this means that $\gamma$ = 5/3. The dimensionless parameter that distinguishes between these two regimes is the Péclet number

$$P_e = \frac{|\dot{R}|R}{\chi}, \tag{3.2}$$

where $\chi$ is the thermal diffusivity of the gas.

This idea about heat transfer is based on a more careful version of this argument by Kamath et al. (1993) and Prosperetti et al. (1988).

If the rate of heat transfer is intermediate between adiabatic and isothermal, the situation is more complicated. Over the years several methods have been proposed that amount to varying $\gamma$ continuously between the isothermal value and the adiabatic value (Kamath et al. (1993), Plesset and Prosperetti (1977), Prosperetti et al. (1988)) depending on the Péclet number. This approach can yield quantitatively incorrect results as shown by Prosperetti and Hao (1999), in large part because energy dissipation from thermal processes is neglected.

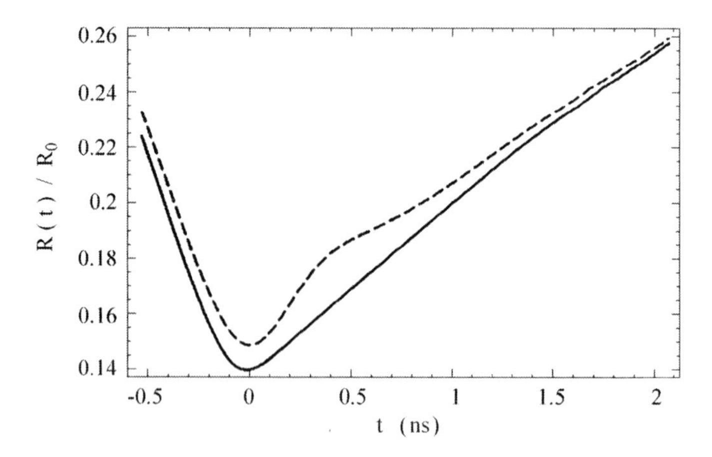


Figure 3.21 Comparison of bubble dynamics with a full simulation of the equations for liquid and gas motion (dashed) and with the Rayleigh–Plesset equation with uniform pressure (solid). The parameters were $P_A = 1.4$ atm, $R_0 = 4$ μm and $f = 26.5$ kHz. The RP equation is not able to reveal the bump after the turning round. (Lin, Storey and Szeri (2002a) © Cambridge University Press.)

### *3.5.3 Extensions to the Rayleigh–Plesset equation*

Brenner et al. (2001) discuss, interestingly, various other forms of the RP equation. If we take into account the back reaction of the sound radiated by the bubble on its own dynamics, this leads to extra terms in the RP equation. The most complete and elegant derivation of this effect is by Prosperetti and Lezzi (1986), Lezzi and Prosperetti (1987), and Prosperetti and Hao (1999).

When $\dot{R}/c \sim 1$, where $c$ is the velocity of sound in the liquid, sound radiation is important and must be taken into account. A standard way of doing this was invented by Keller and Kolodner (1956), Keller and Miksis (1980), and leads to the Keller equation (Brenner et al. (1995) and Comment on Brenner et al. (1995) by Putterman and Roberts (1998), Prosperetti and Lezzi (1986)).

Other forms of the RP equation are by Herring (1941), Kirkwood and Bethe (1942), Trilling (1952), Gilmore (1952), Flynn (1975a,b), Lastman and Wentzell (1981, 1982).

A useful form of the RP equation in the context of SL is by Barber et al. (1997a), Löfsted et al. (1995), but also see the later criticism of this equation by Putterman et al. (2001).

$$R\ddot{R} + \frac{3}{2}\dot{R}^2 = \frac{1}{\rho}\left[(P_g - P_0 - P(t)) - 4\eta\frac{\dot{R}}{R} - \frac{2\sigma}{R} + \frac{R}{c}\frac{\mathrm{d}}{\mathrm{d}t}(p_g)\right] \tag{3.3}$$

When comparing theoretical predictions based on the RP equation to SL experiments, a principal source of modelling error lies in the fact that the approximations leading to the RP equation breakdown at the bubble collapse. The speed of sound may not be constant, an effective polytropic exponent may be used in Eq. (3.1) and the Péclet number in Eq. (3.2) has been used by Hilgenfeldt et al. (1999). Heat transfer has been carried out numerically by Vuong and Szeri (1996).

In view of all these difficulties, it is surprising that solutions to the RP type equations still seem accurate for the mechanics of an SL bubble and for many of its accompanying effects. Recently Lin et al. (2002a) achieved a better understanding why RP type equations are so successful. Their examination reveals that these solutions are quite accurate even in the case of significant inertially driven spatial inhomogeneities in the pressure field, and even when wave-like motion of the gas inside the bubble is present. This extends the utility of the RP equation into the regime where the Mach number for the gas $M_g = \dot{R}/c_g$ ($c_g$ being the speed of sound in the gas) is no longer small. It is shown that the relevant condition is not $|M_g| < 1$, but $|\epsilon_p| < 1$, where

$$\epsilon_p \equiv \frac{R\ddot{R}\rho_{gas}}{\Gamma p(r=0,t)},$$ where $\Gamma$ is the polytropic exponent,

i.e. what is relevant is the bubble wall *acceleration*. So even in the SL regime, Lin et al. (2002a) found good agreement when comparing their full gas dynamical partial differential equation (PDE) simulations with the solutions to the Rayleigh–Plesset ordinary differential equation (ODE) with the assumption of a uniform pressure inside, as shown in Fig. 3.21.

### 3.5.4 *The bubble's response to weak and strong driving*

First consider small oscillations about an ambient radius $R_0$. A straightforward calculation (Young (1999)) shows that such a bubble oscillates at the resonant frequency

$$2\pi f_0 = \sqrt{\frac{1}{\rho R_0^2}\left(3\gamma P_0 + (3\gamma - 1)\frac{2\sigma}{R_0}\right)}$$

A typical bubble in the SL regime has $\ddot{R}_0 \sim 5$ μm, corresponding to a resonant frequency of $f_0 \sim 0.5$ MHz, much faster than driving frequency $f \sim 20$ kHz.

Figure 3.22 from Brenner et al. (2002) shows solutions to the modified RP Eq. (3.3) at forcing pressures $P_A/P_0 = 1.0, 1.1, 1.2, 1.3$. At low forcing the bubble undergoes almost sinusoidal oscillations of relatively small amplitude, with a period equal to that of the external forcing frequency $f$. Here, the oscillations are essentially "quasistatic", because the resonant frequency is so much larger than $f$. The oscillatory pressure forcing is balanced by the gas pressure (Hilgenfeldt et al. (1998a), Löfstedt et al. (1993)), with inertia, surface tension and viscosity playing a negligible role. At a critical pressure amplitude around $P_A \sim P_0$, such quasistatic oscillations are no longer possible, resulting in a nonlinear response of the bubble. The critical amplitude depends slightly on $R_0$, and is called the (dynamical) Blake threshold (Blake (1949), Hilgenfeldt et al. (1998a)).

In the SBSL regime, the solution to Eq. (3.3) can be divided into three stages:

1. *Expansion*. During the negative half-cycle of the driving $-P\sin\omega t$, the applied tension makes the bubble expand. Since $f \ll f_0$, the expansion continues until the applied pressure becomes positive. The timescale of this regime is thus set by the period of the driving pressure wave and is typically ~20 μs for SL experiments. This is sufficient to increase the bubble radius by as much as a factor of 10.

2. *Collapse*. When the driving pressure changes sign, the expanded bubble is "released" and collapses inertially over a very short time (~1 ns for SBSL bubbles).

The solution during collapse is well described by the classical solution of Lord

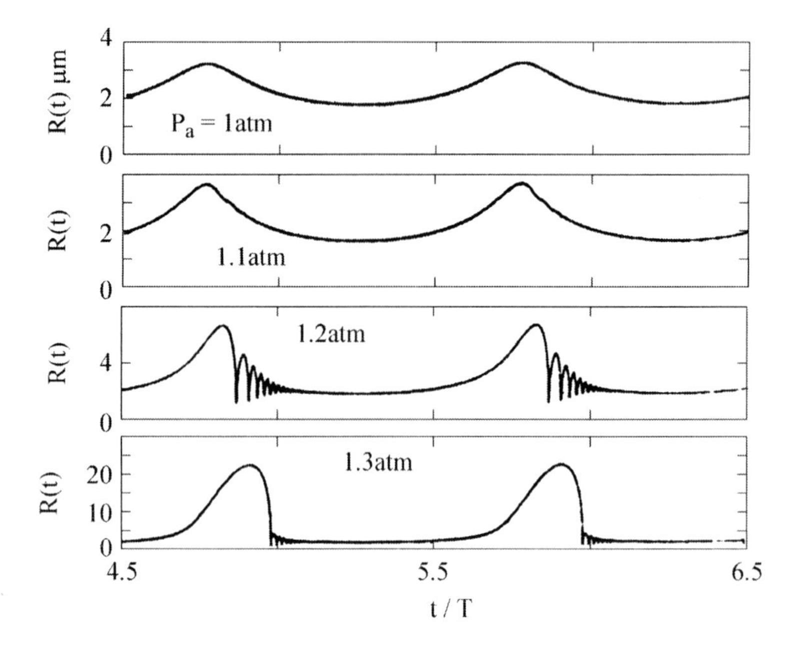


Figure 3.22 Solutions to the modified Rayleigh–Plesset equation at forcing pressures $P_A/P_0 = 1.0$, 1.1, 1.2, 1.3. The ambient bubble radius is $R_0 = 2$ μm, the frequency $f = 1/T = 26.5$ kHz. (Brenner et al. (2002).)

Rayleigh, to which we turn in some detail below. SBSL light emission occurs at the end of the collapse.

3. *Afterbounces.* After the collapse, the bubble spends the remaining half of the cycle oscillating about its ambient radius at roughly its resonant frequency $f_0 \gg f$, giving rise to characteristic afterbounces.

It is worthwhile at this point to comment on the roles of surface tension and viscosity in setting the properties of the solution. The surface tension term is dynamically important when it is as large as the external forcing pressure, implying that $\sigma/R \sim P_A$. This occurs when the bubble radius is below

$$R < R_{st} = \frac{\sigma}{P_A}$$

For water, this corresponds to a radius of ~0.7 $\mu m/(P_A/\text{bar})$. We will see below that this length scale plays an important role in determining the stability of the solutions to the RP equation with respect to both dissolution and breakup.

Viscous effects are important when the viscous damping timescale is of the order of the timescale of the bubble motion, roughly $(\eta/\rho)/R_0^2 \sim f_0$. For water, this timescale condition is never satisfied, though for more viscous fields it can be important.

Moss et al. (2000), in an elegant paper, consider the theoretical predictions of the afterbounces. The Rayleigh–Plesset Equation (RPE) for strongly driven bubbles, such as sonoluminescing bubbles, produces large amplitude rebounds that often last until the next acoustic cycle of the periodic driving pressure. This is in contrast to the experimental data,

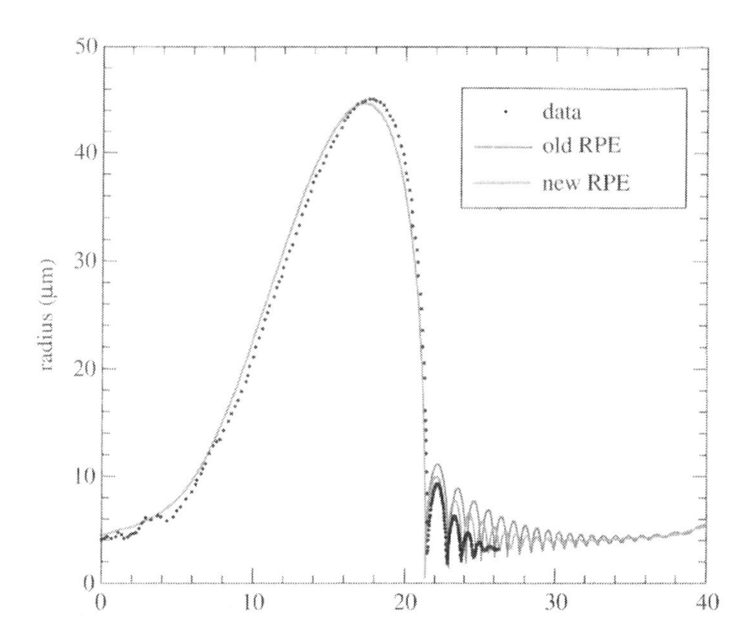


Figure 3.23 Time dependence of the radius of a sonoluminescing air bubble. The data is from Löfstedt et al. (1993) (black dots). Reproduced with permission from Moss et al. (2000) *Proc Roy Soc Lond A* **456**, 2983. Credit is given to the University of California, Lawrence Livermore National Laboratory and the Department of Energy under whose auspices the work was performed. (SEE ALSO COLOR PLATES)

for example from Löfstedt et al. (1993) Fig. 3.3, which shows rapidly damped rebounds. This is shown by the black dots and black line in Fig. 3.23 for an air bubble of equilibrium radius $R_0 = 4.5$ µm, driving pressure $P_A = 1.35$ bar and driving frequency = 26.5 kHz. The traditional RPE assumes that the damping of the rebounds is caused by the compressibility of the *liquid*. This is shown by the blue line labeled old RPE in Fig. 3.23. Moss et al. (2000) show that appreciable damping also comes from the most compressible part of the system, which is the *gas*. This damping arises from the inhomogeneous pressure field within a rapidly collapsing or expanding bubble. Taking this into account generates a new modification of the RPE that better reproduces the extreme damping that is observed experimentally. This is shown by the red line in Fig. 3.23. Figure 3.24 is an expanded view of the rebounds in Fig. 3.23.

### 3.5.5 *The Rayleigh collapse*

Now we turn to the behavior of the bubble radius near the collapse. As we have emphasized above, this is the regime where we have reason to doubt the RP description's accuracy, since the velocity of the bubble wall will grow to be of the order of the sound velocity in the liquid. The approach to the collapsed state, however, can be captured very well by the equation, and is given by the classical solution of Lord Rayleigh.

Lord Rayleigh (1917) imagined a bubble with very large inertia, so that the gas pressure, surface tension and viscosity are all negligible – in other words the collapse of a void. The equation for the wall motion of the bubble/void is then $R\ddot{R} + 3/2\dot{R}^2 = 0$, which

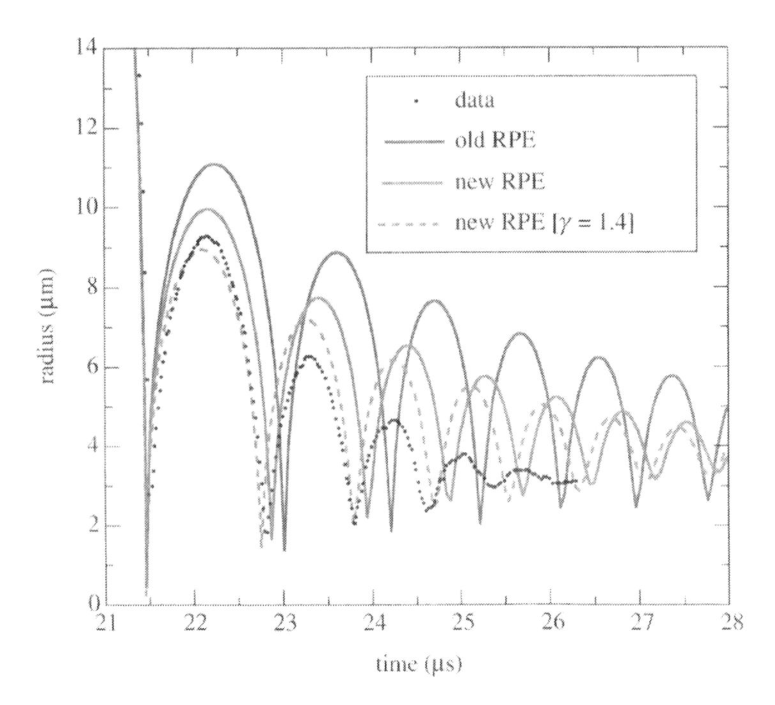


Figure 3.24 Expanded view of the rebounds in Fig. 3.23. Reproduced with permission from Moss et al. (2000) *Proc Roy Soc Lond A* **456**, 2983. Credit is given to the University of California, Lawrence Livermore National Laboratory and the Department of Energy under whose auspices the work was performed. (SEE ALSO COLOR PLATES)

can be directly integrated. The solution is of the form $R(t) = R_0((t_* - t)/t_*)^{\alpha}$ with $\alpha = 2/5$ with the remarkable feature of a divergent bubble wall velocity as $t$ approaches the time $t_*$ of total collapse. Lord Rayleigh pointed out that this singularity is responsible for cavitation damage. It is also the central hydrodynamic feature responsible for the rapid and strong energy focusing that leads to sonoluminescence.

Clearly, something must stop the velocity from diverging. The collapse rate is eventually so fast that the heat does not have time to escape from the bubble. The pressure in the gas obeys the adiabatic equation of state, which diverges like $P_g \sim R^{-\Gamma} \propto (t_* - t)^{-2}$ (for a monatomic ideal gas) which is stronger than the inertial acceleration. This effect is therefore capable of stopping the collapse ($\Gamma$ is the polytropic exponent, remember).

When the bubble is strongly compressed, there is another important modification to the equation of state, the hard core van der Waals forces, see Eq. (3.1). At sufficiently high gas densities there are strong deviations from the ideal gas law. Although this model is crude, it captures the salient effect, and stops the Rayleigh collapse.

Although the gas pressure can halt the Rayleigh collapse, it turns out that the most important effect in halting the collapse involves sound radiation in the liquid during the last stages of collapse; its pressure contribution (the last term in Eq. (3.3)) grows as $(t_* - t)^{-13/5}$ (page 196 in Hilgenfeldt et al. (1998a)), and overwhelms the other terms. This hints at the prime role of sound emission for energy loss during bubble collapse. Up to 50% of the kinetic energy in the collapse may end up as a radiated pressure wave (Gompf and Pecha (2000)).

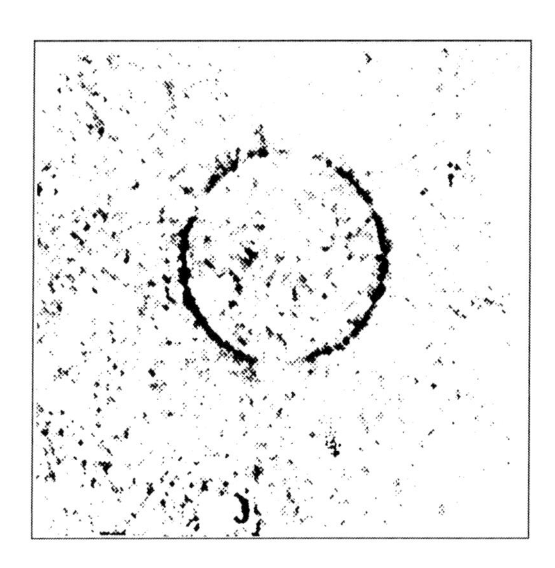

Figure 3.25 Image of shock wave from a sonoluminescing bubble 480 ns after collapse. (Holzfuss et al. (1998a).)

### *3.5.6 Sound emission and shock waves from a bubble*

The last term in Eq. (3.3) represents damping due to radiation of sound into the liquid by the bubble's motion. This radiation damping is dealt with by Keller and Miksis (1980) and Prosperetti (1984).

Shock wave emission from a sonoluminescing bubble was first detected by Holzfuss et al. (1998a). They produced single bubble sonoluminescence in water at a driving frequency of 23.5 kHz and a driving amplitude of 1.2 to 1.5 bars. Together with the emitted light pulse, a shock wave was generated in the water at the collapse time (Fig. 3.25). The time-dependent velocity of the outward-traveling shock front was measured by an imaging technique. Near to the bubble center (6–73 μm away), the velocity of the shock was measured to be about 2,000 m/s. Further away, up to 1200 μm from the bubble center, the velocity of the shock front drops to about 1500 m/s. Thus the shock front has a faster speed than the ambient sound velocity in water at 20°C of 1400 m/s. The pressure in the shock and in the bubble was calculated from its velocity, by a Rankine–Hugeniot relation and the Tait equation for water. The shock pressure came to 5500 bars. Thus we can see that this enormous pressure inside the bubble produced a shock front with a faster speed than the ambient sound velocity.

Figure 3.26 (Lauterborn et al. (1999)) shows the shock wave from a sonoluminescing bubble expanding as a perfect ring. The bubble expands at a much slower rate than the speed of the shock wave and can be seen as a tiny black spot in the middle of the ring of the shock wave.

More work was done by Matula et al. (1998), Wang et al. (1999), Pecha and Gompf (2000), Gompf and Pecha (2000), and Weninger et al. (2000a,b), as described below.

Matula et al. (1998) used a piezoelectric hydrophone to measure a pressure pulse with a fast rise time (5.2 ns) and high amplitude (1.7 bar) at a transducer 1 mm distance from the bubble.

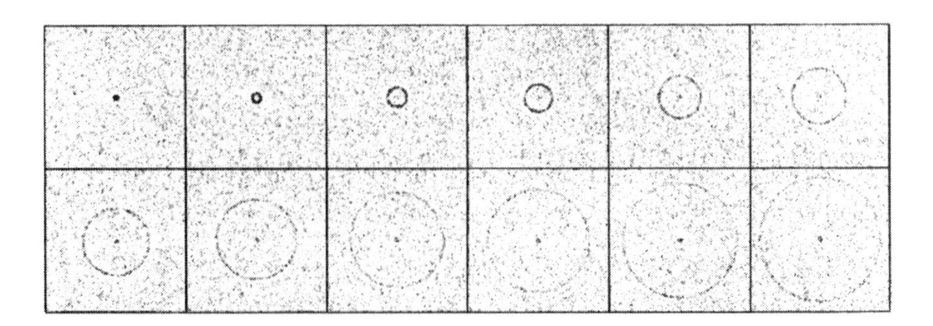

Figure 3.26 Shockwave from a trapped sonoluminescing bubble driven at 21.4 kHz and at a pressure amplitude of 132 kPa. The interframe time is 30 ns. Picture size is 160 μm × 160 μm. (Courtesy of R. Geisler.) (Lauterborn et al. (1999).)

Wang et al. (1999) carried out a systematic study of strength and duration of the pressure pulses as a function of gas concentration, driving pressure and liquid temperature. They demonstrated that a probe 2.5 mm from the bubble, recorded pressure pulses with rise times varying from 5 to 30 ns as the driving pressure and dissolved gas concentration varies. The amplitude of the pressure pulse varied between 1 and 3 bar.

Another study was carried out by Pecha and Gompf (2000) and Gompf and Pecha (2000). They measured pressure amplitudes and rise times consistent with the other measurements, and were able to measure the pressure pulse much closer (within 50 μm) to the bubble. In addition, using a shadowgraph, they were able to visualize the shock wave leaving the bubble, see Fig. 3.27. Pecha and Gompf (2000) found that the shock velocity in the immediate vicinity of the bubble was as fast as 4,000 m/s, much faster than the speed of sound $c = 1{,}430$ m/s under normal conditions, but in good agreement with the calculated value of about 4,000 m/s by Holzfuss et al. (1998a) using the Gilmore model and the Kirkwood–Bethe hypothesis. This high speed shock originates from the strong compression of the fluid around the bubble at collapse. From the nonlinear propagation, the pressure in the vicinity of the bubble can be estimated to be in the range of 40–60 kbar.

It should be emphasized that for large enough $P_A$, the presence of such shocks in the liquid results from the RP dynamics (Brennen (1995)), through the sound radiation term above. No assumption about the motion of the gas inside the bubble is necessary. It has been suggested in the literature (Putterman and Weninger (2000)) that the existence of shocks in the liquid is evidence for shock waves existing inside the bubble at the collapse. Such shocks in the gas would be expected to propagate into the liquid. However, quantitative comparisons by Wang et al. (1999) demonstrate that the strength of the wave in the liquid is accounted for by sound radiation alone, without any consideration of hypothesized shocks in the gas.

Finally, it should be stressed that the emitted shock wave must also be considered when extracting the bubble radius around the collapse from Mie scattering data. Gompf and Pecha (2000) and Pecha and Gompf (2000) have shown that in the last nanoseconds around the minimum radius most of the Mie scattered light gets scattered at the highly compressed water around the bubble (see Fig. 3.13) and not at the bubble surface itself. Previous measurements by Putterman and Weninger (2000) and Weninger et al. (1997a, 2000) neglect this effect, leading to an overestimate of the bubble wall velocity. Instead of a bubble wall velocity larger than Mach 4 or 6,000 m/s obtained in Putterman and Weninger

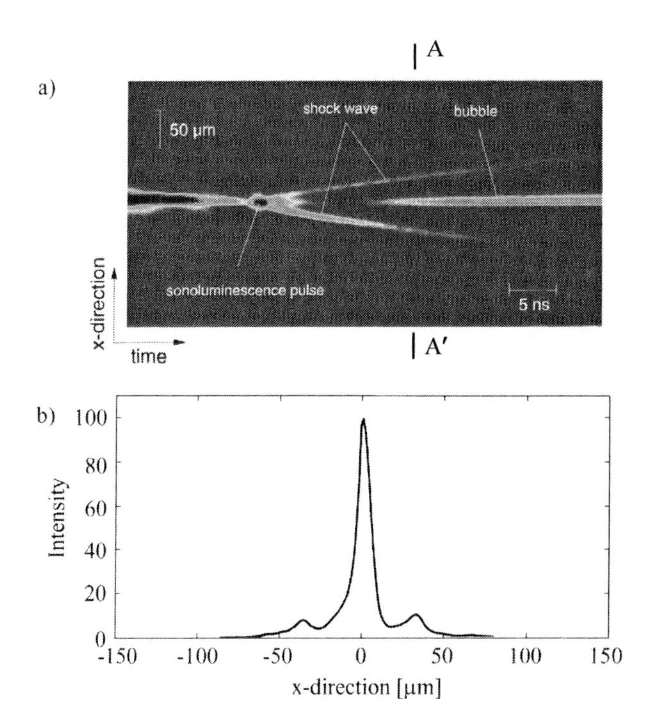


Figure 3.27 (a) Streak image of the SBSL bubble collapse. Laser light is scattered at the bubble as well as at the emitted shock wave. (b) Line scan A–A′ of the streak image. (Pecha and Gompf (2000).) (SEE ALSO COLOR PLATES)

(2000) and Weninger et al. (1997a), Gompf and Pecha (2000) find the more realistic value of 950 m/s by taking the compression effect into account. Weninger et al. (2001) reply to criticisms of Gompf and Pecha (2000) and refer to their earlier work of Weninger et al. (2000a), Vacca et al. (1999) describing a new light scattering technique based on differential measurement and polarization.

Hallaj et al. (1996) is an abstract of measurements of the acoustic emission from a glowing bubble. A broadband, nonfocused transducer was used to record the acoustic emission from single bubbles at various acoustic drive amplitudes in the sonoluminescence and non-sonoluminescence producing regions. The typical acoustic signature included a large amplitude pulse corresponding to the initial collapse followed by smaller amplitude pulses corresponding to the rebounds. The rebounding bubble essentially oscillates at its resonance frequency so that the simple measurement of the time interval between rebounds makes it relatively easy to measure one of the fundamental unknowns of a sonoluminescence bubble – its equilibrium radius.

Other papers on sound emission from bubbles are by Grossman et al. (1997) and Hilgenfeldt et al. (1998b).

### *3.5.7 Simple model of uniform van der Waals gas without heat and mass exchange*

The simplest model to assume is an adiabatic equation of state for the bubble interior (Löfstedt et al. (1993), Barber et al. (1997a)),

$$P_{\text{gas}}(t) = \left(P_0 + \frac{2\sigma}{R_0}\right) \frac{(R_0^3 - h^3)^{\Gamma}}{(R(t)^3 - h^3)^{\Gamma}} \tag{3.4}$$

and the corresponding temperature equation

$$T(t) = T_0 \frac{(R_0^3 - h^3)^{\Gamma - 1}}{(R(t)^3 - h^3)^{\Gamma - 1}}$$

equivalent to Eq. (3.1) when replacing $\gamma$ by $\Gamma = C_p/C_v$, the ratio of the specific heats. The detailed values of $\Gamma$ during the period of collapse are in Prosperetti (1977c). Equation (3.4) supplements the RP equation and permits its solution.

One obvious problem with Eq. (3.4) is that an adiabatic bubble motion is assumed with no heat exchanged between the bubble and the exterior. As pointed out in Sec. 3.5.2, Péclet number estimates via Eq. (3.2) show that there is almost unrestricted heat exchange for most of the oscillation cycle, and that the motion is isothermal. Only at bubble collapse is the Péclet number greater than one. This means that most of the time the ratio of the specific heats $\Gamma$ has to be replaced by an isothermal exponent $\gamma = 1$. Near the cavitation collapse there is a change to the adiabatic value $\gamma \rightarrow \Gamma$. Roughly speaking, this transition will occur when the Péclet number is of order unity.

### *3.5.8 Simple model of uniform van der Waals gas with heat and mass exchange*

Chu and Leung (1997) solved the hydrodynamic equations for the motion of the gas in a bubble in an almost incompressible liquid. They included the effect of the thermal conductivity of the gas, but did not consider the effect of the water vapor in the bubble. They showed that the bubble remains close to isothermal in the expansion stage, and that a cold dense layer of air is formed at the bubble wall in the contraction stage.

Kyuichi Yasui (1997) included water vapor exchange with the bubble's exterior due to condensation and evaporation; heat exchange with the bubble's exterior modified by an energy flux depending on the compression, the temperature gradient, and condensation and evaporation; that a thin layer of water around the bubble can be heated; and following Kamath et al. (1993), 25 chemical reactions of the water vapor are included. (It is crucial that the net effect of these reactions is that they consume thermal energy, i.e. they are endothermic).

Yasui's results are shown in Fig. 3.28. At the bubble maximum, the bubble consists almost entirely of water vapor. Even at collapse, the bubble still retains some of the water (~1% of the total bubble contents). Because of the invading water vapor and the endothermic chemical reactions, the maximum temperature inside the bubble is reduced to about 10,000 K.

In the last 120 ps before collapse Yasui's model gives us a reduction of the thermal energy of 1.4 nJ through chemical reactions and of 0.6 nJ through thermal conduction. The loss through photon emission is extremely small, only about 0.2 pJ.

Yasui (1997) assumed that the transport of mass through the boundary layer was *condensation-limited*, rather than diffusion-limited (mass diffusion was not explicitly modeled, and thus assumed instantaneous). However, Storey and Szeri (2000) showed that

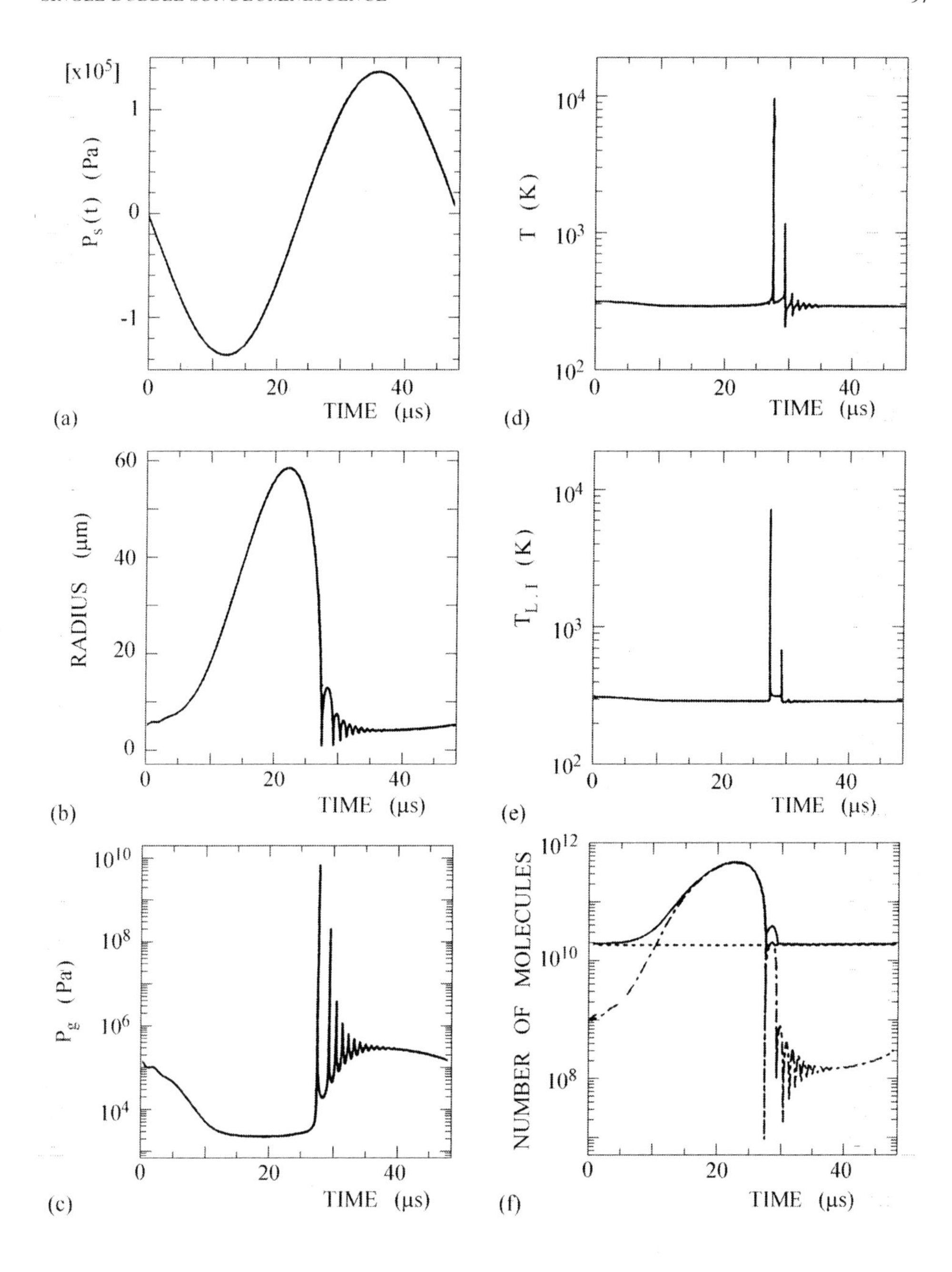


Figure 3.28 Calculated results under the condition of SBSL (Holt and Gaitan (1996)) for one acoustic cycle. The time axes (the horizontal axes) are the same for all the figures. (*a*) The pressure of the acoustic field [$p_s(t)$] employed in the calculation. (*b*) The bubble radius ($R$). (*c*) The pressure inside the bubble ($p_g$) with logarithmic vertical axis. (*d*) The temperature inside the bubble ($T$) with logarithmic vertical axis. (*e*) The liquid temperature at bubble wall ($T_{L,i}$) with the same logarithmic vertical axis with that in (*d*). (*f*) The number of molecules in the bubble with logarithmic vertical axis. The line shows the total number of molecules in the bubble ($n_t$), the dotted line is the number of argon molecules ($n_{Ar}$), and the dash-dotted line is that of vapor molecules ($n_{H_2O}$). (Yasui (1997).)

transport under SBSL conditions is diffusion-limited. Therefore, Toegel et al. (2000b, 2002) took the opposite view to Yasui and developed a simple *diffusion-limited* model for water vapor exchange between bubble and liquid, using a boundary layer approximation. They took into consideration only the most important endothermic process

$$H_2O + 5.1\ \text{eV} \leftrightarrow OH + H$$

Analyzing the reactive thermodynamics within the dense collapsed bubble, they demonstrated that the excluded volume of the nonideal gas results in the pronounced suppression of the particle-producing endothermic reactions. Thus, sufficiently high temperatures for considerable bremsstrahlung emission can be achieved.

Yasui (1997) uses a boundary layer as well, but in addition assumes a temperature jump between the outer edge of this boundary layer and the liquid. This temperature slip is more usually associated with low density systems, and is absent in other work like Toegel et al. (2000b) or Storey and Szeri (2000), but present in Young (1976).

### 3.5.9 The parameter range of single bubble sonoluminescence

Experiments have shown that apart from the parameters implicit in the RP equation, other quantities of crucial importance to SBSL are the concentration of gas $c_\infty$ dissolved in the liquid, the temperature of the liquid, the type of liquid, and the type of gas. We have to understand how these parameter dependencies occur.

Various physical constraints limit the parameter range in which SL can be observed to emit light. The bubble must be forced strongly enough for a cavitation event to occur during each cycle of the drive; the bubble must not break into pieces, which roughly translates into the requirement that viscous processes and surface tension are strong enough to limit the growth of bubble shape instabilities. For the consistent, stable light emission of SBSL, the number of gas molecules inside the bubble must neither increase or decrease over a complete cycle of oscillation. This requirement is what sets the ambient radius $R_0$ of the bubble; it involves a subtle interplay between diffusive processes exchanging gas between the bubble interior and the outside liquid, and chemical reactions. And finally, it is necessary that the Bjerknes forces holding the bubble trapped in the flask are strong enough to ensure that the center of the bubble does not move appreciably.

This section presents the current understanding of each of these effects, and assesses the extent to which the theoretical predictions agree with experiments.

#### 3.5.9.1 The Blake threshold

It is useful to mention here an experimental observation (Toegel and Lohse (2003)). This is that when starting with air bubbles, relative gas concentrations of typically 20%–40% are required for SBSL, whereas for argon bubbles one requires only 0.2%–0.4%. Moreover, for air bubbles the pressure regime where light emission occurs is preceded by a regime where the ambient radius of the bubble $R_0$ considerably shrinks with increasing acoustical pressure — up to a critical pressure amplitude, where the bubble is hardly detectable anymore. Beyond this threshold a sudden growth in $R_0$ takes place and light emission sets in. To explain this, see Toegel and Lohse (2003).

Regardless of the exact mechanism of SL, it is abundantly clear that the light results from energy focusing during a rapid bubble collapse. Therefore, the bubble must be forced

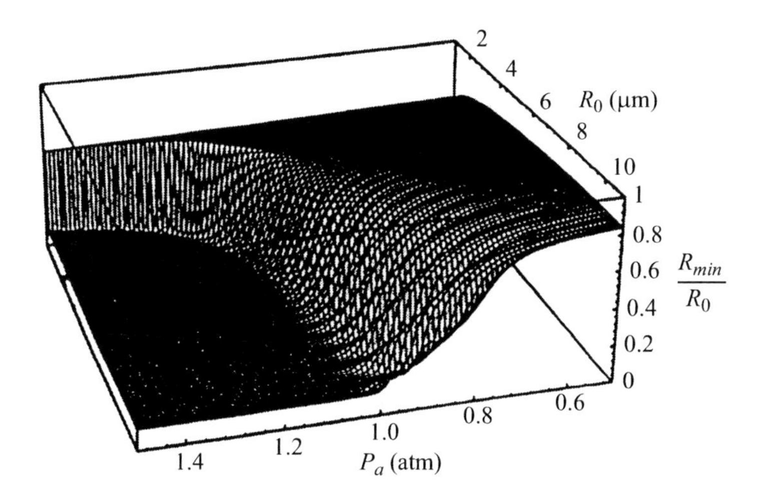


Figure 3.29 Compression ratio $R_{min}/R_0$ as a function of Pa and $R_0$. The two regimes of bubble dynamics are clearly visible: weakly oscillating bubbles for small $R_0$ and small $P_a$, strong collapses to the hard-core radius for large $R_0$ and large $P_a$. (Hilgenfeldt et al. (1998a).)

strongly enough to induce a cavitation event of sufficient violence — in essence, the Rayleigh collapse of Sec. 3.5.5 must be fully established. Whether this happens depends on both the ambient bubble size (mass of gas inside the bubble) and the forcing pressure. Figure 3.29 shows the minimum radius during a cycle of the drive as a function of the forcing pressure $P_A$ and ambient radius $R_0$ at a frequency of 26.5 kHz. There is an abrupt transition $R_0$ ($P_A$) where the onset of the Rayleigh collapse occurs and the gas inside the bubbles becomes strongly compressed, leading to heating. Therefore, SL can only occur above the threshold.

Blake (1949) showed that the critical ambient radius $R_0^c$ is related to the pressure under isothermal bubble movement (Brennen (1995)) by

$$R_0^c = \frac{2}{3}\frac{\sigma}{P_A - P_0}$$

The threshold in $R_0 - P_A$ space separating gently oscillating from violently collapsing bubbles is therefore known as the *Blake threshold.*

Further constraints set an upper limit on the bubble radius in order to ensure the bubble's stability. First we turn to diffusive stability.

#### 3.5.9.2 Diffusive stability

Lohse (2003) summarizes the conditions under which stable single bubble sonoluminescence can occur as

(1) The collapse must be strong enough, that is, above the threshold for the Rayleigh collapse (Sec. 3.5.5) to occur.

(2) The bubble must be spherical and stable.

(3) The bubble must be diffusively stable.

(4) The constituents of the bubble must be chemically stable.

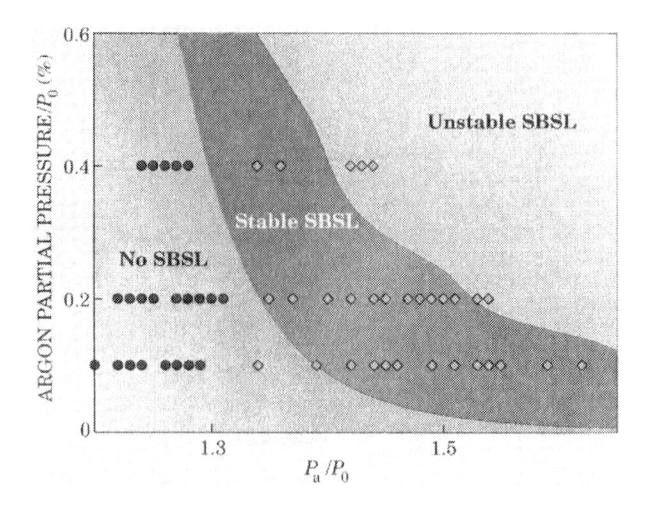


Figure 3.30a Phase diagram of single-bubble sonoluminescence (SBSL). The horizontal axis is the driving-pressure amplitude $P_a$, normalized by the ambient pressure $P_0$. The vertical axis is the argon partial pressure, also normalized by $P_0$. Stable SBSL is predicted to occur only in the red region. At lower forcing or Ar partial pressures, no SBSL occurs. Above the red region, the SBSL is unstable: The bubble grows by diffusion, and the phase and intensity of the emitted light aren't constant. The experimental data (dots in the respective color of the region) are in reasonable agreement with the predictions. Diagram adapted from Brenner et al. (2002)). (Lohse (2003).) (SEE ALSO COLOR PLATES)

Applying these criteria, Lohse (2003) presents Fig. 3.30a for the phase diagram of sonoluminescence as quantitatively calculated.

Since bubble dynamics and energy focusing during collapse are sensitively dependent on $R_0$, it is crucial for a stable SBSL bubble to maintain the same ambient radius, i.e. not to exchange any net mass with its surroundings. The gas exchange between the bubble and the liquid is affected by diffusion of gas through the liquid, and by the conservation of liquid volume in the spherical shell of liquid around the oscillating bubble when it is pushed in or out when the bubble contracts or expands, causing advection of the dissolved gas.

The typical model used for this process starts with the transport equation for the mass concentration $c(r, t)$ (mass/volume) of gas around a spherical bubble,

$$\partial_t c + v\partial_r c = D\frac{1}{r^2}\partial_r(r^2\partial_r c)$$

where $D$ is the gas diffusion coefficient in water. The velocity of the liquid is $u(r, t) = R^2\dot{R}/r^2$ at a distance $r$ from the center of the bubble, with $R(t)$ entering by solution of the RP equation. The gas in the bubble is assumed to remain in equilibrium with that in the liquid at the boundary of the bubble wall; hence the gas concentration at the bubble wall is given by Henry's law

$$c(R, t) = c_0 p_g(R, t)/P_A$$

where $c_0$ is the saturation concentration.

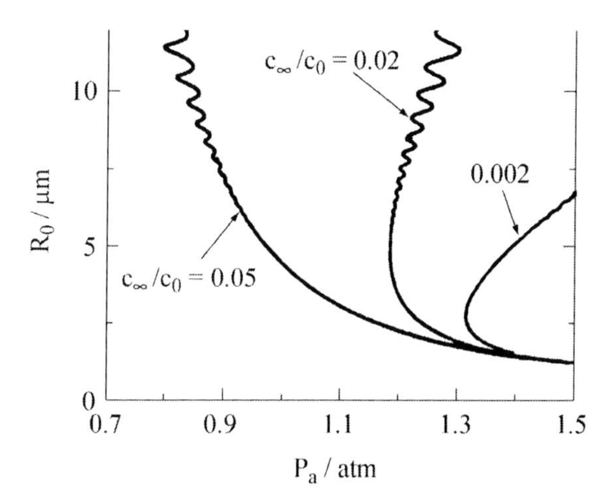


Figure 3.30b Bifurcation diagrams in the $R_0$–$P_a$ parameter space. The regimes with positive slope are stable. Gas concentrations are $c_\infty/c_0 = 0.002$ (right), $c_\infty/c_0 = 0.02$ (middle), and $c_\infty/c_0 = 0.5$ (left). To the left of the curves the bubbles shrink and finally dissolve, to the right of them they grow by rectified diffusion. (Brenner et al. (2002).)

Finally, the gas concentration far from the bubble is given by the ambient concentration $c_\infty$,

$$c(\infty, t) = c_\infty$$

The mass loss or gain of the bubble is then proportional to the concentration gradient at the bubble wall,

$$\dot{m} = 4\pi R^2 D \partial_r c|_{R(t)}$$

An excellent substantial account of bubble diffusion is given by Brenner et al. (2002). From this account, Fig 3.30b shows the famous equilibrium bubble states diagram in $R_0$ – $P_A$ parameter space for three fixed concentrations ($c_\infty/c_0 = 0.5, 0.02, 0.002$).

Figure 3.30b shows that there are no diffusively stable SL bubbles for *large gas concentrations*, where all equilibria are unstable. For small concentrations $c_\infty/c_0$ the situation is quite different. There are stable equilibria at large $P_A$ and small $R_0$. Only in this region for *very low gas concentration* $c_\infty/c_0 \sim 0.001$–$0.02$ (depending on $P_A$) is the bubble diffusively stable, and only then is stable SBSL possible.

Löfstedt et al. (1995) realized that diffusive equilibrium in strongly forced bubbles is only possible at very small gas concentrations, in agreement with experiments using argon or other rare gas bubbles: However, this disagrees with the results obtained for *air* bubbles, where stability is achieved at roughly 100 times larger gas concentrations. Recall that Gaitan (1990) needed to degas to only about 40% of saturation – if he had had to go to 100 times lower concentrations, he may never have discovered SBSL!

Akhatov et al. (1997) have studied the influence of surface tension on the dynamics of small bubbles, taking into account rectified diffusion and the resonancelike response of small bubbles to strong acoustic pressure amplitudes. This theory provides an explanation for the existence of small, stably oscillating bubbles that have been observed in experiments on sonoluminescence.

### 3.5.10 Shape stability

The theoretical diffusive equilibrium curves stretch to far larger ambient radii $R_0$ than those observed for SBSL bubbles. There must be another requirement limiting the size (or the total mass content) of the bubble. It is now well established that this limit is set by the onset of instabilities in the shape of the oscillating bubble. The analysis of shape stability is a classical problem in bubble dynamics, pioneered by Plesset (1954, 1956), Birkhoff (1954), Strube (1971) and Prosperetti (1977a). In this section we present the application of these ideas to single bubble sonoluminescence.

#### 3.5.10.1 Dynamical equations

To analyze the stability of the radial solution $R(t)$, consider a small distortion of the spherical interface $R(t)$,

$$r = R(t) + a_n(t)Y_n(0, \varnothing)$$

where $Y$ is a spherical harmonic of degree $n$. The goal is to determine the dynamics $a_n(t)$ for each mode. The derivation of Plesset (1954) follows the same spirit as the derivation of the Rayleigh–Plesset equation, which it recovers to zeroth order in $a_n$. A potential flow outside the bubble is constructed to satisfy the boundary condition that the velocity at the bubble wall is $\dot{R} + \dot{a}_n Y_n$. This potential is then used in Bernoulli's equation to determine the pressure in the liquid at the bubble wall. If viscous effects are neglected, applying the pressure jump condition across the interface yields a dynamical equation for the distortion amplitude $a_n(t)$,

$$\ddot{a}_n = B_n(t)\dot{a}_n - A_n(t)a_n = 0 \tag{3.5}$$

where $B_n(t) = 3\dot{R}/R$ and $A_n(t) = (n-1)\frac{\ddot{R}}{R} - \frac{\beta_n \sigma}{\rho R^3}$,
where $\beta_n = (n-1)(n+1)(n+2)$.

The stability of the spherical bubble then depends on whether solutions to Eq. (3.5) grow or shrink with time. It is already apparent that Eq. (3.5) has the form of a parametrically driven oscillator equation (the Hill equation), with the radial dynamics $R(t)$ governing the periodic driving.

A more accurate stability analysis requires taking account of viscosity and other dissipative processes. Viscosity, treated by Prosperetti (1977a), poses difficulties because viscous stresses produce vorticity in the neighborhood of the bubble wall, which spreads convectively through the liquid. Once created, the vorticity acts back on the dynamics of $a_n(t)$. This interaction is nonlocal in time, and so the problem requires solving integro-differential equations for the vorticity in the liquid, coupled with the shape oscillations. Details can be found in Prosperetti (1977a), Hilgenfeldt et al. (1996) and Hao and Prosperetti (1999). Here we simply summarize the results of a simple "boundary layer" approximation, which assumes that the vorticity is localized in a thin region round the bubble. It was again Prosperetti (1977b) who first realized the usefulness of this approximation. If $\delta$ is the boundary layer thickness, the prefactors of Eq. (3.5) are modified to

$$A_n(t) = (n-1)\frac{\ddot{R}}{R} - \frac{\beta_n\delta}{\rho R^3} + \frac{2\nu\dot{R}}{R^3}\left[-\beta_n + n(n-1)(n+2)\frac{1}{1+2\frac{\delta}{R}}\right] \tag{3.6}$$

$$B_n(t) = \frac{3\dot{R}}{R} + \frac{2\nu}{R^2}\left[-\beta_n + \frac{n(n+2)^2}{1+2\frac{\delta}{R}}\right] \tag{3.7}$$

where $\nu$ is the kinematic viscosity of the liquid. The viscous contribution to $A_n(t)$ is not important since the ratio between the third and the second term of the RHS in Eq. (3.6) is typically $\nu\rho R_0\omega/\sigma \lesssim 10^{-2}$. However in the second term of the RHS of Eq. (3.7) it introduces a damping factor which causes exponential damping of the shape modulations. The amount of damping strongly depends on both the boundary layer thickness $\delta$ and on $n$. Brenner et al. (1995) and Hilgenfeldt et al. (1996) choose $\delta$ to be the minimum of the oscillatory boundary layer thickness $\sqrt{\nu/\omega}$ and the wavelength of the shape oscillation $R/(2n)$. Wang et al. (2003) calculated the evolution equations for an ellipsoidal bubble derived by aspherical acoustical driving. They showed that the bubble can either grow rapidly, leading to the bubble's breakdown, or can decay gradually, making the bubble spherical. Other choices have been investigated in the literature (Augsdörfer et al. (2000)) and Hao and Prosperetti (1999) and will be commented on below.

The Hill equation Eq. (3.5) (Weisstein (2003) page 1381) is driven by the strongly nonlinear RP dynamics $R(t)$, therefore, in contrast to the monofrequent driving of the prototypical Mathieu equation (Weisstein (2003) page 1383), instabilities in $a_n$ can be excited on the many different time scales of the bubble oscillation. In particular, three types of instability can be distinguished: the *parametric instability* (over time scales of the oscillation period), the *afterbounce instability* (over time scales of the bubble afterbounces), and the *Rayleigh–Taylor instability* (over time scales of the Rayleigh collapse).

Reddy and Szeri (2002) considered a bubble with a time-dependent radius, translating unsteadily in a flow. This situation can be brought, for example, by forcing with an acoustic traveling wave. The equations governing the amplitudes of shape modes were derived using domain perturbation theory, following a classical paper by Plesset (1954). Contrary perhaps to intuition, the results show that driving at the natural frequency of volume oscillations is not necessarily the ideal forcing to engender a shape instability. Moreover, several radial oscillations can have a stabilizing effect on shape oscillations. The results suggest the possibility of destroying bubbles selectively by size.

### 3.5.10.2 Parametric instability

The parametric shape instability, as defined in the last paragraph of Sec. 3.5.10.1, acts over the relatively long time scale $T_d = 2\pi/\omega$ (time period of the driving oscillation). If the non-spherical perturbations of bubble shape show net growth over one oscillation period, they will overwhelm the bubble over many periods. This argument neglects possible (nonlinear) saturation effects not contained in the linear approximation (Eq. 3.5) that could inhibit further growth of the perturbations.

In the relevant parameter regime for parametric instability, $R(t)$ and thus also $A_n(t)$ and $B_n(t)$ are strictly periodic in time with the period $T_d$. Thus the stability of the Hill equation Eq. (3.5) can be rigorously analyzed (Nayfeh and Mook (1979)). Instability occurs when the magnitude of the maximum eigenvalue of the Floquet transition matrix $F_n(T_d)$ of Eq. (3.5) is greater than one. The Floquet transition matrix is defined as the propagator of the perturbation vector over one period,

$$\begin{pmatrix} a_n(T_d) \\ \dot{a}_n(T_d) \end{pmatrix} = F_n(T) \begin{pmatrix} a_n(0) \\ \dot{a}_n(0) \end{pmatrix}$$

By numerically computing the eigenvalues of the Floquet transition matrix one can map out the phase diagram of parametric stability, i.e. identify parametrically stable and unstable regions.

In the SL parameter range of $P_A \approx 1.2$ to 1.5 atm, and a typical frequency $f = 26.5$ kHz, calculations with the boundary layer approximation suggest that parametric instability sets in for ambient radii in excess of $R_0^{PI} \approx 4-5\ \mu$m, with only a weak dependence on $P_A$. Refined boundary layer models like those of Prosperetti and Hao (1999) or Augsdörfer et al. (2000) take into account higher order terms in $\dot{R}/c$, heat losses, or the varying gas density in the bubble on collapse. These models find upper stability bounds for $R_0$ about half a micron larger, because the additional effects result in less violent oscillation dynamics and smaller values of the (destabilizing) bubble acceleration.

Storey (2001) calculated the parametric spherical stability of a single sonoluminescing bubble with detailed modeling of the gas dynamics in the bubble interior.

The most accurate model used one-dimensional *direct numerical simulation (DNS)* to solve the complete Navier–Stokes equation of a multispecies gas including phase change, heat transfer, mass transfer and chemical reactions. This DNS method computed the gas pressure and produced radial oscillations that agreed well with experiments.

The earlier traditional model assumed that the pressure inside the bubble was uniform and that the bubble underwent a polytropic process (i.e. isothermal or adiabatic). This assumption directly related the gas pressure to the volume and makes the nonlinear ordinary differential equation solvable: this equation, we will remind the reader, is the *Rayleigh–Plesset equation* (RPE).

The two methods model the liquid in the same way: the difference is that the *DNS* solves the full equations governing the gas, whereas the *RPE* uses a uniform pressure and polytropic assumption.

In Fig. 3.30c, Storey (2001) plots the radial oscillations for the two models (the *DNS and traditional models*) in order to highlight the practical difference in the two approaches. Afterbounces from the *DNS* model are found to damp much more rapidly when compared to the traditional *RPE* model with an isothermal assumption. The extra damping in the DNS case is not surprising, since the full simulation accounts for transport and dissipation of heat and mass in the bubble, a source of loss for the energy of the collapse.

The traditional linear stability analysis using the *RPE* model predicts a threshold for single bubble sonoluminescence at a much lower pressure threshold than experimental observations. Whereas the more accurate *DNS* model of the radial dynamics predicts a

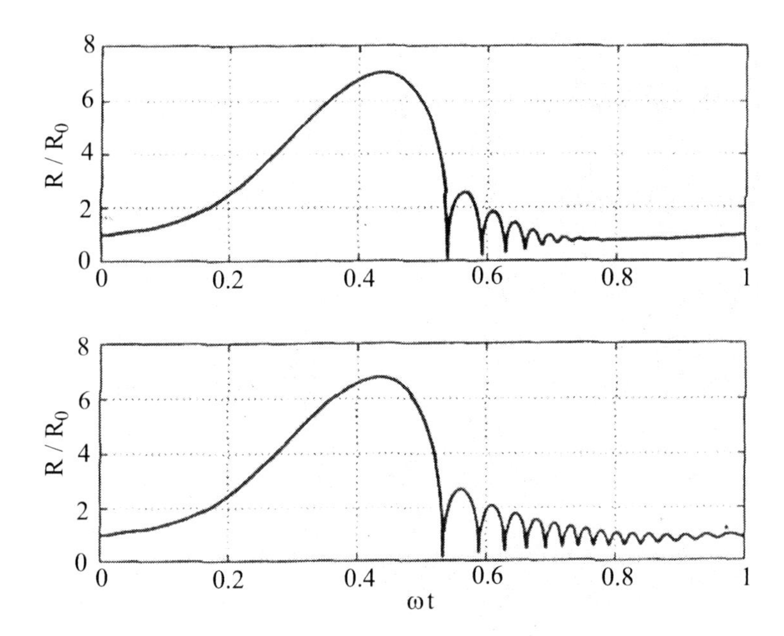


Figure 3.30c Radial dynamics from a DNS model and a traditional RPE model. The conditions are a 4.5-micron bubble forced with a 1.3-atm. pressure amplitude at 32.8 kHz. The top figure is from DNS's, and the bottom figure is from the RPE. The radius is scaled by the ambient radius, and time is scaled by the frequency of the forcing. (Storey (2001).)

threshold for single bubble sonoluminescence that is in excellent agreement with experimental data.

### 3.5.10.3 Afterbounce instability

During the afterbounces the bubble oscillates close to its *resonant frequency* (see Sec. 3.5.4) on a timescale $\tau_0 = 1/f_0 \sim \sqrt{\rho R^2/3P_0} \sim 0.3$ μs. It turns out that the characteristic timescale for shape oscillations about the spherical bubble is very close to this resonant timescale, which is $\sqrt{\rho R_0^3/(\gamma\beta_n)} = 1$ μs$/\sqrt{\beta_n}$. For the $n = 2$ mode, $\sqrt{\beta_n} \approx 3$, so the timescales are very close. The near coincidence of these two timescales is the root cause of the parametric instability (which exhibits maximum growth when the timescale of the forcing is of order of the timescale of the natural oscillation frequency). Under the right circumstances this instability can be so violent that the bubble is destroyed during a single cycle of afterbounces. An experimental example is shown in Fig. 3.31, taken from Matula (1999). The distortion can grow so much that the bubble breaks apart during the afterbounce period. The growth of instabilities during the afterbounce phase has been directly observed by Gaitan and Holt (1999).

The afterbounce instabilities must be triggered by noise: a good way to analyze this dependence is to model the thermal noise through coupling a Langevin type force to the dynamical Eq. (3.5) for the shape distortion (Augsdörfer et al. (2000) and Yuan et al. (2001)), with a magnitude adjusted to satisfy the fluctuation–dissipation theorem.

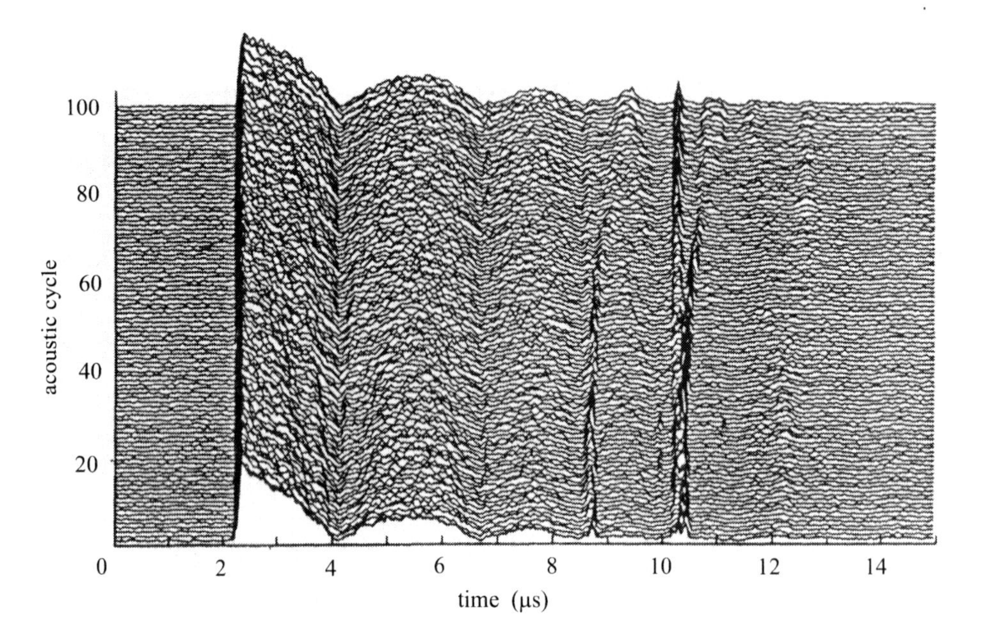


Figure 3.31 The afterbounces of a bubble driven below the luminescence threshold. Parametric instabilities can be inferred from the light-scattered signal. Note the large signal *spikes* near 9 and 10.5 μs in this figure. Direct imaging of the bubble shows non-spherical bubble shapes. (Reproduced with permission from Matula 1999 *Phil Trans Roy Soc Lond A* **357**, 225.)

#### 3.5.10.4 Rayleigh–Taylor instability

The Rayleigh–Taylor (RT) shape instability occurs when a lighter fluid is accelerated into a heavier fluid, the classical example being the interface between two layers of liquid, the lower one being lighter, and with buoyancy as the accelerating force. Löfstedt et al. (1995) discussed a gas bubble in a liquid, where the exponential growth of the angular surface perturbations characterize the planar Rayleigh–Taylor instability of Rayleigh (1883) and Taylor (1950). This was accompanied by effects related to the spherical geometry as described by Plesset and Mitchell (1956) and Plesset (1977). Following these references, Löfstedt et al. (1995) wrote the radius of the bubble as

$$R(t) = \bar{R}(t) + \sum_{n=2}^{\infty} a_n(t) Y_n(\Theta, \Phi) \tag{3.8}$$

where the $Y_n$ are spherical harmonics. Requiring that pressure and velocity be continuous at the surface of the bubble, yields to linear order in $a(t)$,

$$\ddot{a}_n + \frac{3\dot{R}}{R}\dot{a}_n - \left[(n-1)\frac{\ddot{R}}{R}\right]a_n = 0 \tag{3.9}$$

In the equation above, surface tension and viscosity, which tend always to dampen the corrugations of the bubble surface, have been neglected. To investigate the susceptibility of the bubble to corrugations, it is helpful to consider separately three regimes of the sonoluminescence bubble motion. During the expansion of the bubble to $R_m$, the growth is quite accurately linear. The corresponding solution for $a(t)$ decreases as $1/R^2$ from its initial value. During the collapse from the maximum, the bubble obeys

$$\dot{R}^2 = \frac{2}{3}\frac{P_0}{\rho_l}\left[\frac{R_m^3}{R^3} - 1\right] \tag{3.10}$$

which is Rayleigh's solution for a collapsing vacuous cavity, Rayleigh (1917). For this motion equation Eq. (3.9) can be rewritten in the form below (Plesset and Mitchell (1956) and Prosperetti (1977)).

$$y(1-y)a_n''(y) + \left(\frac{1}{3} - \frac{5}{6}y\right)a_n'(y) - \frac{(n-1)}{6}a_n(y) = 0 \tag{3.11}$$

where $y = R_m^3/R^3$. The asymptotic solution of this hypergeometric equation, as $y \to \infty$, is $a_n \sim (R_m/R)^{1/4}$. Löfstedt et al. (1995) go on to say that since the Rayleigh expression Eq. (3.10) is true to radii of the order $R_0$, before the gas pressure affects the collapse, the enchancement of the surface corrugation is only a factor of about $10^{1/4} \approx 1.8$. Alternatively, Eq. (3.9) can be written in a Hamiltonian form

$$H = \frac{p_n^2}{2R^3\rho_l} - \frac{(n-1)}{2}R^2\ddot{R}a_n^2\rho_l \tag{3.12}$$

where $p_n$ is the momentum canonical to $a_n$. Since $\ddot{R}$ is negative during the collapse, the motion in the "potential well" of the Hamiltonian is stable, and the magnification of corrugations is due to the scaling of the effective mass and the potential through $R$ in the spherical geometry. Following the collapse the bubble is virtually motionless at $R_0$ for one half of the acoustic cycle. Applying Eq. (3.9) yields a linear growth of $a(t)$ during that time. Including corrections due to surface tension $\sigma_l$ and viscosity $\eta_l$ shows that the more accurate solution during that part of the motion where the bubble radius is constant is (Prosperetti (1977a)

$$a_n(t) = a_0\cos(\omega_\sigma t + \delta_0)e^{-\alpha t}$$

where

$$\omega_\sigma^2 = \frac{(n-1)(n+1)(n+2)\sigma_l}{R_0^3\rho_l}$$

and

$$a = \frac{(n+2)(2n+1)\eta_l}{R_0^2\rho_l}$$

In particular the time constant for a quadrupolar perturbation to decay for a bubble motionless at $R_0 = 4$ μm is less than 1 μsec. Löfstedt et al. (1995) conclude that the instabilities associated with the collapse of the bubble to its minimum cannot be analyzed within this simplified framework, since even the hydrodynamics is violated at the minimum. They infer that during any given cycle, i.e. from flash to flash, spherical modulations do not build up. The upper threshold of sonoluminescence was due to either some other hydrodynamic instability or to an event occurring at the moment of collapse, which is not described by hydrodynamics. For SL bubbles, this shape instability acts over the extremely short timescales of the final stages of Rayleigh cavitation collapse. Here the bubble interface *decelerates* in preparation for the re-expansion, leading to an extremely large relative acceleration of the gas with respect to the water in excess of $10^{12}$ g! This deceleration only occurs for a short time of nanoseconds; it is roughly the time a sound wave of speed $c_g$ needs to cross a fully collapsed SBSL bubble of radius $R \sim h$, where $h$ is the van der Waals hard core radius. For the RT instability to be effective it must destroy the bubble during this time period. The competition between large magnitude and short duration of the accelerating force determines the stability threshold.

Wu and Roberts (1998) suggested that the Rayleigh–Taylor shape instability of the bubble surface is responsible for the extinction of the sonoluminescence when the amplitude of the acoustic field driving the bubble oscillation exceeds a certain threshold. They also showed that Brenner et al. (1995) and Hilgenfeldt et al. (1996) underestimated the potency of viscosity in quenching shape oscillations.

Lin et al. (2002b) remark that in a classical paper Plesset (1954) determined conditions under which a bubble changing in volume maintained a spherical shape. The stability analysis was further developed by Prosperetti (1977a) to include the effects of viscosity on the evolving shape modes. Lin et al. (2002b) then modified the theory to include *the changing density of the bubble contents*. This is important in violent collapses where the densities of the gas and vapor within a bubble may approach the density of the liquid outside. This exerts a stabilizing influence on the Rayleigh–Taylor mechanism of shape instability of spherical bubbles.

Lin et al. (2002b) state that the assumption of a *uniform density field* in Hilgenfeldt et al. (1996) is of course naive, as the relatively cold wall during a violent collapse results in an increased gas density in the vicinity. This new theory compares well with experimental data; the Rayleigh–Taylor instability providing an extinction threshold for violently collapsing bubbles.

In their analysis, Lin et al. (2002b) mention that the boundary layer thickness $\delta$ is given by

$$\delta = \min\left(\sqrt{\nu}/\omega,\ R/2n\right)$$

where $\nu$ = viscosity of liquid
$\omega$ = annular frequency of driving sound
$R$ = radius of bubble
$n$ = $n$th harmonic

This is from Brenner et al. (1995) and Hao and Prosperetti (1999a).

The quantity $R/2n$ acts as a cut off justified on the basis of a quasistatic argument for small bubbles (Hilgenfeldt et al. (1996a,b)).

For an argon bubble in water of viscosity $\nu = 0.001\ \text{kg m}^{-1}\ \text{s}^{-1}$ under a driving frequency of 32.8 kHz

$$\delta = 2.2 \times 10^{-6}\ \text{m}$$

It is interesting to compare this with the boundary layer thickness calculated by the author (1976, 1999 pages 328 and 300). Here

$$\delta = n\lambda$$

where $\lambda$ is the mean free path and $n$ is about 3.

This gives for argon

$$\delta = 3 \times 10^{-7}\ \text{m}$$

which is not too far different.

Bogoyavlenskiy (2000) analyzed the shape stability for SL bubbles. Building on the results of Hilgenfeldt et al. (1998a), he derived the inverse function of $R(t)$ at the Rayleigh collapse and then considered the time as the parameter which can be eliminated, arriving at $a_n(R)$. This analytical treatment works when surface tension is neglected and the boundary layer is vanishingly thin ($\delta \to 0$); good assumptions in the last violent stages of collapse. The result is that the shape distortion weakly diverges as $a_n(R) \sim R^{-\xi}$ with an exponent $\xi$ in the range between 0 and $\frac{1}{4}$ that generally depends on the mode number $n$.

3.5.10.5 Interplay of diffusive equilibria and shape instabilities

The interplay of diffusive equilibria and shape instabilities has been explained by Hilgenfeldt et al. (1996). The conditions for diffusive and shape stability must be fulfilled simultaneously for stable SL. Outside of this parameter regime, bubbles do not necessarily perish, but can undergo dynamical processes like rectified diffusion that can allow for unsteady SL light emission at a weaker level ("unstable SL").

For low (fixed) forcing pressure $P_A \approx 1.1$ atm and low gas concentration (e.g. $c_\infty/c_0 =$ 2%), bubbles dissolve (remember, $c_\infty$ = ambient concentration, $c_0$ = saturation concentration). When the gas concentration is sufficiently large (e.g. $c_\infty/c_0 = 50\%$) rectified diffusion can overcome the tendency for dissolution, and growing bubbles are possible if $R_0$ lies above the unstable equilibrium line. Rectification then continues until shape instabilities limit the amount of growth, see Fig. 3.32. When $R_0$ reaches the boundary for shape instability, a microbubble pinches off, decreasing the bubble volume (and thus $R_0$). If the remaining bubble is still large enough i.e. above the unstable equilibrium line in Fig. 3.32, the process will repeat. For these low forcing pressures, the allowed size of the bubble after the pinch-off is very restricted. If the pinched-off microbubble is too large, the remaining bubble dissolves.

No diffusively stable regime exists for these low driving pressures, and the bubbles are not driven strongly enough to show SL. For relatively large forcing pressure ($P_A \approx 1.3$ atm) the situation is quite different. For low enough gas concentration (e.g. $c_\infty/c_0 = 0.2\%$) in Fig. 3.33), bubbles can grow (or shrink) and approach a *stable* diffusive equilibrium. Since at large $P_A$ the Rayleigh collapse occurs, we have stable SBSL here, with a well-defined stable $R_0$ following from $P_A$ and the gas concentration. For larger gas concentrations, large enough bubbles will again grow up to the threshold of parametric shape instability where microbubbles pinch off. Here in contrast to the smaller $P_A$ regime the remaining bubble is very likely to end up in a regime where it can grow again.

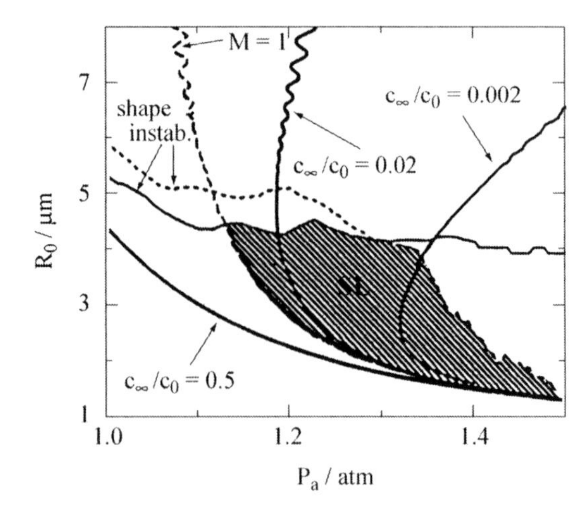


Figure 3.32 The figure shows the discussed effects for argon all together: The $M = 1$ curve (long dashed) gives a characterization when the strong bubble collapse and thus the heating begins. The bubble grows thanks to rectified diffusion right of the diffusive stability curves (thick lines, shown for $c_\infty/c_0 = 0.5$, 0.02, and 0.002, left to right). The thin solid line is the onset of the parametric instability, the short dashed line that of the afterbounce instability, and the dotted line that of the Rayleigh–Taylor instability. These three lines are calculated within the simplified theory of Hilgenfeldt et al. (1996), which slightly underestimate the shape stability, as discussed in the text. (Brenner et al. (2002).)

The characteristic slow growth of $R_0$ (over the timescales of rectified diffusion) and sudden breakdown (at microbubble pinch-off) is reflected in other experimentally observed parameters, such as the maximum radius $R_{max}$. The momentum of the pinched-off microbubble also gives the remaining bubble a recoil. As this repeats again and again on the diffusive timescale of ~0.1 s, the bubble seems to "dance", as originally observed by Gaitan (1990) and later by Barber et al. (1995). If the bubbles in this regime are large enough (close to the instability line), they will also emit SL even as they undergo rectified diffusion, leading to the same pattern of slow increase and sudden breakdown of the light signal. The phase of light emission (with respect to the driving signal) behaves in the same way. Experimental measurements of the *phase* are shown in Fig. 3.33 clearly showing the growth/pinch-off dynamics.

Hilgenfeldt et al. (1996) quote a personal communication from Gompf (1996) to explain the "ageing" of water in the SL container in case the container is not gas tight. By "ageing" it is meant that stable SL, and finally also unstable SL, becomes impossible with "old" water. The reason is that external air diffuses into the water, dissolves, and $c_\infty/c_0$ ($c_\infty$ = ambient concentration, $c_0$ = saturation concentration) increases. Consequently, the original stable bubble is pushed into the unstable region and starts to "dance", shedding off microbubbles. The dancing frequency becomes higher and higher and finally the bubble dissolves, after a too large pinch off. Bubbles may be reseeded, but will also die very soon.

A good summary article on bubble stability is by Gaitan (1999).

Toegel and Lohse (2003) presented a detailed model for the phase diagrams for single bubble sonoluminescence. The model employed was based on a set of ordinary differential equations and accounts for the bubble hydrodynamics, heat exchange, phase change of water vapor, chemical reactions of the various gas species in the bubble ($N_2$, $O_2$, and $H_2O$ being the most important of these), and diffusion/dissolution of the reaction products in the

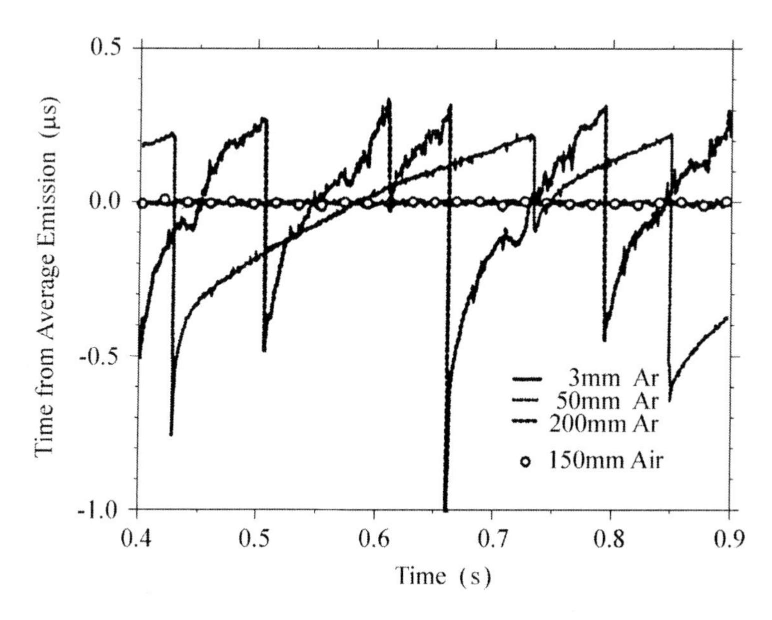


Figure 3.33 Figure taken from Barber et al. (1995). It shows the experimental result for the phase of light emission $\phi_s(\bar{t})$ for three different argon concentrations $c_\infty/c_0 = 0.00395$, $c_\infty/c_0 = 0.0658$, and $c_\infty/c_0 = 0.26$, corresponding to a gas pressure overhead of $p_\infty = 3$ mmHg, 50 mmHg, and 200 mmHg, respectively. Also shown is the relative phase of light emission for air bubbles: Stable SL is achieved for much higher concentration $c_\infty^{\mathrm{air}}/c_0 = 0.2$, corresponding to 150 mmHg. Hilgenfeldt et al. (1996) interpreted the drift in the phase of light emission as a result of the bubble growth through rectified diffusion, which is followed by a pinch-off of a microbubble when the bubble is running into the shape instability. (Barber et al. (1995).)

liquid. The results of the model are compared in detail to various phase diagram data from recent experimental work. Very good agreement was found. The onset of SBSL was found to be hysteretic. When starting with air, typical temperatures before onset were 5,500 K and 15,000 K thereafter. In the light emitting regime the bubbles were found to nearly entirely consist of argon.

### 3.5.11 *Mixture segregation within sonoluminescing bubbles*

Inside a sonoluminescing gas bubble in water, there will always be a *mixture* of gas and water vapor. To simplify the problem, Storey and Szeri (1999) considered the case of a mixture of the two monatomic inert gases, helium and argon. For this model, they assumed that no water vapor was present in the model. This avoids the complications caused by evaporation and condensation at the bubble interface. Due to *thermal diffusion* there will be a tendency for the gases to segregate due to the large temperature gradient. This phenomenon was first a theoretical conjecture of the Chapman–Enskog theory and later confirmed by experimentation (Chapman and Cowling (1900)). In a rare gas mixture, molecules of the gas with the larger molecular mass will be driven to the cooler regions and molecules of the gas with the smaller molecular mass will be driven to the hotter regions. In sonoluminescence, this thermal diffusion can be expected to play a part during the collapse of the bubble when the center is extremely hot and the bubble wall is relatively cool.

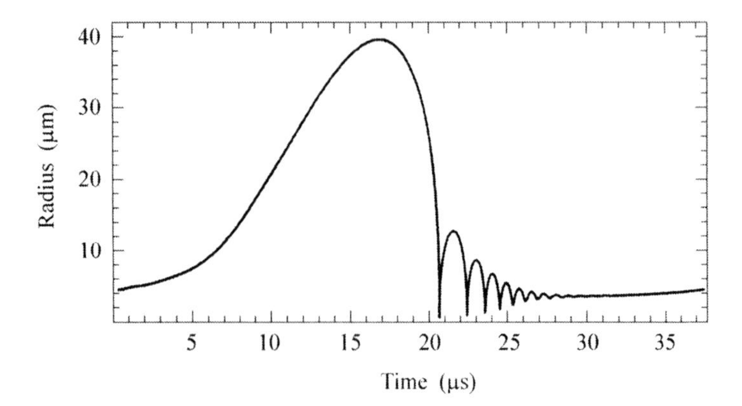


Figure 3.34 Radial response of the bubble to one cycle of the applied acoustic field. The bubble is He–Ar (10% mass He) forced at a pressure amplitude of 1.3 (Storey and Szeri (1999) © Cambridge University Press.)

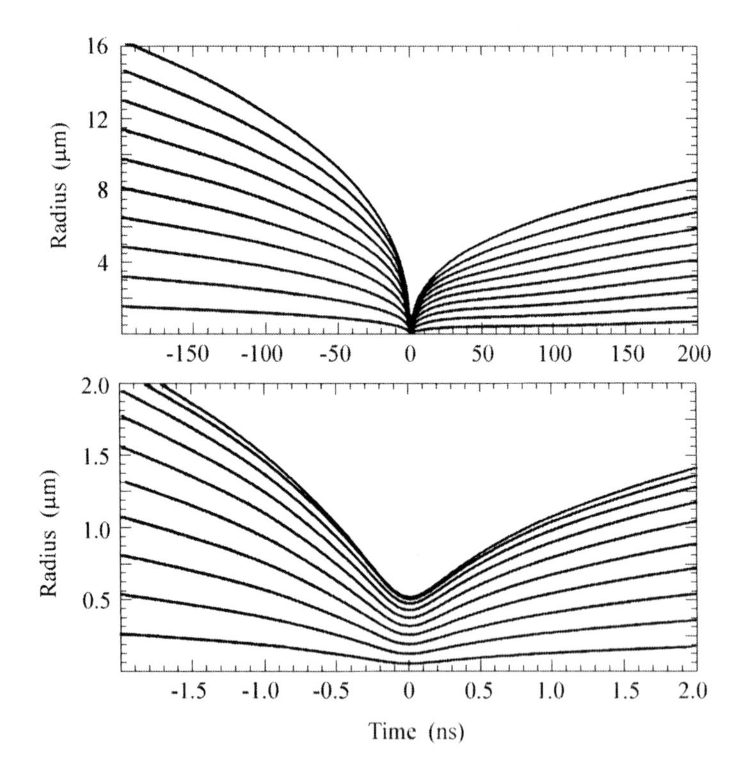


Figure 3.35 Radial response of the bubble to the applied acoustic field at the main collapse on two time scales. The marker particles were evenly spaced in the reference configuration. The outermost marker particle shows the evolving position of the interface. The bubble is He–Ar (10% mass He) forced at a pressure amplitude of 1.3. (Storey and Szeri (1999) © Cambridge University Press.)

To analyze the problem, Storey and Szeri (1999) used a van der Waals type model for the equation of state (Vuong et al. (1999)) to avoid the shortcomings of less sophisticated equations of state. To solve the governing partial differential equations of the gas,

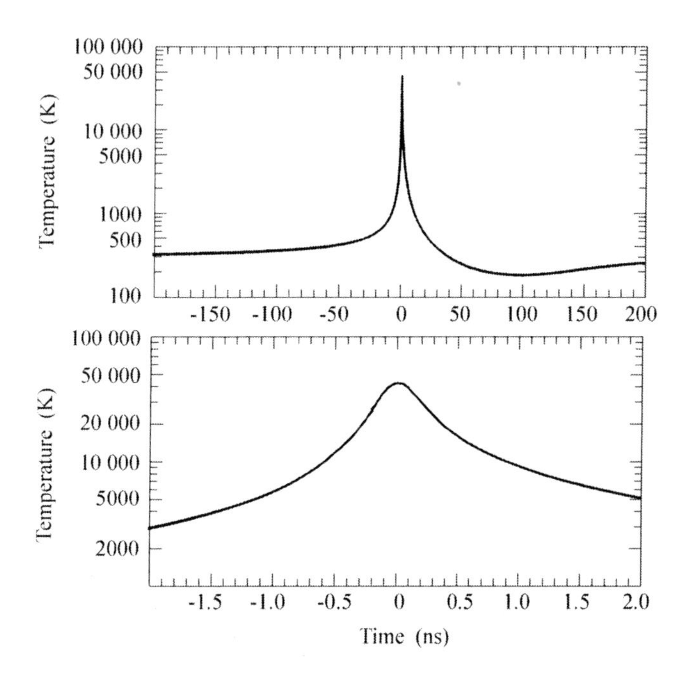


Figure 3.36 Temperature of the center of the bubble as a function of time at the main collapse. The bubble is He–Ar (10% mass He) forced with a pressure amplitude of 1.3. (Storey and Szeri (1999) © Cambridge University Press.)

Chebyshev polynominals were used. They assumed that no transport occurs across the bubble surface. They also neglected heat transfer by radiation and made no attempt to model the light emission.

Storey and Szeri (1999) investigated a *single bubble* SL case with the following parameters: liquid pressure $P_l = 101$ kPa, initial radius $R_0 = 4.5$ μm, initial temperature $T_0 = 300$ K, pressure of gas in bubble $P_0 = P_l + 2\sigma / R_0$, sound pressure amplitude ratio $P_A = P_{max} / P_0 = 1.3$ and sound frequency $f = 26.5$ kHz. The bubble contained a *helium–argon* mixture composed of 10% helium by mass (52.58% helium on a mole basis). The radial response of this bubble to one cycle of the applied acoustic field is shown on Fig. 3.34. Attention is then focused on the first main collapse of the bubble which occurs at approximately 20.6 ns from the beginning of the cycle. The next five figures show a beautiful sequence of events. Figure 3.35 shows the evolving radial position of selected marker particles in the bubble interior on two timescales which differ by two orders of magnitude. Time is shifted in this figure to put zero at the point of minimum radius. On the 100 ns timescale the collapse is quite sharp but on the 1 ns timescale the bubble smoothly reaches the minimum radius and expands.

Figure 3.36 shows the temperature history of the bubble center on the same two time scales as Fig. 3.35. The temperature rises an order of magnitude on a 1 ns time scale to a maximum of approximately 42,000 K. Figures 3.35 and 3.36 show the bubble motion at minimum radius and the temperature peak on a time scale of several hundred picoseconds.

Figure 3.37 shows the helium mole fraction at the bubble wall and center of the bubble as a function of time. The same two time scales from the previous two figures are used here. Close examination reveals that compositional inhomogeneities develop much more slowly

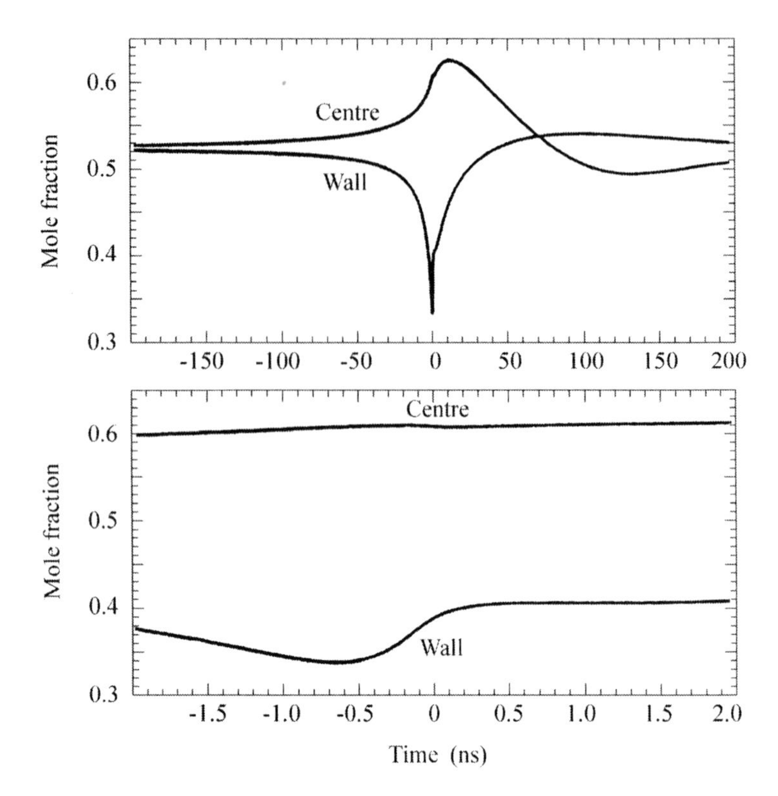


Figure 3.37 Composition history of marker particles at the center and at the bubble wall as a function of time. The wall particle reaches the lowest mole fraction of helium while the center reaches the maximum. The bubble is He–Ar (10% mass He) forced with a pressure amplitude of 1.3. (Storey and Szeri (1999) © Cambridge University Press.)

than the peak dynamic and thermal fields. The species segregation is driven by the slow build up and release of heat throughout the collapse and not by the short burst of energy supplied to the bubble contents at the point of minimum radius. Note that at the bubble wall the mole fraction of helium decreases to a minimum slightly before $t = 0$ then increases. In contrast, at the center, helium continues to accumulate for some time after the collapse and then eventually the center becomes slightly argon rich during the expansion.

Another view of the composition field is provided in Fig. 3.38. In this figure are shown snapshots of the composition field within the bubble at intervals of 2 ns with the gray scale indicating the mole fraction of helium. Black indicates argon rich and white indicates helium rich. Upon close inspection it will be seen that the argon is heavily concentrated in a thin region near the wall up to the time of minimum radius. As the bubble expands, the helium continues to move to the center while the sharp argon "shell" relaxes by diffusion.

Storey and Szeri (1999) also considered the *multibubble sonoluminescence* of the same helium/argon mixture and obtained a maximum temperature at the bubble center of 22,000 K.

Yasui (2000) undertook computer simulations of bubble oscillations under multi-bubble sonoluminescence conditions taking into account the segregation of water vapor and noble gas inside a collapsing bubble. The MBSL was performed in 20°C water for helium, argon and xenon. The ambient bubble radius was 4 μm. The maximum bubble temperature

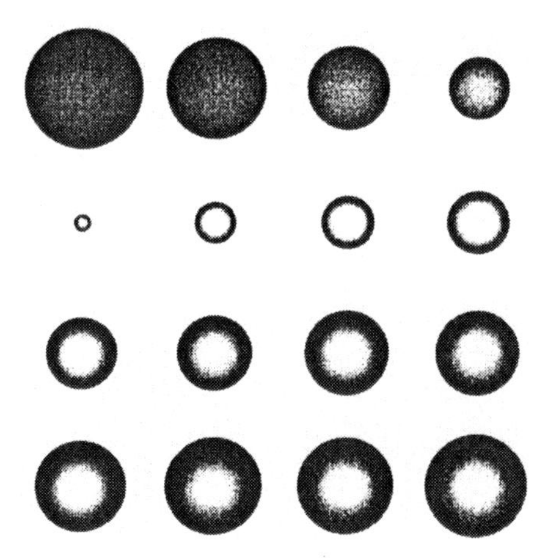

Figure 3.38 Composition field and radius of the bubble around the time of minimum radius. The gray scale is set so that pure black is the minimum in mole fraction of helium that occurred in this case and pure white is the maximum. A uniformly gray disk would indicate uniform composition. The frames are 2 ns apart from −8 to 24 ns from left to right, then down. The bubble is He–Ar (10% mass He) forced with a pressure amplitude of 1.3. (Storey and Szeri (1999) © Cambridge University Press.)

at the collapse occurred at the relatively low acoustic amplitude of about 1.5 bar. The maximum temperatures were 17,000 K for helium, and 24,000 K for argon and xenon.

Matula et al. (1996, 1997) describe experiments in microgravity in two abstracts.

## 3.6 INFLUENCE OF ARGON ON STABLE SONOLUMINESCENCE

Hiller et al. (1994) found that the sonoluminescence from a nitrogen bubble is increased thirty times by the addition of 1% argon (Fig. 3.39). Note that air is 1% argon. This is why air bubbles in water glow so easily (Brenner et al. (1996)). Hilgenfeldt et al. (1996), Brenner et al. (1996), Lohse et al. (1997) and Brenner et al. (1999) suggest that the high temperatures generated by the bubble collapse dissociate the oxygen and nitrogen producing O and N radicals which react with the H and O radicals formed from the dissociation of water vapor. Rearrangement of the radicals will lead to the formation of NO, OH, NH, which eventually dissolve in water to form $HNO_2$ and $HNO_3$ among other products. These reaction products pass into the surrounding liquid and are dissolved. This chemical process deprives the gas in the bubble of its reactive components. The only gases that can remain in the bubble are those which at high temperatures do not react with the liquid vapor i.e. the rare gases. Experimental verification of this dissociation hypothesis has been made by Matula and Crum (1998) and Ketterling and Apfel (1998). Didenko and Suslick (2002) measured the yields of nitrite ions, hydroxyl radicals and photons in a collapsing bubble. They showed that endothermic sonochemical reactions are a major limitation on the conditions produced during cavitation.

Brenner et al. (1995, 1996), Hilgenfeldt et al. (1996), Lohse and Hilgenfeldt (1997) and Lohse et al. (1997) discuss the complicated phase diagrams for sonoluminescing

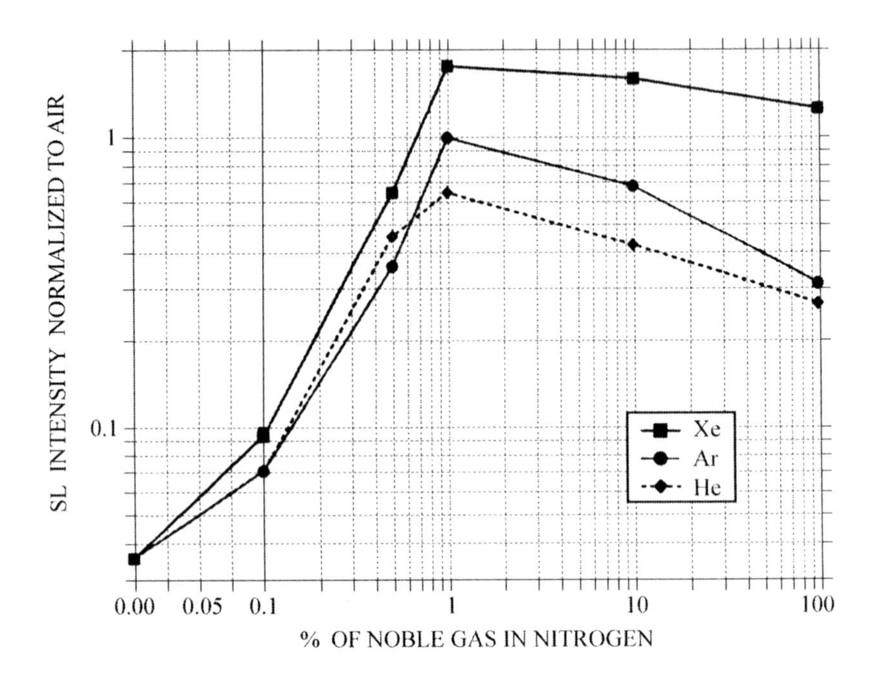


Figure 3.39 Intensity of light emission from a sonoluminescing bubble in water as a function of the percentage (mole fraction) of noble gas mixed with nitrogen. The gas mixture was dissolved into water at a pressure head of 150 mm. The data are normalized to the light emission of an air bubble in 24°C purified water with a resistance greater than 5 MΩ cm dissolved at 150 mm. Such an air bubble emits about $2 \times 10^5$ photons per flash. (Reprinted with permission from Hiller et al. *Science* **266**, 248. Copyright 1994 American Association for the Advancement of Science.)

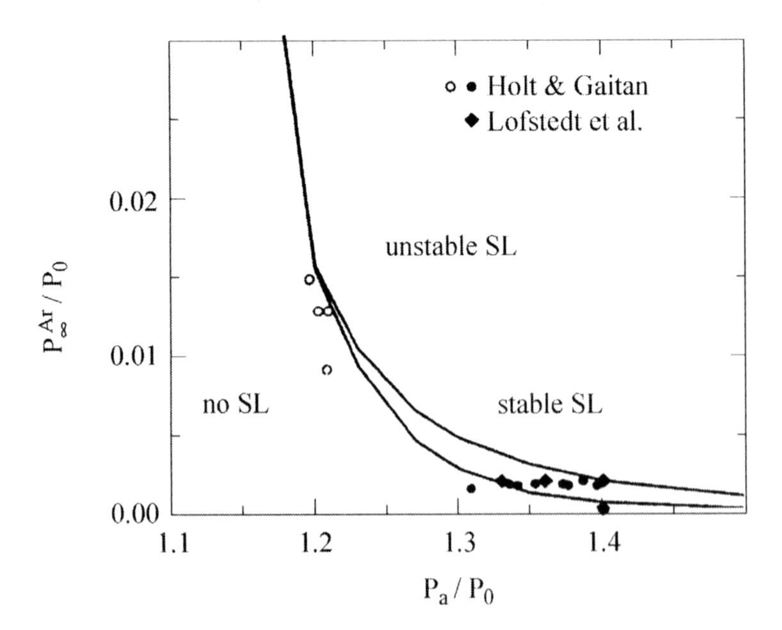


Figure 3.40 Phase diagram for pure argon bubble in the $p_\infty^{Ar}/P_0$ versus $P_a/P_0$ parameter space. (Lohse et al. (1997).)

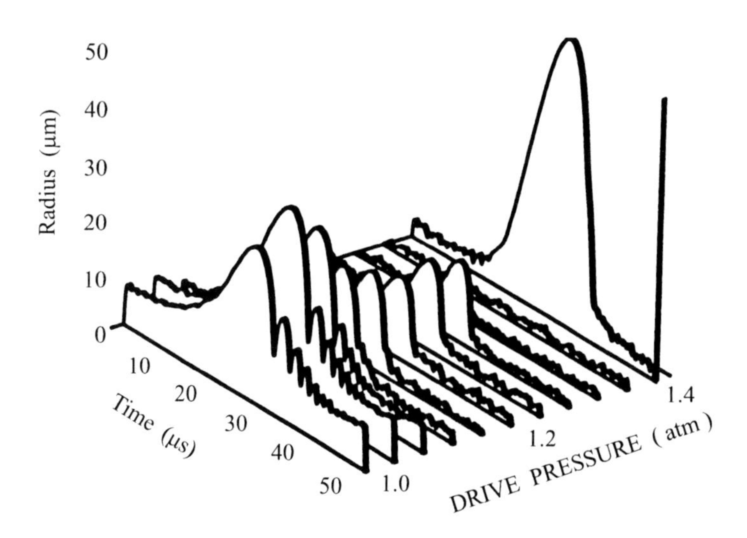


Figure 3.41 This figure shows the transition to SL for a bubble filled with an initial 0.1% xenon in nitrogen gas mixture at a partial pressure of 150 mmHg. According to the dissociation hypothesis this corresponds to $p_{\infty}^{Xe}/P_0 = 0.0002$ for a pure sonoluminescing xenon bubble. From the theoretical phase diagram one concludes that bubbles driven at $P_a$ = 1.3 atm dissolve for these low concentrations, whereas bubbles at $P_A$ = 1.4 atm show stable SL, just as seen in this experimental figure. (Löfstedt et al. (1995).)

bubbles. Figure 3.40 is a phase diagram for pure argon bubbles where $P_0$ = ambient pressure (usually 1 atm), $p_{\infty}^{Ar}$ = partial pressure of argon and $P_A$ = forcing pressure.

Stable SL is only possible in a very small window of argon concentration. The experimental data points refer to observed stable SL (filled symbols) or stable non-SL bubbles (open symbols) and are from Löfstedt et al. (1995) (diamonds) and Holt and Gaitan (1996) (circles) and show good agreement with the theory.

This is shown another way in Fig. 3.41 by Löfstedt et al. (1995). Holt and Gaitan (1996) and Gaitan and Holt (1999) were the first to measure detailed phase diagrams as a function of the ambient radius $R_0$ and the forcing pressure $P_A$ for different dissolved gas concentrations.

Recently, a comprehensive experimental study of bubble stability diagrams as a function of forcing pressure and rare gas concentration has been undertaken by Ketterling and Apfel (1998, 2000a,b,c). By performing experiments over a wide range of parameters, they showed that when the bubble is stable and light-emitting, it indeed lies on a positively sloped equilibrium curve in the $R_0$–$P_A$ plane, corresponding to the partial pressure of argon in the mixture. This occurs regardless of how much nitrogen is present. Three representative phase diagrams are shown in Figs. 3.42, 3.43 and 3.44. Figure 3.42 shows the phase diagram of air in water at $c_{\infty}/c_0$ = 0.2. The solid line represents the (unstable) diffusive equilibrium for the mixture, and the dashed line is the (stable) diffusive equilibrium for $c_{\infty}/c_0$ = 0.002 (the argon in the mixture). As the forcing pressure is increased the bubble follows a stable equilibrium line above the unstable diffusive equilibrium (where it starts emitting light). Figure 3.43 demonstrates that, without molecular gases present, a pure argon bubble at $c_{\infty}/c_0$ = 0.0026 follows its stable diffusive equilibrium. No other stable configurations are possible. On the other hand Fig. 3.44 shows that a pure nitrogen bubble at $c_{\infty}/c_0$ = 0.1 follows a stable equilibrium but cannot reach a stable sonoluminescing state at higher $P_A$.

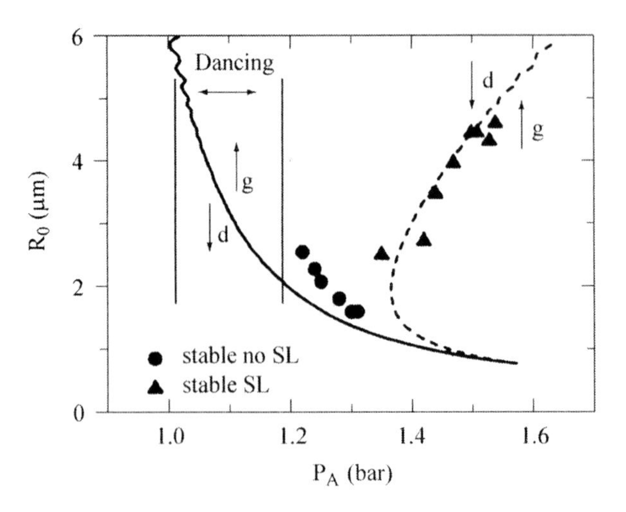


Figure 3.42 Phase diagram for air saturated in water to 20% yielding a final argon concentration of 0.2%. Each data point represents the $P_A$ and $R_0$ found from a single $R(t)$ curve and is indicated to be luminescing and/or stable. The curves in the plot are lines of diffusive equilibrium for a given gas concentration $C = 20\%$ (solid line) and $C = 0.2\%$ (dashed line). The range of $P$ where dancing was observed is indicated, as are regions of bubble growth (g) and dissolution (d) relative to each equilibrium curve. The stable no SL points (circles) correspond to a *stable* chemical equilibrium which would lie above the $C = 20\%$ curve if plotted. (Ketterling and Apfel (1998).)

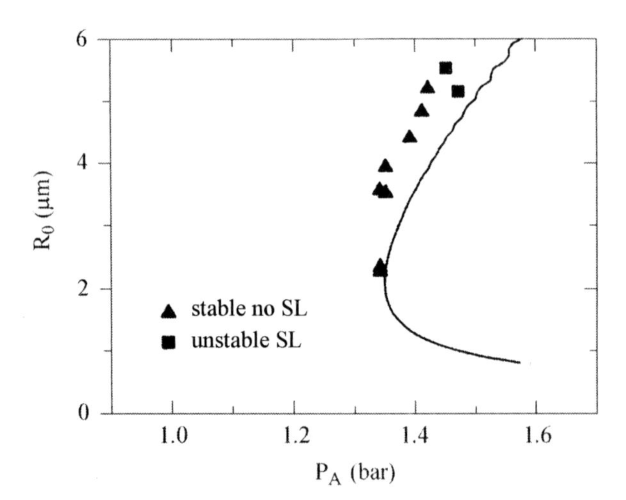


Figure 3.43 Phase diagram for pure argon saturated to 0.26%. The diffusive equilibrium curve is for $C = 0.26\%$ (solid line). With only argon present, only stable SL was observed. No bubble (stable or unstable) could be trapped below $P_A \approx 1.3$ bar and no dancing was observed. (Ketterling and Apfel (1998).)

Very recently, Simon et al. (2001) measured the phase diagrams of sonoluminescing bubbles at very low air concentration. They employed a new experimental technique to obtain the parameters of the bubble dynamics ($P_A$ and $R_0$) which is based on the information provided by the timing of the light flash in the acoustic period. Also Simon et al. (2001)

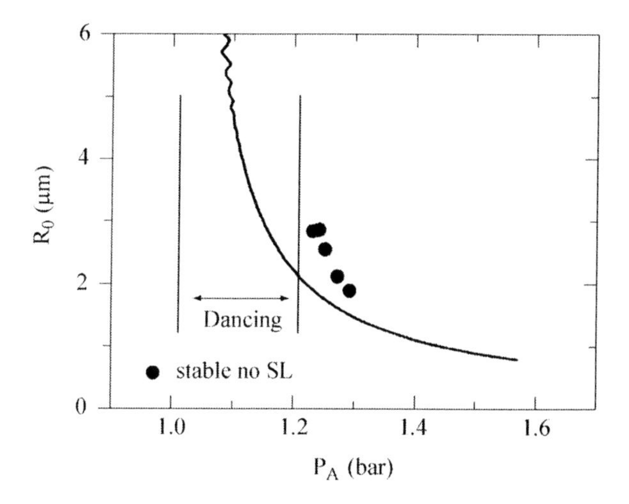


Figure 3.44 Phase diagram for pure nitrogen saturated to 10%. The diffusive equilibrium curve is for $C = 10\%$ (solid line). With no argon present, no SL was observed. Above $P_A \approx 1.3$ bar, the bubble dissolved. (Ketterling and Apfel (1998).)

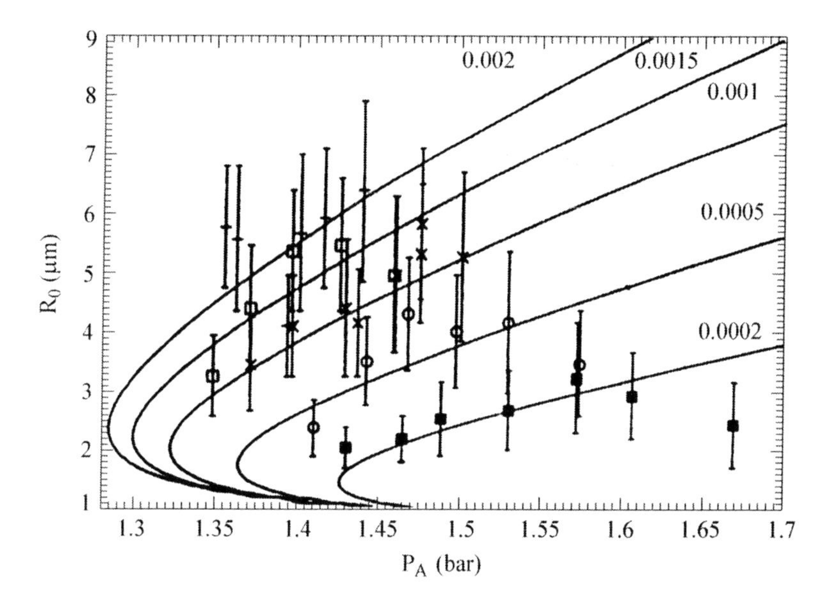


Figure 3.45 Experimental data and stable diffusive equilibrium curves with $C_i/C_0^{(N)} = 0.01 \times C_i/C_0$ in the $R_0$, $P_A$ plane. The symbols correspond to the measurements at $C_i/C_0 = 0.2$ (pluses), 0.15 (open boxes), 0.1 (crosses), 0.05 (circles), 0.02 (filled boxes). (Simon et al. (2001).)

find that the light emitting bubbles follow the stable diffusive equilibrium curve based only on the argon concentration, as predicted by the theory of Lohse et al. (1997). See Fig. 3.45 by Simon et al. (2001).

Storey and Szeri (2002) used a simplified model (Storey and Szeri (2001)) of the inter-species mass diffusion, heat transfer, chemistry and gas exchange with the liquid.

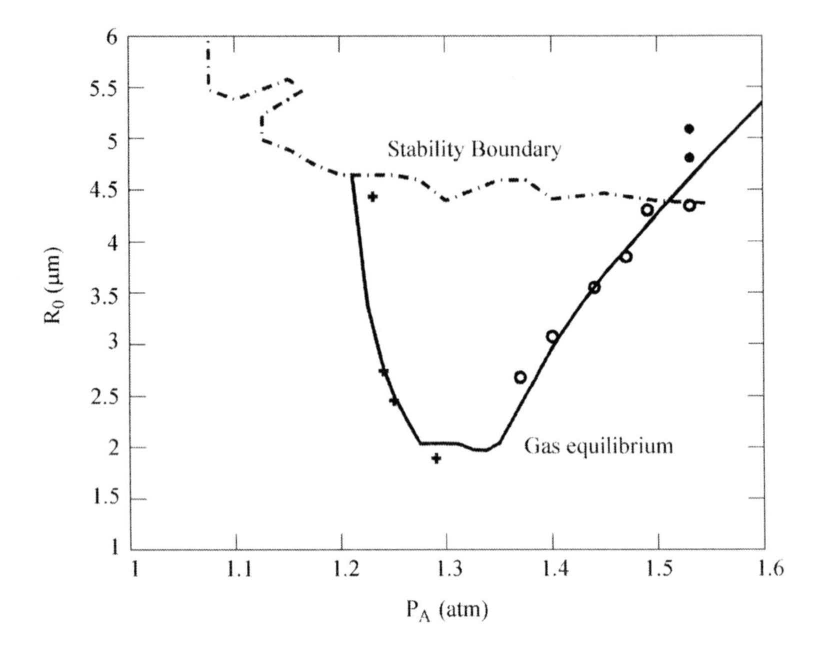


Figure 3.46 A diagram of parameter space, showing the chemical-diffusive equilibrium (solid curve) and shape stability boundary (dashed curve). The points are experimental measurements (Ketterling and Apfel (2000a)), open circles denote stable glowing bubbles, crosshairs are stable non-glowing bubbles, and stars are unstable glowing bubbles. (Storey and Szeri (2002).)

Since the location of the stability boundary in parameter space is very sensitive to detail of the radial dynamics, a complete Navier–Stokes model is used to compute the shape stability boundary (Storey (2001)). Storey and Szeri (2002) considered a driving frequency of 32.8 kHz and air dissolved at 20% saturation in water as the pressure amplitude is varied as in Fig. 3.46. The solid curve is the shape stability boundary, below which bubbles are spherically stable. The experimental points from Ketterling and Apfel (2000a) are also shown. The agreement between the model of Storey and Szeri (2002a) and the experiment is excellent. The result confirms that the dissociation hypothesis is correct. Most importantly, the dissociation hypothesis is supported in a calculation by Storey and Szeri (2002) with internal peak temperatures of only ~7,000 K and large amounts of trapped vapor.

In Fig. 3.47 is a depiction of the chemical composition of the bubble as the pressure amplitude is varied. The mole fraction of the dominant species (argon and nitrogen) at the beginning of each acoustic cycle is shown, as obtained by the simulation of Storey and Szeri (2002). The transition from a primarily nitrogen (i.e. air) to an almost pure argon bubble is clearly captured as the driving pressure increases. In the center figure the mole fraction of water vapor and OH are shown at the time of collapse, indicating that more water is trapped at higher driving by the mechanism previously described by Storey and Szeri (2000). Finally, the peak temperatures are shown in the bottom figure. Note from Fig. 3.46 that the light emission in the experiments commences as the driving pressure exceeds ~1.35 atm in the experiments.

Storey and Szeri (2002) suggest that emission from water and its reaction products should receive consideration in emission models, whether from molecular excited states or electron–atom bremsstrahlung (Didenko et al. (2000b), Young et al. (2001), Hammer and

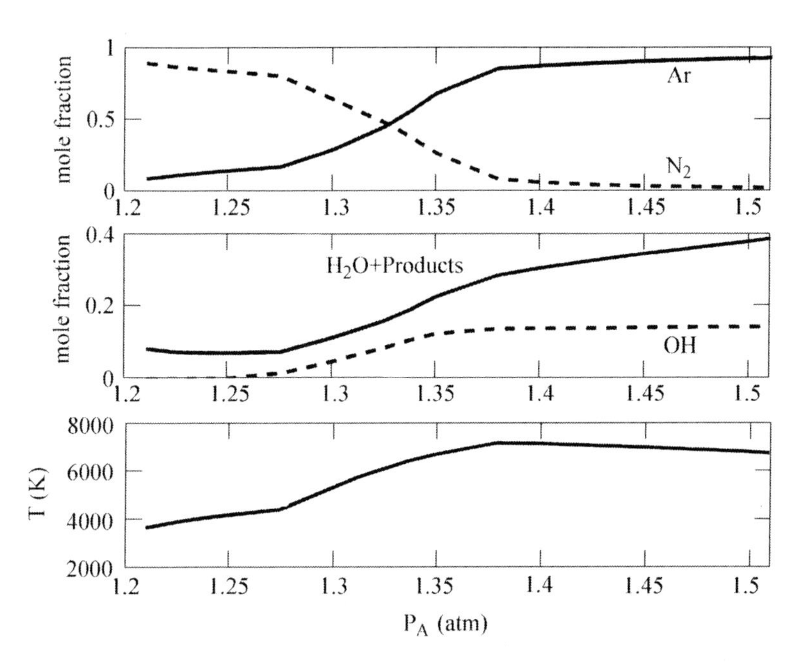


Figure 3.47 Bubble composition and temperature as the pressure amplitude is varied, for bubbles on the solid curve of Fig. 3.46. In the top figure is the repeating mole fraction at the start of each cycle. In the center figure are the fractions of water plus all its dissociated products, and OH at collapse. In the bottom figure is the peak temperature at collapse. (Storey and Szeri (2002).)

Frommhold (2001)). Water is present in considerable quantities at the collapse, and the temperatures seem consistently lower than those assumed by bremsstrahlung models that neglect vapor.

Asai and Watanabe (2000) theoretically estimated the existence of a hysteresis loop on the graph of single bubble sonoluminescence intensity with respect to sound pressure. This prediction is confirmed experimentally, and the effect of argon rectification in the bubble is discussed.

The ultimate proof of the argon rectification hypothesis would be the detection of the chemical species leaving the bubble. Lepoint and Lepoint-Mullie (1998, 1999) exchanged the *water* in the experiment for *Weissler's reagent*. *Weissler's reagent* is an aqueous NaI solution containing $CCl_4$ and starch. The oscillating bubble produced chlorine radicals which oxidized iodide to iodine, giving the distinct blue color of the iodine–starch reaction. Thus chemical reactions in and around a single bubble were conclusively demonstrated. Lepoint et al. (1999) found that the bubble triggered the reaction even at driving pressures below the single bubble sonoluminescence threshold as the temperature there can already be sufficiently large. A thread of blue color was observed to emerge from the bubble, usually in either an up- or downwards direction, Fig. 3.48.

Graça and Kojima (2002) measured the light intensities from single bubble sonoluminescence from mixtures of pairs of rare gases dissolved in water. The light from *He–Ar or He–Xe mixtures* was measured as a function of the mixing ratio, to search for the effect of species segregation during the violent bubble collapse. The results are difficult to interpret.

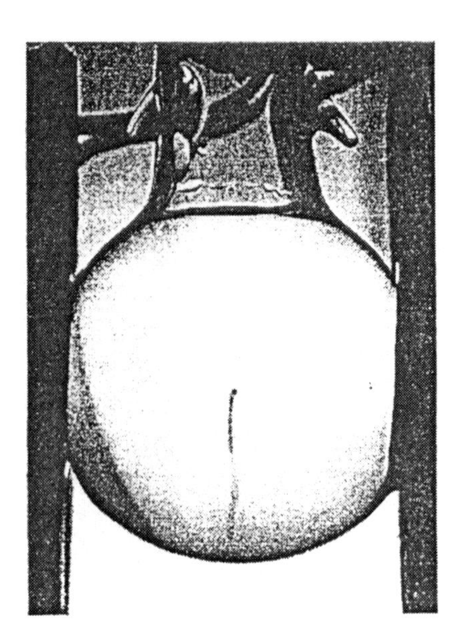

Figure 3.48 A single bubble in *Weissler's reagent.*
The single bubble activity produces chlorine radicals from the $CCl_4$ in the Weissler's reagent. These chlorine radicals oxidize iodine ions to molecular iodine, giving the blue filament from the iodine–starch reaction. (Lepoint and Lepoint-Mullie (1999).)

## 3.7 AMBIENT RADIUS

The ambient radius of the bubble is not an adjustable parameter as the system chooses it dynamically (Brenner et al. (1999), Löfstedt et al. (1995) and Barber et al. (1997a)). These papers give wide discussions of this subject. Barber et al. (1997a) show the wonderful waterfall plots in Fig. 3.49a, and Cheeke (1997) reproduces Fig. 3.49b from Barber et al. (1997a).

Section 3.4 describes how the bubble radius is measured.

A parameter of interest is the sphericity of the collapsing bubble. A spherical bubble is the most stable. As the bubble becomes less spherical it becomes more unstable. If the bubble is not spherical, more sonoluminescence will be emitted in one direction than in another.

These correlations were measured with multiple photodetectors as a function of the angle between their line of sight (Weninger et al. (1996), Barber et al. (1997a)). The angular dependence of the intensity of the sonoluminescence is characterized by the correlation

$$\Delta Q_{\mathrm{AB}} = \frac{1}{\bar{Q}_{\mathrm{A}}\bar{Q}_{\mathrm{B}}}\langle[Q_{\mathrm{A}}(i)-\bar{Q}_{\mathrm{A}}][Q_{\mathrm{B}}(i)-\bar{Q}_{\mathrm{B}}]\rangle_i$$

as a function of the angle $\theta_{\mathrm{AB}}$ between detectors A and B. $Q_{\mathrm{A}}(i)$ is the charge recorded in detector A on the $i$th flash, $\bar{Q}_{\mathrm{A}}$ is the running average of $Q(i)$ and $\langle\ \rangle_i$ denotes an average over $i$.

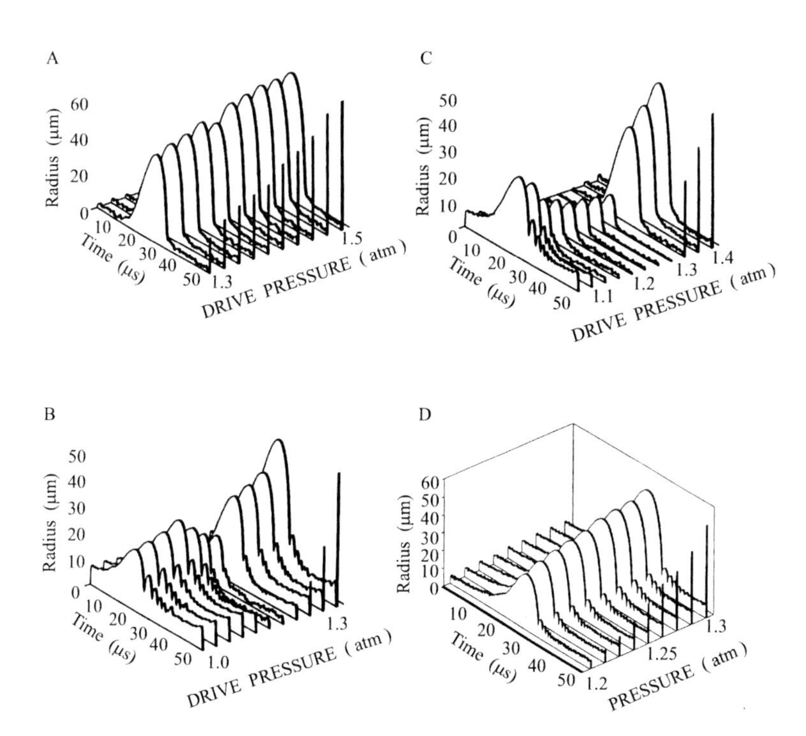


Figure 3.49a Waterfall plots. (Reprinted from *Physics Reports* **281**, 65, Barber et al. (1997a), with permission from Elsevier Science.)

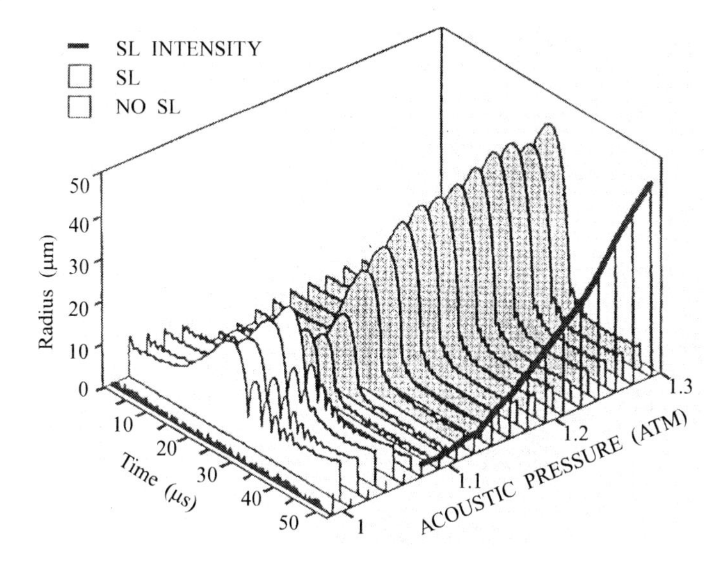


Figure 3.49b Bubble radius versus time for about one cycle of the imposed sound field as a function of increasing drive level. The shaded area represents the light-emitting region. The relative intensity of emitted light as a function of drive level is indicated by the continuous-line ramp. For the unshaded region, the bubble is trapped but no light is emitted. At drive levels below the unshaded region the bubble dissolves over a long time (~ 1 s). The lowest amplitude sweep (no bubble present) indicates the noise level. (Cheeke (1997).)

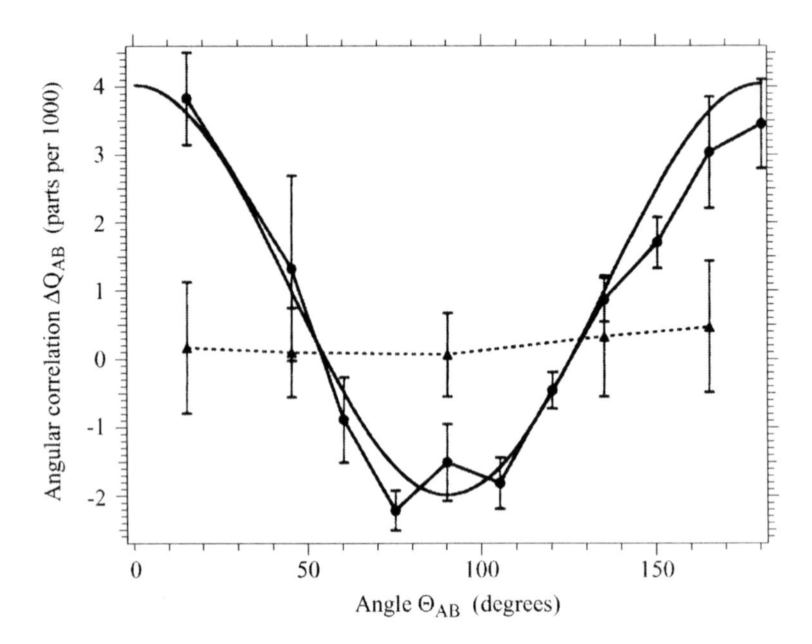


Figure 3.50 Correlation of light intensity between two phototubes subtending an angle $\theta_{AB}$ with respect to a sonoluminescing bubble. The solid line corresponds to an SL bubble whose flash to flash intensity has a large variation. The flash to flash fluctuations for the dotted line are much less and are furthermore consistent with Poisson statistics. Note the appearance of a negative correlation at 90°. The maximum dipole observed is about 10 parts per thousand. If the dipole is due to refraction of light at the gas–fluid interface of the bubble then the ellipticity of the bubble in the state with large fluctuations is about 20%. The sine wave fit is $0.001(1 + 3\cos\theta)$. (Reprinted from *Physics Reports* **281**, 65, Barber et al. (1997a), with permission from Elsevier Science.)

The solid line in Fig. 3.50 displays the angle dependent correlation. It can be attributed to a dipole component in the detected photon field.

Madrazo et al. (1998) calculated the shape and size of sonoluminescing single bubbles from a theory for the sonoluminescent light flash diffraction occurring at the gas–liquid interface of a single bubble. Their results maintain that the bubble surface is actually an *ellipsoid of eccentricity very close to 0.2 and major radius between 1.5 and 2 μm.* They found excellent agreement between these theoretical results and the experimental data by Weninger et al. (1996).

## 3.8 SPECTRA FROM SINGLE BUBBLE SONOLUMINESCENCE

Hiller et al. (1992, 1994, 1998), Hiller and Putterman (1995) and Gaitan et al. (1996) in Barber et al. (1997a) show various spectra of the rare gases, nitrogen doped with the rare gases, xenon in various organic liquids, and ethane in water. Figures 3.51 and 3.52 are from these papers.

Young et al. (2001) discovered spectra peaks in extremely dim SBSL in water, with the driving pressure very close to the forcing pressure for SBSL. The SL was so weak that the photons had to be collected over several days. Figure 3.53 shows how as the forcing

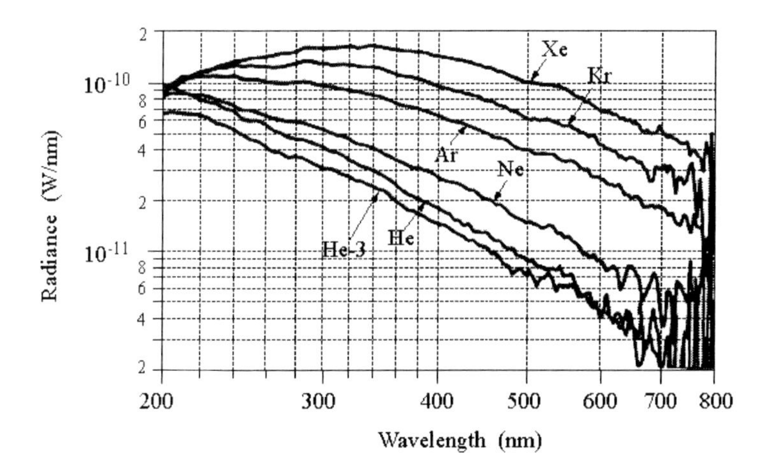


Figure 3.51 Room temperature spectra of various noble gases in a cylindrical "supracil" resonator. No transmission corrections have been used. The gases have been dissolved at 3 mm. (Reprinted from *Physics Reports* **281**, 65, Barber et al. (1997a), with permission from Elsevier Science.)

pressure is increased, the OH-line disappears behind the enhanced continuum contribution to the spectrum. Didenko et al. (2000a,b) discovered spectra of SBSL in organic liquids, see Fig. 3.54. These tend to need higher driving to show SBSL because the vapor molecules have more rotational and vibrational degrees of freedom, leading to a smaller temperature increase at bubble collapse. In order to achieve SBSL with organic fluids, Didenko et al. (2000b) chose their fluids according to the two criteria: (i) a small liquid vapor pressure to reduce the vapor concentration in the bubble, (ii) the dissociation products of the vapor should dissolve in the liquid so that they do not accumulate in the liquid, which would again lower the polytropic exponent (Ashokkumar et al. (2000)). Didenko et al. (2000b) thus found adiponitrile to be an optimal liquid, and indeed they could observe very pronounced CN lines in its SBSL spectrum. The observation of spectral lines is a way of finding out what is going on in the bubble in SL, for example, what chemical reactions are occurring in the bubble. Also, Suslick and collaborators, Suslick et al. (1986), Flint and Suslick (1991), Suslick et al. (1999), Didenko et al. (1999) have used the widths and intensities of spectral lines in multibubble sonoluminescence to deduce the temperature of cavitation.

Matula et al. (1995) measured the multibubble and single bubble sonoluminescence in a 0.1 M sodium chloride solution at a pressure amplitude of approximately 1.2 bars and a frequency of 32 kHz. Figure 3.55 shows the result.

Matula et al. (1995), Matula and Roy (1997) and Crum (1994) explain the difference between single bubble and multibubble sonoluminescence spectra as follows. In single bubble sonoluminescence, symmetric bubble collapse leads to shock wave formation in the gas in the bubble, whereas in multibubble sonoluminescence, asymmetric bubble collapse leads to liquid jets that penetrate the hot bubble interior and result principally in incandescence of the host liquid.

However, doubt has been cast on this simple explanation, because it has now been established that although shock waves occur in the liquid, they do not occur in the gas in the bubble. See Section 4.7.

Kordomenos et al. (1999) made an apparatus to detect *microwave emission* from a sonoluminescing single bubble. For a xenon-doped bubble, which emits the greatest amount

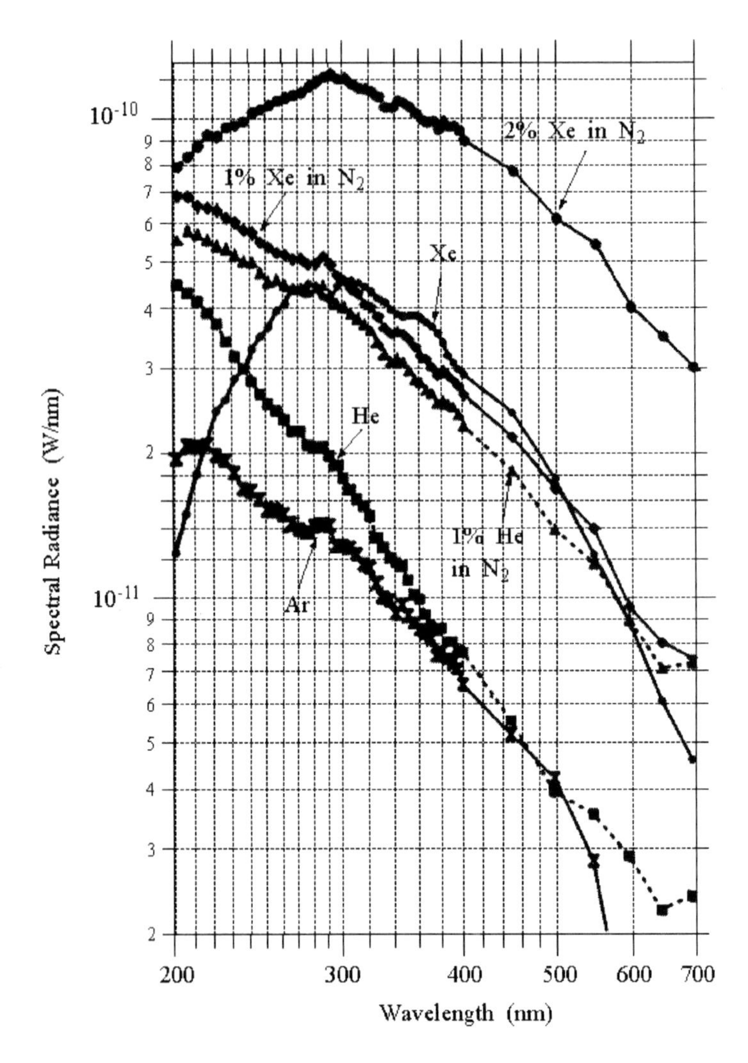


Figure 3.52 Spectrum of SL for various gas mixtures dissolved into water at 150 mm and $T = 24°C$. Note that although helium is dimmer than xenon it has a greater spectral density in the ultraviolet. (Reprinted with permission from Hiller et al. *Science* **266**, 281. Copyright 1994 American Association for the Advancement of Science.)

of light, no microwave emission was observed in a band of frequencies ranging from 1.65 GHz to 2.35 GHz. The sensitivity of the experiment was such that signals greater than 1 nW would have been detected. However, Hammer and Frommhold (2002b) show that this upper bound is compatible with the radiation processes that they think generate significant emission at optical frequencies, electron neutral and electron ion bremsstrahlung. Hammer and Frommhold (2002b) argue that, almost independently of the specific assumptions concerning the hydrodynamics or the nature of the radiative processes, microwave emission from single bubble sonoluminescence exceeding that upper bound can hardly be expected.

Recently, Simon and Levinsen (2003b) measured the spectra from single sonoluminescing bubbles in water at different argon concentrations and sound excitation levels. All the experimental conditions were either measured directly or were derived from measured

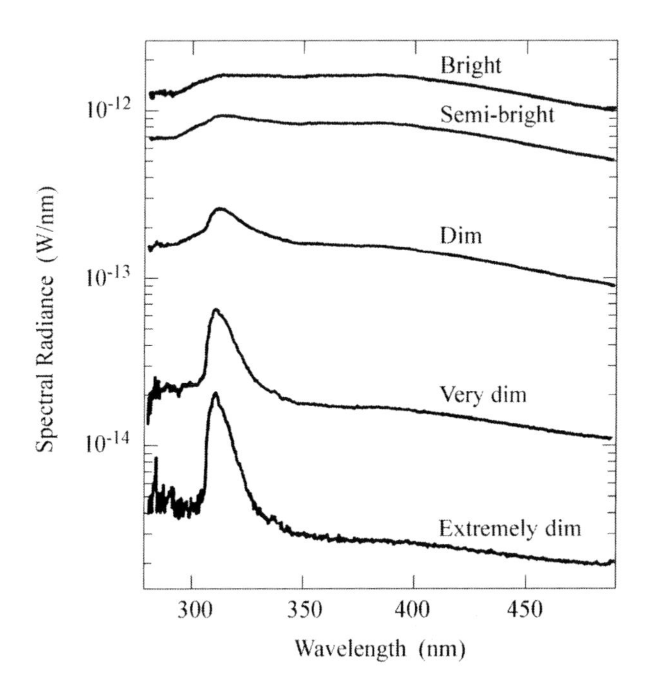


Figure 3.53 Average intensity dependence of argon single-bubble sonoluminescence spectral peaks. Spectra are shown for five different levels of overall brightness. The sonoluminescence was produced using mixtures of deionized water and argon gas with a partial pressure of ~150 torr at 25°C. (Young et al. (2001).)

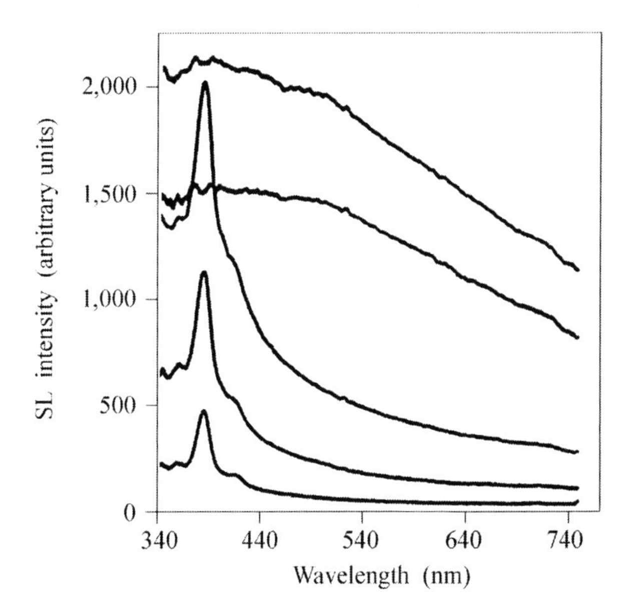


Figure 3.54 Single-bubble sonoluminescence spectra of adiponitrile. Acoustic pressure increases from bottom to top, from 1.7 to 1.9 bar. (Reprinted with permission from *Nature*, Didenko et al. **407**, 877. Copyright (2000) Macmillan Magazines Limited.)

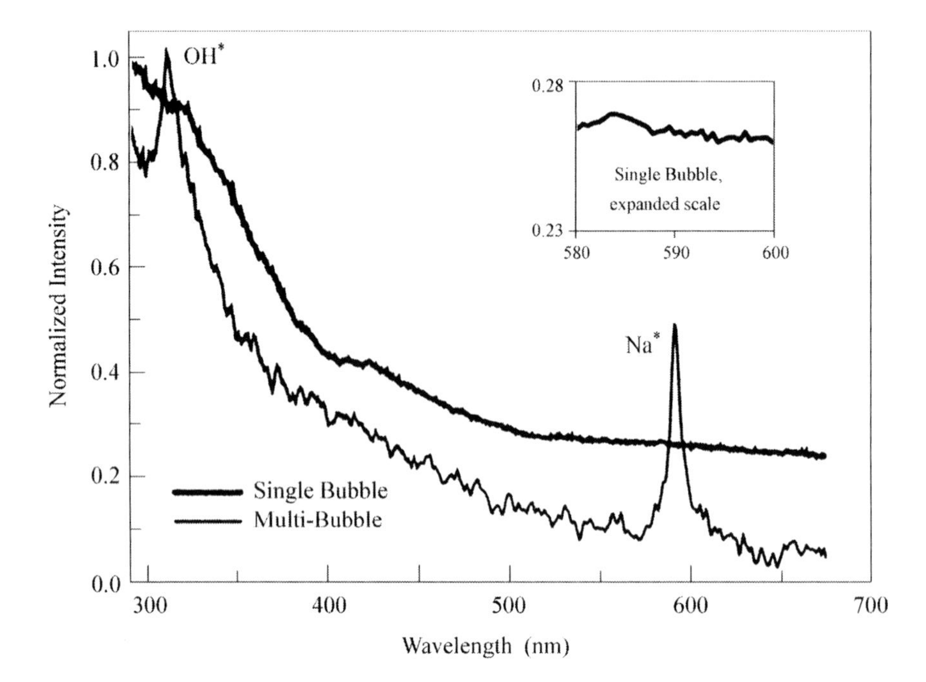


Figure 3.55 Comparison of the background subtracted spectra of MBSL and SBSL in a 0.1 M sodium chloride solution. Each spectrum was normalized to its highest intensity. Absolute radiance comparisons cannot be made due to differences in light gathering techniques. Note the prominence (in MBSL) and absence (in SBSL) of the sodium emission line near 589 nm. The inset illustrates the absence of structure associated with SBSL around 589 nm. The signal level in this region is a factor of 3 above background. (Matula et al. (1995).)

quantities for each spectrum, thus enabling the parametric dependence of the spectra to be analyzed. To characterize the data in a given wavelength interval, the shape of the spectra was fitted with the Planck function. The effective temperatures obtained from these fits were in the range 12,000–18,000 K, and were almost independent of the expansion ratio $R_{max}/R_0$. The spectra intensity normalized by the volume of the bubble increased with the expansion ratio as a power law. The effective temperatures *decreased* with the pressure amplitude each argon concentration, while the light intensity as measured with a photomultiplier tube, *increased*. Simon and Levinsen (2003b) suggest that these observations arise because the increased energy input due to the higher pressure amplitudes results in an increased number of less energetic photons as compared to the case of a lower excitation level.

## 3.9 SONOLUMINESCENCE PULSE WIDTH

Using time-correlated single photon counting, Hiller et al. (1998) measured pulse widths from under 40 to over 350 ps for mixtures of various gases in water. Figure 3.56 shows the SL pulse from 1.8 mg/l $O_2$ with 1.2 bar driving pressure at 20 kHz, from Gompf et al. (1997). The acoustic period will be 50 $\mu$s. Thus the pulse width is over $10^6$ times smaller than the period of the sound field. Note that the fastest visible transition ($3 \rightarrow 2$) of the

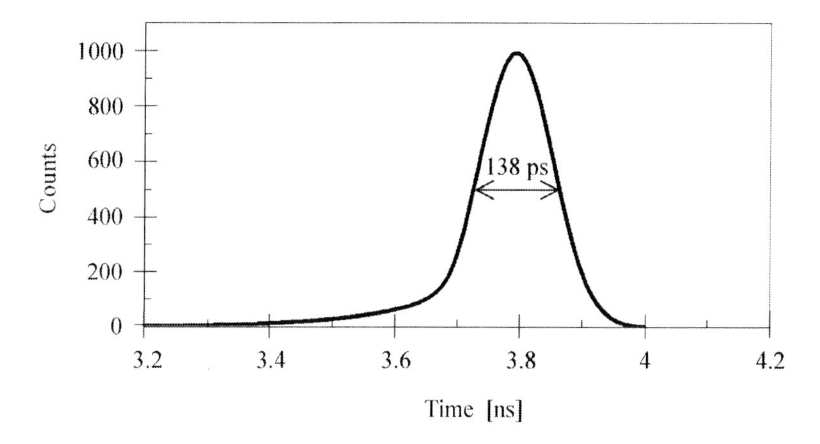


Figure 3.56 SL pulse after subtraction of the reflections and deconvolution. The pulse shape is slightly asymmetric and its FWHM at 1.2 atm driving pressure and a gas concentration of 1.8 mg/l $O_2$ at 22°C is 138 ps (±10 ps). (Gompf et al. (1997).)

hydrogen atom is over 100 times slower. Contributions to this topic have been made by Barber et al. (1991, 1992, 1997a,b), Moran et al. (1995), Gompf et al. (1997), Matula et al. (1997) and Hiller et al. (1998).

## 3.10 EFFECT OF REDUCTION OF AMBIENT PRESSURE ON SINGLE BUBBLE SONOLUMINESCENCE

Dan et al. (1999, 2000) measured this for a 25% by weight degassed water/glycerine mixture with an acoustic drive amplitude $P_A$ = 1.29 atm and a frequency of around 17.5 kHz. Figure 3.57 shows that decreasing the ambient pressure over the liquid $P_0$ from

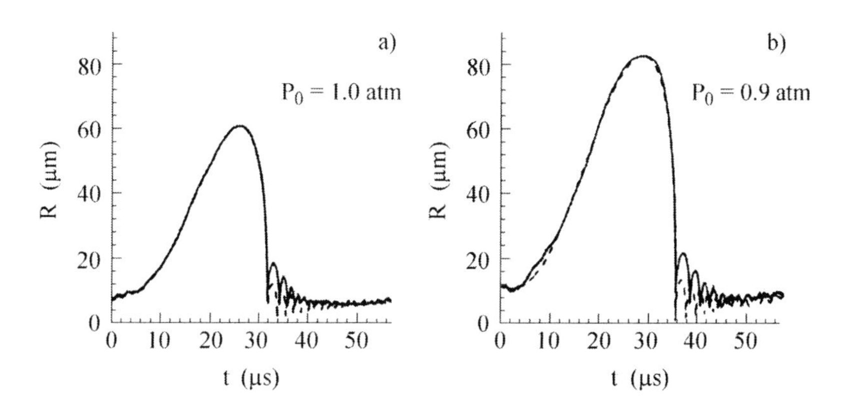


Figure 3.57 The experimental results (broken lines) and the fits (solid lines) for two values of ambient pressure, $P_0$. In (*a*) the acoustic pressure amplitude, $P_a$, and $R_0$ are used as fitting parameters ($R_0$ = 7.3 μm, $P_a$ = 1.29 atm); in (*b*) we use $P_a$ = 1.29 atm, and fit only $R_0$ = 9.0 μm. The liquid viscosity is 0.021 cm²/s, and surface tension is 69.4 dyn/cm. (Dan et al. (1999).)

1.00 atm to 0.90 atm increases $R_0$ from 7.3 μm to 9.0 μm, and $R_{max}$ from 60 μm to 82 μm. Thus $R_{max}/R_0$ increases from 8.2 to 9.1, leading to a stronger collapse, and an increase in SL intensity of about seven times!

## 3.11 SINGLE-BUBBLE SONOLUMINESCENCE IN MICROGRAVITY

Matula (2000) took his SBSL apparatus on board NASA's parabolic research aircraft in order to determine the effect of microgravity on SL. The aircraft performs parabolic trajectories that allow the participants and experiments brief periods (≈25 s) of microgravity (near 0 g) and hypergravity (near 2 g) (Fig. 3.58).

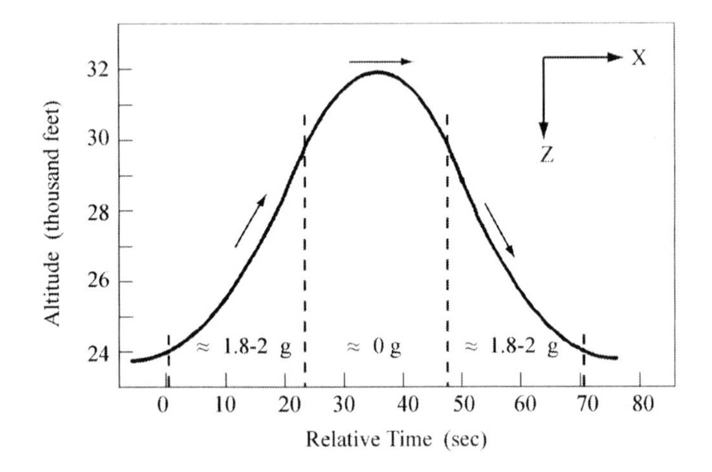


Figure 3.58 The parabolic trajectory of NASA's research aircraft. (Matula (2000).)

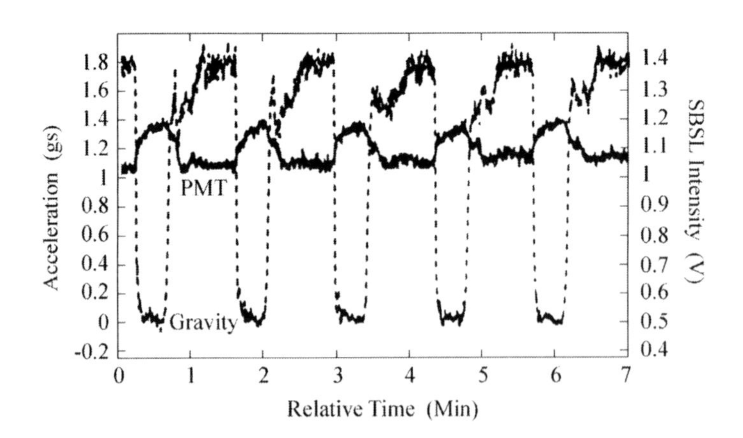


Figure 3.59 The change in sonoluminescence intensity (—) can be correlated directly with the change in gravity (- - -). Note that the intensity increases as soon as the gravitational acceleration begins to decrease, followed by a continued increase in intensity during the microgravity phase. The cell used in these studies was a 100 ml round-bottom spherical cell operated at 31 kHz. (Matula (2000).)

Figure 3.59 shows that the SL increases by about 20% in microgravity. The microgravity affects the buoyancy and this leads to the increase in SL.

## 3.12 EVIDENCE FOR NUCLEAR REACTIONS IN SONOLUMINESCENCE

Taleyarkhan et al. (2002a) claim to have achieved the conditions for D + D nuclear fusion in a multibubble sonoluminescence experiment.

A deuteron D is a proton and a neutron.

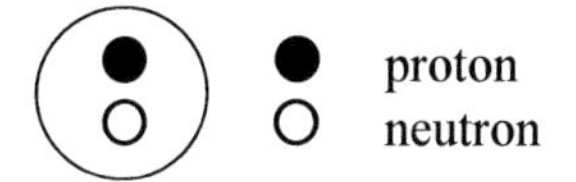


The fusion of 2 deuterons can take 2 equally likely paths:

$$D + D = P + T$$

or

$$D + D = N + {}^{3}He$$

where T is a tritium nucleus (proton + neutron + neutron) and $^{3}He$ is a helium-3 nucleus. Neutrons, N, would be capable of being detected.

Taleyarkhan et al. (2002a) cavitated deuterated acetone ($CD_3COCD_3$) at 19.3 kHz with a large acoustic pressure amplitude of 15 bars over the long time of 7 hours. They reported the production of sonoluminescence, tritium (detected by a scintillation counter), and high-energy neutrons. Since very high temperatures ($\geq 10^6$ K) are needed to produce D + D fusion these results have produced scepticism (Highfield (2002)). They will certainly need to be repeated.

Becchetti (2002) and Pennicott (2002) have commented on the above results.

Shapira and Saltmarsh (2002) have repeated the experiment of Taleyarkhan et al. (2002a) in the same laboratory with the same equipment except for substituting a larger scintillation detector and a more sophisticated data acquisition system. They could find no evidence for 2.5-MeV neutron emission correlated with sonoluminescence from collapsing bubbles. And they state that any neutron emission that might occur is at least 4 orders of magnitude too small to explain the tritium production reported in Taleyarkhan et al. (2002a) as being due to D–D fusion.

Taleyarkhan et al. (2002b) point out, among other things, that because Shapira and Saltmarsh's (2002) detector did not fit within the experimental enclosure, it was placed outside, beyond some shielding. They also say that Shapira and Saltmarsh's (2002) detector threshold was set too high to capture many of the lower-energy neutrons. Dumè (2004) also comments.

The alleged production of nuclear fusion from sound waves in a liquid is sometimes called *sonofusion*.

## 3.13 SINGLE CAVITATION BUBBLE LUMINESCENCE

Although *single cavitation bubble luminescence* is not sonoluminescence, as it is not produced by sound but by a laser, it is included here for completeness.

Ohl (2002) has studied single cavitation bubble luminescence for varying degrees of asphericity of the bubble. Single bubbles were generated by focusing an intense and short infrared light pulse from a Q-switched Nd:YAG laser tightly into a water filled cuvette. The degree of asphericity was controlled by an adjustable rigid boundary near the bubble. Temporal, single and multiple light emission events occurred during a time interval of 80 ns. The luminescence duration increased with increasing asphericity. Spatially, the emissions from nonspherically collapsing bubbles displayed a pronounced halo around the central spot several times larger than the luminescence from a spherical collapse. Spectrally, the ratio of the line to the continuum emission of the sodium doublet was enhanced, whereas the total emitted energy decreased by four orders of magnitude as compared to the spherical collapse. These findings pointed towards emission not only from the bubble interior but also from the liquid surrounding the bubble.

## 3.14 HARMONIC ENHANCEMENT OF SINGLE BUBBLE SONOLUMINESCENCE

Lu et al. (2003) and Prosperetti and Lu (2001) discussed ways of enhancing sonoluminescence emission. Increasing the sound field amplitude does not work, since at high driving amplitudes, the spherical shape becomes unstable, which leads to fragmentation and ultimate destruction of the bubble. Another approach to achieve the same objective has been the use of a lower-frequency drive which, however, has proved equally unsuccessful due to the accumulation of water vapor inside the bubble.

To advance further, some investigators have experimented with multi-drive systems. Holzfuss et al. (1998b) used a cylindrical cell to obtain single bubble sonoluminescence and varied the relative phase shift between the *fundamental (or first harmonic) and the second harmonic* of the driving frequency. They reported an increase of up to 300% of the maximum light output. Similar results were obtained by Hargreaves and Matula (2000). Krefting et al. (2002) used a *first harmonic mode of 25 kHz and the second harmonic at 50 kHz*, and enhanced the brightness by 250%. Morago et al. (2000) used the *first harmonic and the tenth harmonic* and obtained an enhancement of more than 250%. Seeley (1999) did not find any enhancement with the same type of experiment as in Holzfuss's work.

Other studies have been made by Ohsaka and Trinh (2000), Ketterling and Apfel (2000b) and Chen et al. (2002).

Thomas et al. (2002) used a spherical glass flask with two 27.6 kHz piezoelectric transducers for driving the single bubble at the center of the flask, and eight piezoelectric transducers with center frequency of 700 kHz operating in the pulse mode and *focusing* on the sonoluminescing bubble. They obtained a 90% gain in the light intensity. Ogi et al. (2003) also used a high frequency pulse. They produced sonoluminescence at 40 kHz in a rectangular epoxy cell filled with distilled water and 25% glycerine. A polyvinylidene fluoride point-focusing transducer was driven by a 700 W pulse generator to create an acoustic pulse on the sonoluminescing bubble. The center frequency of the pulse was 10 MHz and the duration was 0.15 μs. The pulse was triggered every 100 cycles of the 40 kHz standing wave. Only the pulse that arrived at the bubble at the growing stage could increase the brightness. The brightness was increased by 300%.

Also, a series of experiments have been performed by Umemura and his group on the enhancement of sonochemical reactions by imposing the second harmonic on the fundamental ultrasonic wave. Although sonoluminescence was not involved, these references are noted here. They are Umemura and Kawabata (1993), Umemura et al. (1994, 1995, 1996, 1997) and Kawabata and Umemura (1996a,b).

Lu et al. (2003), in an elegant paper demonstrating a deep understanding of bubble behavior, made a systematic theoretical investigation of the matter to see if it would be possible to further increase the light emission. They considered a sound field of two harmonics to maximize the peak temperature of the bubble, under the condition that the spherical stability of the bubble was preserved. The two frequencies were 26.5 kHz and 53 kHz for an argon bubble in water. The fundamental frequency $\omega/2\pi$ was chosen as 26.5 kHz because this is of the order of the frequency used in much of the experimental work, e.g. Barber et al. (1997a). Lu et al. (2003) considered 3 harmonic amplitudes ($p_1, p_2, p_3$) so that the acoustic pressure $P_A$ is

$$P_A = p_1 \cos \omega t + p_2 \cos(2\omega t) + q_2 \sin(2\omega t)$$

They chose the values of the individual amplitudes $p_1$, $p_2$, $p_3$ in such a way as to maximize the bubble compression ratio $R_{max}/R_{min}$. Using a mathematical method and taking into account the effects of water vapor diffusion, conductive heat loss, and chemical reactions of the water vapor, they were able to calculate the maximum bubble temperatures for single-frequency drive, and multi-frequency drive for various bubble equilibrium radii $R_0$. One example will be given:

| | Single-frequency drive | | | | Multi-frequency drive | | | | | |
|---|---|---|---|---|---|---|---|---|---|---|
| $R_0$ μ | $R_{max}/R_{min}$ | $p_1$ kPa | $c_\infty/c_{sat}$ | $T_{max}$ K | $R_{max}/R_{min}$ | $p_1$ kPa | $p_2$ kPa | $q_2$ kPa | $c_\infty/c_{sat}$ | $T_{max}$ K |
| 3.0 | 267 | 243 | $2.86 \times 10^{-5}$ | 21.500 | 633 | 16.3 | 1,306 | 108 | $3.7 \times 10^{-6}$ | 319,000 |

$c_\infty$ = gas concentration in the water
$c_{sat}$ = saturation concentration at $P_\infty$
$P_\infty$ = pressure in the water far from the bubble
1 kPa = $10^{-2}$ atm

Regardless of the equilibrium size of the bubble, the time dependence of the bubble radius is markedly different between one- and two-frequency driving. Figure 3.60a shows this for $R_0$ = 5 μm, where the dashed and solid lines are for the single- and two-frequency drive, respectively. The bubble core temperature and the argon mole fraction are also shown in this figure.

Although this analysis assumes bubble stability, and neglects ionization; provided the dissolved argon gas is kept extremely low in order to maintain diffusive stability, the trend to higher temperatures for two-frequency driving is clear, and further work is to be encouraged.

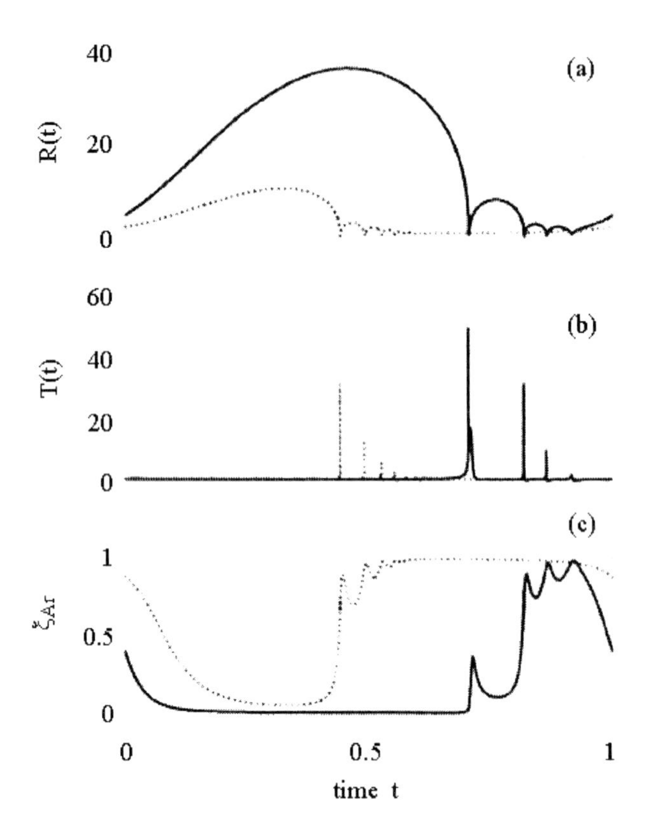


Figure 3.60 (*a*) Time dependence of the radius (normalized by its equilibrium value $R_0 = 5$ mm) for single- (dashed line) and two-frequency (full line) drive. (*b*) Corresponding temperature $T(t)$ normalized by $T_0 = 293.15$ K. (*c*) Argon mole fraction $\xi_{Ar} = C_\infty / C_{sat}$. (Lu et al. (2003).)

## 3.15 PERIOD DOUBLING IN SONOLUMINESCENCE

Holt et al. (1994) studied *period doubling, chaos and quasiperiodicity* in single bubble sonoluminescence.

Dam et al. (2002, 2003), in comprehensive papers, remark that *period doubling* and chaos are common phenomena in nonlinear systems. Single bubble sonoluminescence represents an extreme example of such systems. It therefore comes as no surprise that shape instabilities and *period doubling* to chaos have been observed in the system. Early attempts to explain *period doubling* in multibubble sonoluminescence and single bubble sonoluminescence centered on spherically symmetrical collapse e.g. Holzfuss and Lauterborn (1989). This gives evidence that shape instabilities and *period doubling* in single bubble sonoluminescence are linked, giving additional evidence that stable single bubble sonoluminescence does not necessarily imply perfect spherical bubble collapse, even if the extinction threshold is linked to the onset of shape distortions.

Dam et al. (2002, 2003) used a 6.5 cm diameter glass sphere with a pair of piezoelectric transducers glued to opposite sides with epoxy resin. A drive frequency of 25,100 Hz was maintained. A bubble was generated by a computer controlled blast of air

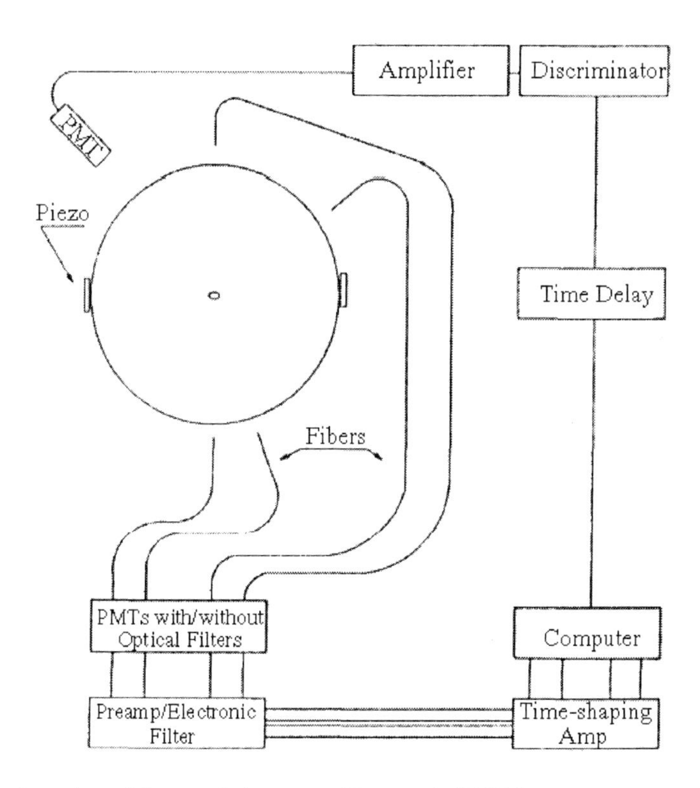


Figure 3.61 Overview of the correlation setup. (Dam et al. (2003).)

onto the free surface. The computer was programmed to automatically redo this if bubble extinction occurred (or none was generated), and to automatically search for stable light emission above a preset intensity limit. Figure 3.61 shows the correlation setup consisting of four quartz fibers of 1 mm diameter leading the emitted light to four photomultiplier tubes. The fibers were placed at well defined longitudes in the equatorial plane of the sphere pointing at the bubble from a distance of about 4.5 cm. The observations were made both with and without narrow band optical filters. The symmetry broken states were seen in all cases even using a 650±40 nm filter. This fact can be used to distinguish between different theories for the light emission. Prior to the measurements reported in Dam et al. (2002, 2003), theoretical attempts to explain observations of *period doubling* bifurcation phenomena in single bubble sonoluminescence were centered on radially bifurcated collapses. The experiments reported in Dam et al. (2002, 2003) show equivocally that the observations are primarily a result of breaking the spherical symmetry in the bubble collapse. *Period doubling* will at most show up as secondary effects in the total light output, if at all.

Levinsen et al. (2003) continued their study of *period doubling* for the case of states that were stable with their spherical symmetry broken. Observations were made using seven detectors distributed in the equatorial plane of the bubble. Contrary to earlier experiments by Holt et al. (1994), where *period doubling* was observed in the time intervals between flashes but *not* in the pulse heights, they observed *period doubling* in pulse heights, but *no* corresponding *period doubling* was seen in the time intervals. In parameter space the *period doubling* was observed below the $n = 2$ shape instability boundary line where extinction was shown to have taken place.

## 3.16 VORTEX RINGS AND PARTICLE DRIFT IN SONOLUMINESCENCE

These contributions are by Longuet-Higgins.

Longuet-Higgins (1996) remarks that drops of water falling on a plane surface are known to produce *vortex rings* which carry vertical momentum away from the surface. A similar phenomenon should also occur in a cavity which collapses asymmetrically. An inward jet will also produce *vortex rings* on impact with the opposite wall of the cavity. It is suggested that the phenomenon may be taking place on each cycle of collapse of a single *sonoluminescing* bubble. The transport of jet momentum away from the bubble by the *vortex rings* should help to restore the bubble, on rebound, to a spherical shape.

Longuet-Higgins (1997) next considered *the drift of fluid particles* induced by a bubble oscillating with motions typical of a single *sonoluminescent* bubble. In the first calculation the bubble is assumed to remain spherical and to oscillate both radially, with a smaller amplitude. The paths of marked particles were traced by numerical integration of the velocity field. It was found that the resulting drift motion at large distances from the oscillating sphere is similar to a steady *dipole* flow. The strength of the *dipole*, however, is typically seven orders of magnitude less than the streaming motion observed in the laboratory by Lepoint-Mullie et al. (1997) and Lepoint et al. (1997). A second calculation supposed that the bubble does not remain spherical but collapses asymmetrically, with one side falling inwards towards the other at a greater speed. The resulting impulse again induces a dipole motion in the far field, whose strength may be estimated on certain assumptions. It was found to be quite comparable to that observed.

## 3.17 BOOSTING SINGLE BUBBLE SONOLUMINESCENCE WITH A DROP OF GLYCERINE IN WATER

Young (1965a) in Sec. 2.2.5 found that glycerine (sometimes called glycerol) gave more multibubble sonoluminescence than any other liquid. He studied glycerine–water mixtures and found that the multibubble sonoluminescence increased with glycerine content, and that this indicated a correlation with viscosity.

Harba and Hayashi (2003a) injected a drop of glycerine at the antinode of standing acoustic waves in pure water and found that it diffused nearly isotropically leaving a stable *sonoluminescing bubble.* Their apparatus is shown in Fig. 3.62. The Langevin transducer operated in the (1,1,1) mode at a frequency of 29.5 kHz. The dynamics of a *single sonoluminescing bubble* was measured primarily by Mie light scattering. The bubble was illuminated by a 20 mW laser diode of wavelength 532 nm, and the scattering light was focused onto an optical guide connected to a photomultiplier tube at a scattering angle of 90°. The output of the photomultiplier tube was fed to a digital oscilloscope, which was triggered by the function generator. The status of the bubble and/or its surroundings were also monitored by a video tele-microscope.

The experimental procedure was as follows. A standing acoustic wave was produced in the water in a ($40\times40\times53$ mm$^3$) quartz cell. A drop of glycerine about 1 mm in diameter was injected at an antinode by a dropper i.e. a glass syringe with a rubber bulb. The glycerine drop contained small air bubbles, which turn into seeding bubbles. The glycerine

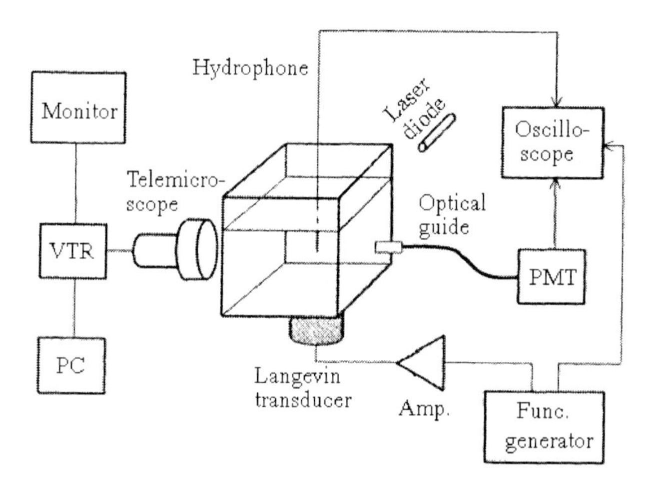


Figure 3.62 Apparatus (Harba and Hayashi (2003a)). © Jap Soc App Phys

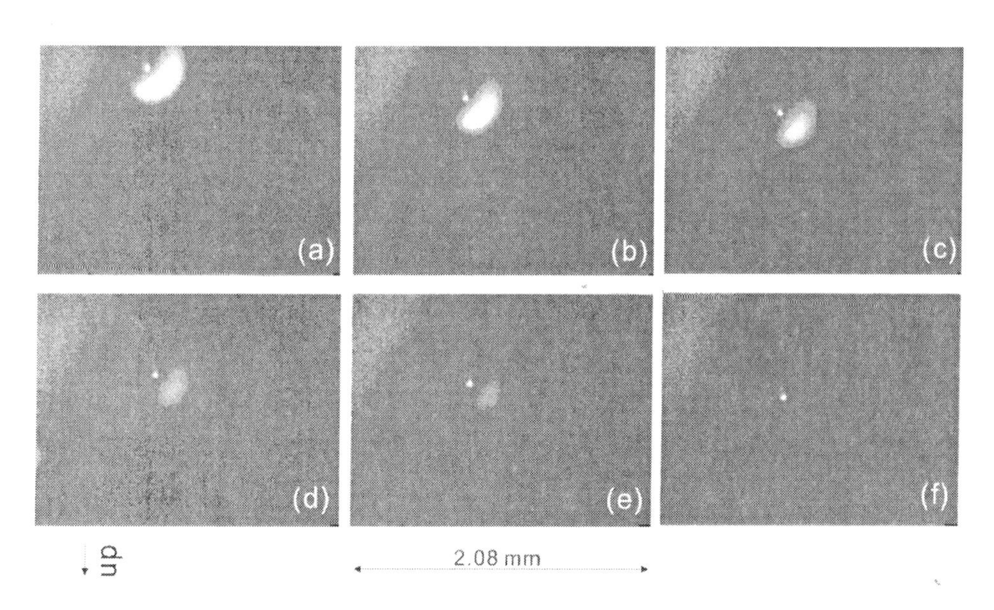


Figure 3.63 Selected frames from film shot at 30 f.p.s. show SBSL bubble from glycerol drop and the process of glycerol diffusion. The frame dimension is 2.08 mm and the arrow direction indicates the upper side of the images. (Harba and Hayashi (2003b).) © Jap Soc App Phys

diffused into the water leaving a *stable sonoluminescing bubble* behind, which was found to be boosted as much as 40% in comparison with that in pure water. Figure 3.63 shows selected frames from film shot at 30 frames per second for a *single sonoluminescing bubble* originating from a drop of glycerine, where the microscope resolution was better than 3 μm. The bubble was *sonoluminescing*. The frames (a), (b), (c), (d), (e) and (f), corresponding to frame numbers 10, 20, 30, 40, 50 and 70, respectively, illustrate how the glycerine diffuses. The glycerine is colored white in the frames because of the difference in refractive index. The bubble in frame (a) was surrounded by a hemispherical glycerine drop. As the bubble approaches the antinode, the glycerine has apparently diffused almost completely as shown in frame (f).

Harba and Hayashi (2003b) injected a *dusty* glycerine drop at the antinode of an acoustic standing wave in water. After the glycerine had diffused into the water, sonoluminescence was produced. Whereas glycerine boosted sonoluminescence (Young (1965a), Harba and Hayashi (2003a)), the dust particles diminished sonoluminescence. With the dust particles the bubble was dim and the sonoluminescence intensity changed from cycle to cycle. The effects of the dust particles were revealed by a telemicroscope, a stroboscope and by Mie scattering. Various types of deviation from the ordinary radius–time curve were observed from Mie scattering. An outstanding example was that in which the scattering intensity alternated between two types of waveform every 10 seconds while the acoustic pressure was kept fixed.

## 3.18 SINGLE BUBBLE SONOLUMINESCENCE IN AIR-SATURATED WATER

Single bubble sonoluminescence experiments are usually conducted in more or less degassed liquids, as above 50% air saturation of the water, shape and diffusion instabilities of the oscillating bubble set in, as described for example by Holt and Gaitan (1996). Krefting et al. (2003) pointed out that with an argon gas fraction of 1% of total air content, diffusively stable bubbles *theoretically* exist even in water which has not been degassed, *though only over a small range*. To test this hypothesis, Krefting et al. (2003) made a careful study using an experimental system with a frequency of about 23 kHz. *Air-saturated water* was used and gave luminescence which was bright and observable with a dark adapted eye, though a photomultiplier tube was used for the measurements. The pressure range of stable bubble oscillations was very narrow and limited by shape instabilities at high and low pressure amplitudes. The bubble radius was measured from direct images recorded with a digital video camera and a long distance microscope, illuminated by a microsecond flashlight. Experimental radius–time curves agreed well with the theoretical model described by Krefting et al. (2002). The experiment showed that single bubble sonoluminescence in *air-saturated water* is possible, provided cavitation bubble sources are absent and a suitable pressure amplitude is chosen.

## REFERENCES

Akhatov I, Gumerov N, Ohl CD, Parlitz U and Lauterborn W 1997 The role of surface tension in stable single-bubble sonoluminescence, *Phys Rev Lett* **78**, 227.

Arakeri VH and Giri A 2001 Optical pulse characteristics of sonoluminescence at low acoustic drive levels, *Phys Rev E* **63**, 066303.

Asai T and Watanabe Y 2000 Existence of hysteresis based on argon rectification in single bubble sonoluminescence, *Jpn J Appl Phys* **39**, 2969.

Ashokkumar M, Crum LA, Frensley CA, Grieser F, Matula TJ, McNamara III WB and Suslick KS 2000 Effect of solutes on single-bubble sonoluminescence in water, *J Phys Chem A* **104**, 8462.

Augsdorfer UH, Evans AK and Oxley DP 2000 Thermal noise and the stability of single sonoluminescing bubbles, *Phys Rev E* **61**, 5278.
Barber BP 1992 PhD thesis, University of California, Los Angeles.
Barber BP and Putterman SJ 1991 Observation of synchronous picosecond sonoluminescence, *Nature* **352**, 318.
Barber BP and Putterman SJ 1992 Light scattering measurements of the repetitive supersonic implosion of a sonoluminescing bubble, *Phys Rev Lett* **69**, 3839.
Barber B, Löfstedt R and Putterman SJ 1991 Sonoluminescence, *J Acoust Soc Am* **89**, S1885 (Abstract).
Barber BP, Hiller R, Arisaka K, Fetterman H and Putterman SJ 1992 Resolving the picosecond characteristics of synchronous sonoluminescence, *J Acoust Soc Am* 91, 3061.
Barber BP, Hiller RA, Löfstedt R, Putterman SJ and Weninger KR 1997a Defining the unknowns of sonoluminescence, *Phys Rep* **281**, 65.
Barber BP, Weninger K and Putterman SJ 1997b Sonoluminescence, *Phil Trans Roy Soc London A* **355**, 641.
Barber BP, Weninger K, Löfstedt R and Putterman SJ 1995 Observation of a new phase of sonoluminescence at low partial pressures, *Phys Rev Lett* **74**, 5276.
Barber BP, Wu CC, Löfstedt R, Roberts PH and Putterman SJ 1994 Sensitivity of sonoluminescence to experimental Parameters, *Phys Rev Lett* **72**, 1380.
Becchetti FD 2002 Evidence for nuclear reactions in imploding bubbles, *Science* **295**, 1850.
Birkhoff G 1954 Note on Taylor instability, *Quart Appl Maths* **12**, 306.
Blake FG 1949 Harvard University Acoustic Research Laboratory, Technical Memorandum 12, 1.
Bogoyavlenskiy VA 2000 Single-bubble sonoluminescence: shape stability analysis of collapse dynamics in a semianalytical approach, *Phys Rev E* **62**, 2158.
Brennen GE 1995 *Cavitation and Bubble Dynamics*, Oxford University Press.
Brenner MP, Hilgenfeldt S, Lohse D and Rosales RR 1996 Acoustic energy storage in single bubble sonoluminescence, *Phys Rev Lett* **77**, 3467.
Brenner MP, Hilgenfeldt S and Lohse D 1999 *The Hydrodynamical/Chemical Approach to Sonoluminescence, in* Sonochemistry and Sonoluminescence eds. Crum LA et al. Kluwer, Dordrecht, 165.
Brenner MP, Hilgenfeldt S and Lohse D 2002 Single-bubble sonoluminescence, *Rev Mod Phys* **74**, 425.
Brenner MP, Lohse D and Dupont TF 1995 Bubble shape oscillations and the onset of sonoluminescence, *Phys Rev E* **75**, 954. (Comment by Putterman SJ and Roberts PH 1998 *Phys Rev Lett* **80**, 3666. Reply 1998 *Phys Rev Lett* **80**, 3668.)
Chapman S and Cowling TG 1990 *The Mathematical Theory of Non-Uniform Gases*, Cambridge University Press.
Cheeke JDN 1997 Single-bubble sonoluminescence: "bubble, bubble, toil and trouble", *Can J Phys* **75**, 77.
Chen W, Chen X, Lu M, Miao G and Wei R 2002 Single-bubble sonoluminescence driven by non-simple-harmonic ultrasounds, *J Acoust Soc Am* **111**, 2632.
Chu M-C and Leung D 1997 Effects of thermal conduction in sonoluminescence, *J Phys Condens Matter* **9**, 3387.
Crum LA 1994 Sonoluminescence, sonochemistry, and sonophysics, *J Acoust Soc Am* **95**, 559.
Crum LA and Cordry S 1994 *Single-bubble Sonoluminescence, Bubble Dynamics and Interface Phenomena*, ed Blake JR et al. Kluwer, Dordrecht, 287.
Dam JS, Levinsen MT and Skogstad M 2002 Period-doubling bifurcations from breaking the spherical symmetry in sonoluminescence: experimental verification, *Phys Rev Lett* **89**, 084303.

Dam JS, Levinsen MT and Skogstad M 2003 Stable nonspherical bubble collapse including period doubling in sonoluminescence, *Phys Rev E* **67**, 026303.
Dan M, Cheeke JDN and Kondić L 1999 Ambient pressure effect on single-bubble sonoluminescence, *Phys Rev Lett* **83**,1870.
Dan M, Cheeke JDN and Kondić L 2000 Dependence of single-bubble sonoluminescence on ambient pressure, *Ultrasonics* **38**, 566.
Dave JV 1969 Scattering of electromagnetic radiation by a large, absorbing sphere, *IBM J Res Dev* **13**, 302.
Delgadino GA and Bonetto FJ 1997 Velocity interferometry technique used to measure the expansion and compression phases of a sonoluminescent bubble, *Phys Rev E* **56**, R6248.
Didenko YT, McNamara III WB and Suslick KS 1999 Temperature of multibubble sonoluminescence in water, *J Phys Chem A* **103**, 10783.
Didenko YT, McNamara III WB and Suslick KS 2000a Effect of noble gases on sonoluminescence temperatures during multibubble cavitation, *Phys Rev Lett* **84**, 777.
Didenko YT, McNamara III WB and Suslick KS 2000b Molecular emission from single-bubble sonoluminescence, *Nature* **407**, 877.
Didenko YT and Suslick KS 2002 The energy efficiency of formation of photons, radicals and ions during single-bubble cavitation, *Nature* **418**, 394.
Dumè B 2004 Bubble fusion: the sequel, *Physics World* **17**, No. 4, 7.
Flint EB and Suslick KS 1991 Sonoluminescence from alkali-metal salt solutions, *J Phys Chem* **95**, 1484.
Flynn HG 1975a Cavitation dynamics. I. A mathematical formulation, *J Acoust Soc Am* **57**, 1379.
Flynn HG 1975b Cavitation dynamics. II. Free pulsations and models for cavitation bubbles, *J Acoust Soc Am* **58**, 1160.
Gaitan DF 1990 An experimental investigation of acoustic cavitation in gaseous liquids, PhD thesis, University of Mississippi.
Gaitan DF 1999 Sonoluminescence and bubble stability, *Physics World* **12** No. 3, 20.
Gaitan DF and Crum LA 1990a Observation of Sonoluminescence from a Single, Stable Cavitation Bubble in a Water/Glycerine Mixture, Frontiers of Nonlinear Acoustics: Proceedings of 12th ISNA eds. Hamilton MF and Blackstock DT, Elsevier, pp 459–463.
Gaitan DF and Crum LA 1990b Sonoluminescence from single bubbles, *J Acoust Soc Am* Supplement 1 Vol 87, S141 (Abstract).
Gaitan DF and Holt RG 1999 Experimental observations of bubble response and light intensity near the threshold for single bubble sonoluminescence in an air–water system, *Phys Rev E* 59, 5495.
Gaitan DF, Atchley AA, Lewis SD, Carlson JT, Maruyama XK, Moran M and Sweider D 1996 Spectra of single-bubble sonoluminescence in water and glycerine–water mixtures, *Phys Rev E* **54**, 525.
Gaitan DF, Crum LA, Church CC and Roy RA 1992 Sonoluminescence and bubble dynamics for a single, stable, cavitation bubble, *J Acoust Soc Am* **91**, 3166.
Gilmore FR 1952 Hydrodynamics Laboratory Report 26-4, California Institute of Technology.
Gompf B 1996 Personal communication.
Gompf B, Gunther R, Nick G, Pecha R and Eisenmenger W 1997 Resolving sonoluminescence pulse width with time-correlated single photon counting, *Phys Rev Lett* **79**, 1405.
Gompf B and Pecha R 2000 Mie scattering from a sonoluminescing bubble with high spatial and temporal resolution, *Phys Rev E* **61**, 5253. (Comment by Weninger KB, Evans PG and Putterman SJ 2001 *Phys Rev E* **64**, 038301.)

Graça J da and Kojima H 2002 Single-bubble sonoluminescence from noble gas mixtures, *Phys Rev E* **66**, 066301.
Greenland PT 1999 Sonoluminescence, *Contemporary Phys* **40**, 11.
Grossmann S, Hilgenfeldt S and Lohse D 1997 Sound radiation of 3-MHz driven gas bubbles, *J Acoust Soc Am* **102**, 1223.
Hallaj IM, Matula TJ, Roy RA and Crum LA 1996 Measurements of the acoustic emission from glowing bubbles, *J Acoust Soc Am* **100**, 2717 (Abstract).
Hammer D and Frommhold L 2001 Topical review, Sonoluminescence: how bubbles glow, *J Mod Opt* **48**, 239.
Hammer D and Frommhold L 2002b Microwave emission of sonoluminescing bubbles, *Phys Rev E* **66**, 017302.
Hansen GM 1985 Mie scattering as a technique for the sizing of air bubbles, *Appl Opt* **24**, 3214.
Hao Y and Prosperetti A 1999a The effect of viscosity on the spherical stability of oscillating gas bubbles, *Phys Fluids* **11**, 1309.
Hao Y and Prosperetti A 1999b The dynamics of vapour bubbles in acoustic pressure fields, *Phys Fluids* **11**, 2008.
Harba N and Hayashi S 2003a A glycerol drop in water boosts single-bubble sonoluminescence, *Jpn J Appl Phys* **42**, 716.
Harba N and Hayashi S 2003b Dust particles counteract the boosting of single-bubble sonoluminescence by glycerols droplets, *Jpn J Appl Phys* **42**, 2971.
Hargreaves K and Matula TJ 2000 The radial motion of a sonoluminescence bubble driven with multiple harmonics, *J Acoust Soc Am* **107**, 1774.
Hayashi S 1999 Finite angular resolution Mie-scattering measurements of sonoluminescing bubbles, *Jpn J Appl Phys* **38**, 6562.
Hayashi S, Uchiyama S and Harba N 2001 Double scattering of light by a sonoluminescing bubble and many small particles, *J Phys Soc Japan* **70**, 3544.
Herring G 1941 Office of Science Research and Development Report 236. (NDRC Report C-4-sr 10-010 Columbia University.)
Highfield Roger 2002 Has the bubble burst on sun-tapping claims? *Daily Telegraph, London*, 13 March 2002, page 23.
Hilgenfeldt S, Brenner MP, Grossman S and Lohse D 1998a, Analysis of Rayleigh–Plesset dynamics for sonoluminescing bubbles, *J Fluid Mech* **365**, 171.
Hilgenfeldt S, Grossman S and Lohse D 1999 Sonoluminescence light emission, *Phys Fluids* **11**, 1318.
Hilgenfeldt S, Lohse D and Brenner MP 1996a Phase diagrams for sonoluminescing bubbles, *Phys Fluids* **8**, 2808.
Hilgenfeldt S, Lohse D and Brenner MP 1996b Erratum: "Phase diagrams for sonoluminescing bubbles" [*Phys Fluids* **8**, 2808 (1996)], *Phys Fluids* **9**, 2462.
Hilgenfeldt S, Lohse D and Zomack M 1998b Response of bubbles to diagnostic ultrasound: a unifying theoretical approach, *Eur Phys J B* Vol 4 Number 2, page 247.
Hiller RA 1995 PhD thesis, University of California, Los Angeles.
Hiller RA and Barber BP 1995 Producing light from a bubble of air, *Sci Am* **272**,78.
Hiller RA and Putterman SJ 1995 Observation of isotope effects in sonoluminescence, *Phys Rev Lett* **75**, 3549.
Hiller RA, Putterman SJ and Barber BP 1992 Spectrum of synchronous picosecond sonoluminescence, *Phys Rev Lett* **69**, 1182.
Hiller RA, Putterman SJ and Weninger KR 1998 Time-resolved spectra of sonoluminescence, *Phys Rev Lett* **80**, 1090.
Hiller RA, Weninger K, Putterman SJ and Barber BP 1994 Effect of noble gas doping in single-bubble sonoluminescence, *Science* **266**, 248.

Holt RG and Crum LA 1992 Acoustically forced oscillations of air bubbles in water: Experimental results, *J Acoust Soc Am* 91, 1924.
Holt RG and Gaitan DF 1996 Observation of stability boundaries in the parameter space of single bubble sonoluminescence, *Phys Rev Lett* 77, 3791.
Holt RG, Gaitan DF, Atchley AA and Holzfuss J 1994 Chaotic sonoluminescence, *Phys Rev Lett* **72**, 1376.
Holzfuss J and Lauterborn W 1989 Liapunov exponents from a time series of acoustic chaos, *Phys Rev A* **35**, 2146.
Holzfuss J, Rüggeberg M and Billo A 1998a Shock wave emissions of a sonoluminescing bubble, *Phys Rev Lett* **81**, 5434.
Holzfuss J, Rüggeberg M and Mettin R 1998b Boosting sonoluminescence, *Phys Rev Lett* **81**, 1961.
van de Hulst HG 1957 *Light Scattering by Small Particles*, Wiley, New York.
Hund M 1969 Untersuchungen zur Einzelblasenkavitation bei 10 Hz in einer wassergefüllten Druckkammer, *Acustica* **21**, 269.
Jeon J, Yang I, Karng S and Kwak H 2000 Radius measurement of a sonoluminescing gas bubble, *Jpn J App Phys* **39**, 1124.
Kaji M, Bergeal N, Asase T and Watanabe Y 2002 Observation of sonoluminescing bubble motion at the rebounding phase, *Jpn J Appl Phys* **41**, 3250.
Kamath V, Prosperetti A and Egolfopoulos FN 1993 A theoretical study of sonoluminescence, *J Acoust Soc Am* **94**, 248.
Kawabata K and Umemura S 1996a Effect of second-harmonic superimposition on efficient induction of sonochemical effect, *Ultrasonics Sonochemistry* **3**, 1.
Kawabata K and Umemura S 1996b Use of second-harmonic superimposition to induce chemical effects of ultrasound, *J Phys Chem* **100**, 18784.
Keller JB and Kolodner II 1956 Damping of underwater explosion bubble oscillations, *J Appl Phys* **27**, 1152.
Keller JB and Miksis M 1980 Bubble oscillations of large amplitude, *J Acoust Soc Am* **68**, 628.
Kerker M 1969 *The Scattering of Light and Other Electromagnetic Radiation*, Academic Press, New York.
Ketterling JA and Apfel RE 1998 Experimental validation of the dissociation hypothesis for single bubble sonoluminescence, *Phys Rev Lett* **81**, 4991.
Ketterling JA and Apfel RE 2000a Extensive experimental mapping of sonoluminescence parameter space. *Phys Rev E* **61**, 3832.
Ketterling JA and Apfel RE 2000b Using phase space diagrams to interpret multiple frequency drive sonoluminescence, *J Acoust Soc Am* **107**, 819.
Ketterling JA and Apfel RE 2000c Shape and extinction thresholds in sonoluminescence parameter space. *J Acoust Soc Am* **107**, L13.
Kirkwood JG and Bethe HA 1942 The pressure wave produced by an underwater explosion, Office of Science and Development Report 558.
Kordomenos JN, Bernard M and Denardo B 1999 Experimental microwave radiometry of a sonoluminescing bubble, *Phys Rev E* **59**, 1781.
Kozuka T, Hatanaka S, Tuziuti T, Yasui K and Mitome H 2000 Observation of a sonoluminescing bubble using a stroboscope, *Jpn J Appl Phys* **39**, 2967.
Kozuka T, Hatanaka S, Yasui K, Tuziuti T and Mitome H 2002 Simultaneous observation of motion and size of a sonoluminescing bubble, *Jpn J Appl Phys* **41**, 3248.
Krefting D, Mettin R and Lauterborn W 2002 Two-frequency driven single-bubble sonoluminescence, *J Acoust Soc Am* **112**, 1918.
Krefting D, Mettin R and Lauterborn W 2003 Single-bubble sonoluminescence in air-saturated water, *Phys Rev Lett* **91**, 174301.

Lastman GJ and Wentzell RA 1981 Comparison of five models of spherical bubble response in an inviscid compressible liquid, *J Acoust Soc Am* **69**, 638.
Lastman GJ and Wentzell RA 1982 On two equations of radial motion of a spherical gas-filled bubble in a compressible liquid, *J Acoust Soc Am* **71**, 835.
Lauterborn W, Kurz T and Ohl CD 1999 Experimental and theoretical bubble dynamics, *Adv Chem Phys* **110**, 295.
Lentz WJ, Atchley AA and Gaitan DF 1995 Mie scattering from a sonoluminescing air bubble in water, *Appl Opt* **34**, 2648.
Lepoint T and Lepoint-Mullie F 1998 *Synthetic Organic Sonochemistry.* Luche JL (ed) Plenum, New York, page 1.
Lepoint T and Lepoint-Mullie F 1999 *Advances in Sonochemistry*, Vol. 5 pages 1–108, JAI Press.
Lepoint T, De Pauw D and Lepoint-Mullie F 1997 Sonoluminescence: An alternative "electrohydromatic" hypothesis, *J Acoust Soc Am* **100**, 2012.
Lepoint T, Lepoint-Mullie F and Henglein A 1999 *Sonochemistry and Sonoluminescence*, Crum LA et al. (eds), Kluwer Academic, Dordrecht, page 285.
Lepoint T, Lepoint-Mullie F, Voglet N, Labouret S, Pétrier C, Avni R and Lugue J 2003 OH/D $A^2\Sigma^+ - X^2\Pi_i$ rovibronic transitions in multibubble sonoluminescence, *Ultrasonics Sonochemistry* **10**, 167.
Lepoint-Mullie F, Lepoint T and Henglein A 1997 Observation of single-bubble sonochemistry: preliminary report, *Ber Bunsenges Phys Chem* (submitted).
Levi BG 2002 Skepticism greets claim of bubble fusion, *Physics Today* **55(4)**, 16.
Levinsen MT, Weppenaar N, Dam JS, Simon G and Skogstad M 2003 Direct observation of period-doubled non-spherical states in single-bubble sonoluminescence, *Phys Rev E* **68**, 035303.
Lezzi A and Prosperetti A 1987 Bubble dynamics in a compressible liquid. Part 2. Second-order theory, *J Fluid Mech* **185**, 289.
Lin H, Storey BD and Szeri AJ 2002a Inertially driven inhomogeneities in violently collapsing bubbles: the validity of the Rayleigh–Plesset equation. *J Fluid Mech* **452**, 145.
Lin H, Storey BD and Szeri AJ 2002b Rayleigh–Taylor instability of violently collapsing bubbles, *Phys Fluids* **14**, 2925.
Löfstedt R, Barber BP and Putterman SJ 1993 Toward a hydrodynamic theory of sonoluminescence, *Phys Fluids* **A5**, 2911.
Löfstedt R, Weninger K, Putterman SJ and Barber BP 1995 Sonoluminescing bubbles and mass diffusion, *Phys Rev E* **51**, 4400.
Lohse D, Brenner MP, Dupont TF, Hilgenfeldt S and Johnston B 1997 Sonoluminescing air bubbles rectify argon, *Phys Rev Lett* **78**,1359.
Lohse D and Hilgenfeldt S 1997 Inert gas accumulation in sonoluminescing bubbles, *J Chem Phys* **107**, 6986.
Lohse D and Hilgenfeldt S 1999 Festkörperprobleme, Seite 215 herausgegeben von B Kramer, Vieweg-Verlag.
Longuet-Higgins MS 1996 Shedding of vortex rings by collapsing cavities, with application to single-bubble sonoluminescence, *J Acoust Soc Am* **100**, 2678 (Abstract).
Longuet-Higgins MS 1997 Particle drift near an oscillating bubble, *Proc Roy Soc London A* 453, 1551.
Louisnard O and Gomez F 2003 Growth by rectified diffusion of strongly acoustically forced gas bubbles in nearly saturated liquids, *Phys Rev E* **67**, 036610.
Lu X, Prosperetti A, Toegel R and Lohse D 2003 Harmonic enhancement of single-bubble sonoluminescence, *Phys Rev E* **67**, 056310.
Madrazo A, García N and Nieto-Vesperinas M 1998 Determination of the size and shape of a sonoluminescent single bubble: Theory on angular correlations of the emitted light, *Phys Rev Lett* **80**, 4590.

Marston PL 1979 Critical angle scattering by a bubble: physical-optics approximation and observations, *J Opt Soc Am* **69**, 1205.
Marston PL 1991 Colors observed when sunlight is scattered by bubble clouds in sea-water, *Appl Opt* **30**, 3479.
Matula TJ 1999 Inertial cavitation and single-bubble sonoluminescence, *Phil Trans Roy Soc Lond A* **357**, 225.
Matula TJ 2000 Single-bubble sonoluminescence in microgravity, *Ultrasonics* **38**, 559.
Matula TJ and Crum LA 1998 Evidence for gas exchange in single-bubble sonoluminescence, *Phys Rev Lett* **80**, 865.
Matula TJ, Hallaj IM, Cleveland RO and Crum LA 1998 The acoustic emissions from single-bubble sonoluminescence, *Phys Rev Lett* **103**, 1377.
Matula TJ and Roy RA 1997 Comparisons of sonoluminescence from single-bubbles and cavitation fields: bridging the gap, *Ultrasonics Sonochemistry* **4**, 61.
Matula TJ, Roy RA and Crum LA 1996 Preliminary experimental observations of the effects of buoyancy on single-bubble sonoluminescence in microgravity and hyper-gravity, *J Acoust Soc Am* 100, 2717.
Matula TJ, Roy RA and Mourad PD 1997 Optical pulse width measurements of sono-luminescence in cavitation-bubble fields, *J Acoust Soc Am* **101**, 1994.
Matula TJ, Roy RA, Mourad PD, McNamara III WB and Suslick KS 1995 Comparison of multibubble and single-bubble sonoluminescence, *Phys Rev Lett* **75**, 2602.
Matula TJ, Swalwell JE, Bezzerides V, Hilmo P, Chittick M and Crum LA 1997 Single-bubble sonoluminescence in microgravity, *J Acoust Soc Am* 102, 1185 (abstract).
Metcalf H 1998 That Flashing Sound, *Science* **279**, 1322.
Mie G 1908 Beiträge zur Optik trüber Medien, speziell kolloidaler Metallösungen, *Ann Phys* **25**, 377. For English translation see Library of Royal Aircraft Establishment, Farnborough, UK, No. 1873.
Mitome H 2001 Micro bubble and sonoluminescence, *Jpn J Appl Phys* **40**, 3480.
Moraga FJ, Taleyarkhan RP, Lahey Jr RT and Bonetto FJ 2000 Role of very-high-frequency excitation in single-bubble sonoluminescence, *Phys Rev E* **62**, 2233.
Moran MJ, Haigh RE, Lowry ME, Sweider DR, Abel GR, Carlson JT, Lewia SD, Atchley AA, Gaitan DF and Maruyame XK 1995 Direct observations of single sono-luminescence pulses, *Nucl Instrum Methods* **B96**, 651.
Moss WC, Clarke DB and Young DA 1999 *Star in a Jar, Sonoluminescence and Sono-chemistry*, eds Crum et al. Kluwer, page 159.
Moss WC, Levatin JL and Szeri AJ 2000 A new damping mechanism in strongly collapsing bubbles, *Proc Roy Soc Lond A* **456**, 2983.
Nayfeh AH and Mook DT 1979 *Nonlinear oscillations*, John Wiley, New York.
Neppiras EA 1980 Acoustic Cavitation, *Physics Reports* **61**, 159.
Nyborg WL 2002 personal communication.
Ogi H, Matsuda A, Wada K and Hirao M 2003 Brightened single-bubble sonolumines-cence by phase-adjusted high-frequency acoustic pulse, *Phys Rev E* **67**, 056301.
Ohl G-D 2002 Probing luminescence from nonspherical bubble collapse, *Phys Fluids* **14**, 2700.
Ohsaka K and Trinh EH 2000 A two-frequency acoustic technique for bubble resonant oscillation studies, *J Acoust Soc Am* **107**, 1346.
Pecha R and Gompf B 2000 Microimplosions: Cavitation collapse and shock wave emission on a nanosecond time scale, *Phys Rev Lett* **84**, 1328.
Pecha R, Gompf B, Nick G, Wang ZQ and Eisenmenger W 1998 Resolving the sonoluminescence pulse shape with a streak camera, *Phys Rev Lett* **81**, 717.

Pennicott Katie 2002 Bubble bursts for "sono-fusion", PhysicsWeb http:physics web.org/article/news/6/7/18, 31/08/02.
Plesset MS 1954 On the stability of fluid flows with spherical symmetry, *J Appl Phys* **25**, 96.
Plesset MS and Mitchell TP 1956 On the stability of the spherical shape of a vapour cavity in a liquid, *Quart Appl Maths* **13**, 419.
Plesset MS and Prosperetti A 1977 Bubble dynamics and cavitation, *Ann Rev Fluid Mech* **9**, 145.
Prosperetti A 1977a Viscous effects on perturbed spherical flows, *Quart Appl Math* **34**, 339.
Prosperetti A 1977c Thermal effects and damping mechanisms in the forced radial oscillations of gas bubbles in liquids, *J Acoust Soc Am* **61**, 17.
Prosperetti A 1984 Physics of acoustic cavitation, *Rendiconti SIF* **XC111**, 145.
Prosperetti A 1991 The thermal behaviour of oscillating bubbles, *J Fluid Mech* **222**, 587.
Prosperetti A 1997b On the stability of spherically symmetric flows, *Atti Accad Naz Lincei, Rendiconti Cl Sci Fis Mat Nat* **62**, 196.
Prosperetti A, Crum LA and Commander KW 1988 Nonlinear bubble dynamics, *J Acoust Soc Am* **83**, 502.
Prosperetti A and Hao Y 1999 Modelling of spherical gas bubble oscillations and sonoluminescence, *Phil Trans Roy Soc London A* **357**, 203.
Prosperetti A and Lezzi A 1986 Bubble dynamics in a compressible liquid. Part 1. First-order theory, *J Fluid Mech* **168**, 457.
Prosperetti A and Lu X 2001 Harmonic Enchancement of Single-Bubble Sonoluminescence, 17th Int Conf on Acoustics (Rome), Paper 6C.05.02.
Putterman SJ 1998 Sonoluminescence: the star in a jar, *Physics World* **11** No 5, 38.
Putterman SJ 1999 Sonoluminescence – the plot thickens, *Physics World* **12** No 8, 18.
Putterman SJ, Evans PG, Vazquez G and Weninger K 2001 Is there a simple theory of sonoluminescence? *Nature* **409**, 782.
Putterman SJ and Roberts PH 1998 Comment on "Bubble shape oscillations and the onset of sonoluminescence", *Phys Rev Lett* **80**, 3666.
Putterman SJ and Weninger KR 2000 Sonoluminescence: How bubbles turn sound into light, *Ann Rev Fluid Mech* **32**, 445.
Rayleigh Lord 1883 Investigation of the character of the equilibrium of an incompressible heavy fluid of variable density, *Proc London Math Soc* **XIV**, 170.
Rayleigh Lord 1917 On the pressure developed in a liquid during the collapse of a spherical cavity, *Phil Mag* **34**, 94.
Reddy AJ and Szeri AJ 2002 Shape instability of unsteadily translating bubbles, *Physics of Fluids* **14**, 2216.
Rozenberg LD 1971 *High Intensity Ulrasonic Fields* Parts IV--VI, Plenum Press.
Seeley FB 1999 Effects of higher-order modes and harmonics in single-bubble sonoluminescence, *J Acoust Soc Am* **105**, 2236.
Shapira D and Saltmarsh M 2002 Nuclear fusion in collapsing bubbles – is it there? An attempt to repeat the observation of nuclear emissions from sonoluminescence, *Phys Rev Lett* **89**, 104302.
Simon G and Levinsen MT 2003a Alternative method to deduce bubble dynamics in single-bubble sonoluminescence experiments, *Phys Rev E* **67**, 026320.
Simon G and Levinsen MT 2003b Parametric dependence of single-bubble sonoluminescence spectra, *Phys Rev E* **68**, 046307.
Simon G, Csabai I, Horváth A and Szalai F 2001 Sonoluminescence and phase diagrams of single bubbles at low dissolved air concentrations, *Phys Rev E* **63**, 026301.
Simon G, Cvitanović P, Levinsen MT, Csabai I and Horváth A 2002 Periodic orbit theory applied to a chaotically oscillating gas bubble in water, *Nonlinearity* **15**, 25.

Sochard S, Wilhelm AM and Delmas H 1997 Modelling of free radicals production in a collapsing gas-vapour bubble, *Ultrasonics Sonochemistry* **4**, 77.

von Sonntag C, Mark G, Tauber A and Schuchmann H 1999 OH Radical formation and dosimetry in the sonolysis of aqueous solutions, *Advances in Sonochemistry* Vol 5, 109, JAI Press.

Storey BD 2001 Shape stability of sonoluminescence bubbles: Comparison of theory to experiments, *Phys Rev E* **64**, 017301.

Storey BD and Szeri AJ 1999 Mixture segregation within sonoluminescence bubbles, *J Fluid Mech* **396**, 203.

Storey BD and Szeri AJ 2000 Water vapour, sonoluminescence and sonochemistry, *Proc Roy Soc London A* **456**, 1685.

Storey BD and Szeri AJ 2001 A reduced model of cavitation physics for use in sonochemistry, *Proc Roy Soc London A* **457**, 1685.

Storey BD and Szeri AJ 2002 Argon rectification and the cause of light emission in single-bubble sonoluminescence, *Phys Rev Lett* **88**, 074301.

Strube HW 1971 Numerische Untersuchungen zur Stabilität nichsphärisch schwingender Blasen, *Acustica* **25**, 289.

Suslick KS, Didenko Y, Fang MM, Hyeon T, Kolbeck KJ, McNamara III WB, Mdleleni MM and Wong M 1999 Acoustic cavitation and its chemical consequences, *Phil Trans Roy Soc London A* **357**, 335.

Suslick KS, Hammerton DA and Cline Jr RE 1986 The sonochemical hot spot, *J Am Chem Soc* **108**, 5641.

Taleyarkhan RP, Block RC, West C and Lahey Jr RT 2002b Comments on Shapira and Saltmarsh (2002), http://www.rpi.edu/~laheyr/SciencePaper.pdf.

Taleyarkhan RP, West CD, Cho JS, Lahey Jr RT, Nigmatulin RI and Block RC 2002a Evidence for nuclear emissions during acoustic cavitation, *Science* **295**, 1868.

Taylor G 1950 The instability of liquid surfaces when accelerated in a direction perpendicular to their planes. I, *Proc Roy Soc London Ser A* **201**, 192.

Temple PR 1970 Sonoluminescence from the gas in a single bubble, MS thesis, University of Vermont, USA.

Thomas J, Forterre Y and Fink M 2002 Boosting sonoluminescence with a high-intensity ultrasonic pulse focused on the bubble by an adaptive array, *Phys Rev Lett* **88**, 074302.

Tian Y, Ketterling JA and Apfel RE 1996 Direct observation of microbubble oscillations, *J Acoust Soc Am* **100**, 3976.

Toegel RB, Gompf B, Pecha R and Lohse D 2000b Does water vapour prevent upscaling sonoluminescence? *Phys Rev Lett* **85**, 3165.

Toegel RB, Hilgenfeldt S and Lohse D 2002 Suppressing dissociation in sonoluminescing bubbles: The effect of excluded volume, *Phys Rev Lett* **88**, 034301.

Toegel RB, Hilgenfeldt S and Lohse D 2000a Squeezing alcohols into sonoluminescing bubbles: The universal role of surfactants, *Phys Rev Lett* **84**, 2509.

Toegel R and Lohse D 2003 Phase diagrams for sonoluminescing bubbles: A comparison between experiment and theory, *J Chem Phys* **118**, 1863.

Trilling L 1952 The collapse and rebound of a gas bubble, *J Appl Phys* **23**, 14.

Umemura S and Kawabata K 1993 Enhancement of Sonochemical Reactions by Second-Harmonic Superimposition, Proc 1993 IEEE Ultrasonics Symposium (IEEE NY 1993) page 917.

Umemura S, Kawabata K and Sasaki K 1994 Theoretical Analysis of Cavitation Enhancement by Second-Harmonic Superimposition, Proc 1994 Ultrasonics Symposium (IEEE NY 1994) page 1843.

Umemura S, Kawabata K and Sasaki K 1995 Reduction of Threshold for Producing Sonodynamic Tissue Damage by Second-Harmonic Superimposition, Proc 1995 Ultrasonics Symposium (IEEE NY 1995) page 1567.

Umemura S, Kawabata K and Sasaki K 1996 Enhancement of sonodynamic tissue damage production by second-harmonic superimposition: Theoretical analysis of its mechanism, *IEEE Trans Ultrasonics, Ferroelectrics and Frequency Control* **43**, 1054.

Umemura S, Kawabata K and Sasaki K 1997 *In vitro* and *in vivo* enhancement of sonodynamically active cavitation by second-harmonic superimposition, *J Acoust Soc Am* **101**, 569.

Vacca G, Morgan RD and Laughlin RB 1999 Differential light scattering: Probing the sonoluminescence collapse, *Phys Rev E* **60**, R6303.

Verraes T, Lepoint-Mullie F, Lepoint T and Longuet-Higgins MS 2000 Experimental study of the liquid flow near a single sonoluminescent bubble, *J Acoust Soc Am* **108**, 117.

Vuong VQ, Szeri AJ and Young DA 1999 Shock formation within sonoluminescence bubbles, *Phys Fluids* **11**, 10.

Wang W, Chen W, Lu M and Wei R 2003 Bubble oscillations driven aspherical ultrasound in liquid, *J Acoust Soc Am* 114, 1898.

Wang ZQ, Pecha R, Gompf B and Eisenmenger W 1999 Single bubble sonoluminescence: Investigations of the emitted pressure wave with a fibre optic probe hydrophone, *Phys Rev E* **59**, 1777.

Weisstein EW 2003 Concise Encyclopedia of Mathematics, Chapman & Hall.

Weninger KR, Barber BP and Putterman SJ 1997 Pulsed Mie scattering measurements of the collapse of a sonoluminescing bubble, *Phys Rev Lett* **78**, 1799.

Weninger KR, Evans PG and Putterman SJ 2000 Time correlated single photon Mie scattering from a sonoluminescing bubble, *Phys Rev E* **61**, R1020.

Weninger KR, Evans PG and Putterman SJ 2001 Comment on "Mie scattering from a sonoluminescing bubble with high spatial and temporal resolution", *Phys Rev E* **64**, 038301.

Weninger KR, Hiller R, Barber BP, Laoste D and Putterman SJ 1995 Sonoluminescence from single bubbles in nonaqueous liquids: New parameter space for sonochemistry, *J Phys Chem* **99**, 14195.

Weninger KR, Putterman SJ and Barber BP 1996 Angular correlations in sonoluminescence: Diagnostic for the sphericity of a collapsing bubble, *Phys Rev E* **54**, R2205.

Wiscombe WJ 1980 Improved Mie scattering algorithms, *Appl Opt* **19**, 1505.

Wu CC and Roberts PH 1998 Bubble shape instability and sonoluminescence, *Phys Lett A* **250**, 131.

Yasui K 1997 Alternative model of single-bubble sonoluminescence, *Phys Rev E* **56**, 6750.

Yasui K 2002 Segregation of vapour and gas in a sonoluminescing bubble, *Ultrasonics* **40**, 643.

Yosioka K and Kawasima Y 1955 Acoustic radiation pressure on a compressible sphere, *Acustica* **5**, 167.

Yosioka K and Omura A 1962 The light emission from a single bubble driven by ultrasound and the spectra of acoustic oscillations, *Proc Annual Meet Acoust Soc Jpn* (1962) pages 125–126 (in Japanese).

Young FR 1965a Sonoluminescence from glycerine–water mixtures, *Nature* **206**, 706.

Young FR 1976 Sonoluminescence from water containing dissolved gases, *J Acoust Soc Am* 60, 100.

Young FR 1999 *Cavitation*, Imperial College Press.

Young JB, Nelson JA and Kang W 2001 Line emission in single-bubble sonoluminescence, *Phys Rev Lett* **86**, 2673.

Yuan L, Ho CY, Chu M-C and Leung PT 2001 Role of gas density in the stability of single-bubble sonoluminescence, *Phys Rev E* **64**, 016317.

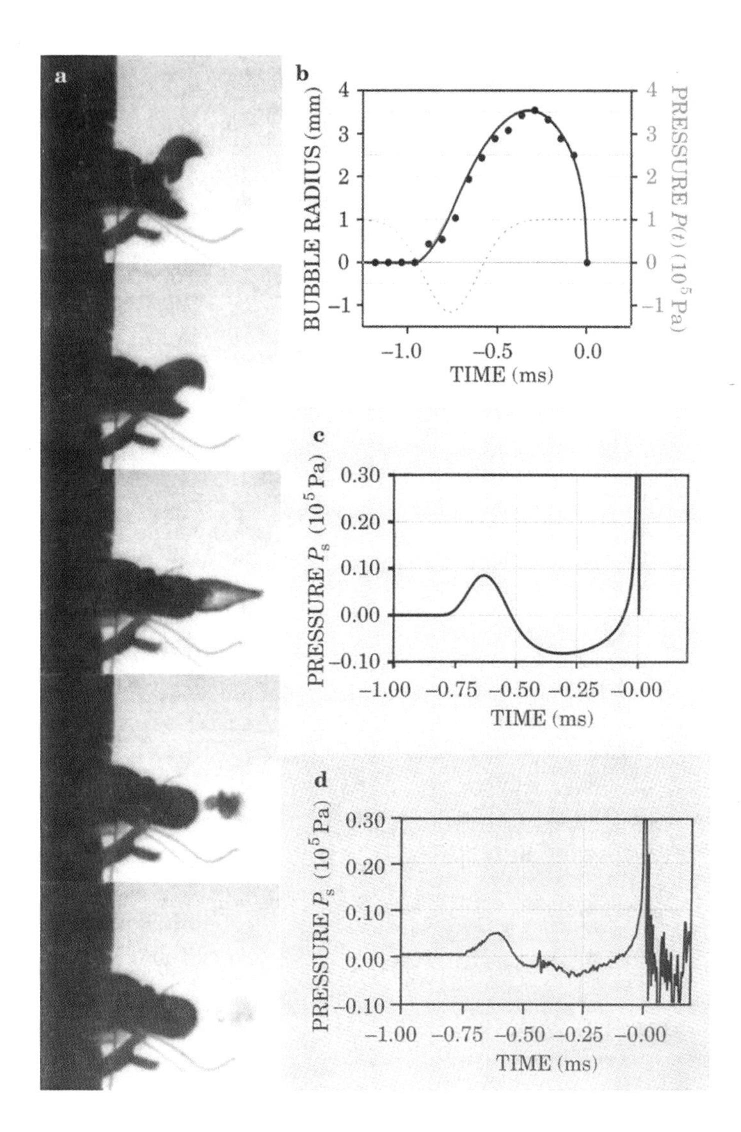


Figure 2.22 *Snapping shrimp* use bubble cavitation to stun and even kill prey. (a) Frames taken every 0.5 ms from a high-speed video recording capture the claw closure of a snapping shrimp. A fast water jet is emitted and generates cavitation. In the last frame, the bubble has collapsed; only microbubbles remain. (b) The measured and calculated bubble radius. The points are the measured data; the solid line shows the results from our Rayleigh–Plesset based theory (equation (1.14)). The dashed line is the assumed pressure reduction $P(t)$ due to the jet. (c) The corresponding sound emission $P_s$ from the bubble. (d) The measured sound emission. From Versluis et al. (2000). (Lohse (2003).)

Figure 2.24 Photograph of the "barometer light", which is the orange line of light generated at the intersection of the mercury meniscus and the wall of the rotating glass cylinder. The rotation speed is $30° s^{-1}$; the cylinder is made from Supracil with a wall thickness of 1 mm, a length of 5 cm, and an outer diameter of 2 cm. The cell is filled with 15 ml of mercury and above the mercury is neon gas at 340 torr. The light is emitted on the side where the wall leaves the mercury. With dimmed ambient lighting this effect is easily seen with the unaided eye, provided that the mercury is properly cleaned. Each black Delrin endcap is joined to the glass with an O-ring seal. We note that the Delrin–mercury interface also emits light. (Reprinted with permission from *Nature*, **391**, 266, Budakian et al. (1998) Copyright Macmillan Magazines Limited.)

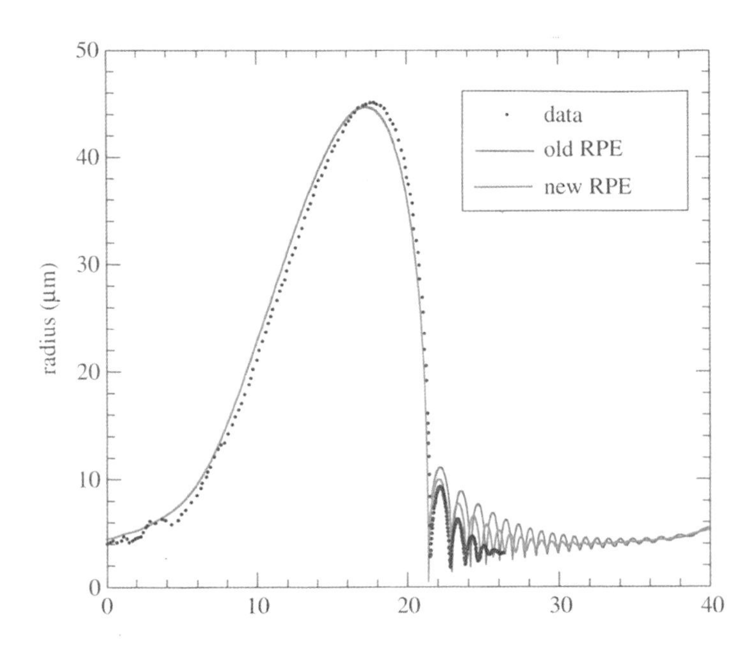


Figure 3.23 Time dependence of the radius of a sonoluminescing air bubble. The data is from Löfstedt et al. (1993) (black dots). Reproduced with permission from Moss et al. (2000) *Proc Roy Soc Lond A* **456**, 2983. Credit is given to the University of California, Lawrence Livermore National Laboratory and the Department of Energy under whose auspices the work was performed.

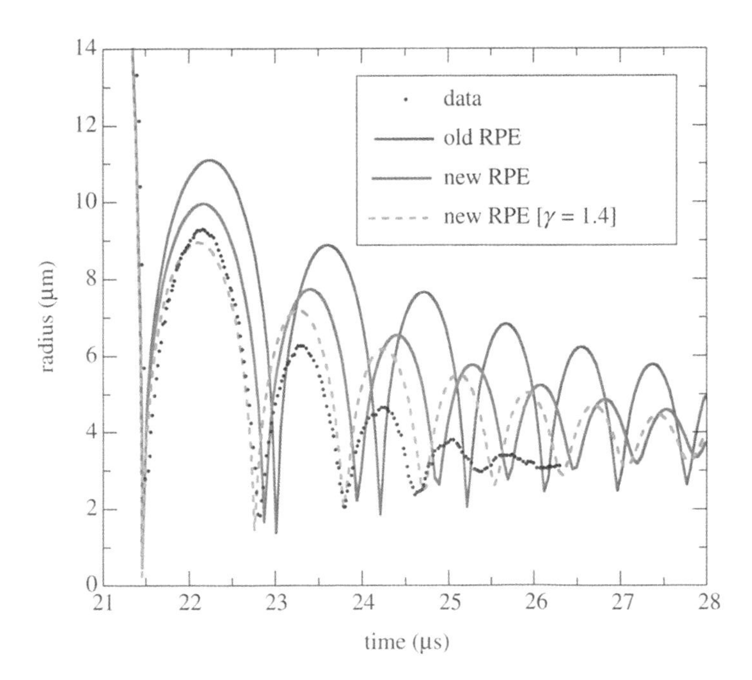


Figure 3.24 Expanded view of the rebounds in Fig. 3.23. Reproduced with permission from Moss et al. (2000) *Proc Roy Soc Lond A* **456**, 2983. Credit is given to the University of California, Lawrence Livermore National Laboratory and the Department of Energy under whose auspices the work was performed.

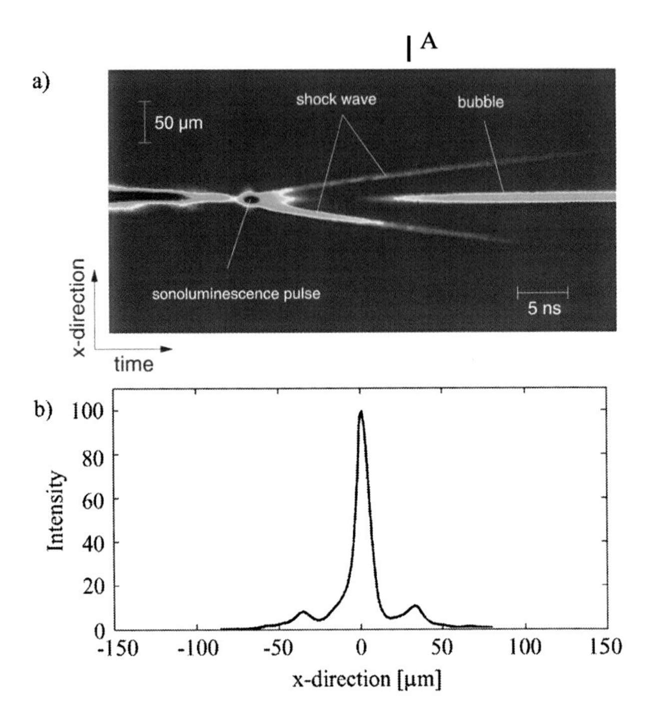


Figure 3.27 (a) Streak image of the SBSL bubble collapse. Laser light is scattered at the bubble as well as at the emitted shock wave. (b) Line scan A–A′ of the streak image. (Pecha and Gompf (2000).)

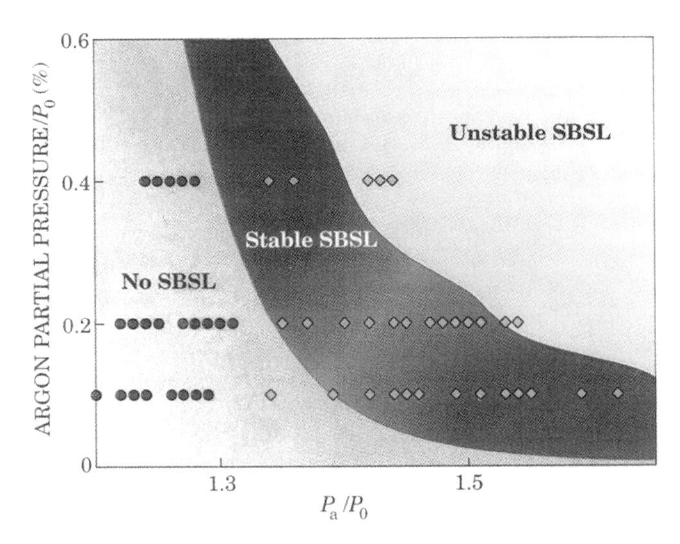


Figure 3.30a Phase diagram of single-bubble sonoluminescence (SBSL). The horizontal axis is the driving-pressure amplitude $P_a$, normalized by the ambient pressure $P_0$. The vertical axis is the argon partial pressure, also normalized by $P_0$. Stable SBSL is predicted to occur only in the red region. At lower forcing or Ar partial pressures, no SBSL occurs. Above the red region, the SBSL is unstable: The bubble grows by diffusion, and the phase and intensity of the emitted light aren't constant. The experimental data (dots in the respective color of the region) are in reasonable agreement with the predictions. Diagram adapted from Brenner et al. (2002)). (Lohse (2003).)

CHAPTER FOUR

# Theories of Sonoluminescence

If a man will begin with certainties, he shall end in doubts.
But if he will be content to start with doubts, he shall end
in certainties.

Francis Bacon *Advancement of Learning*

Great spirits have always encountered violent opposition
from mediocre minds

Albert Einstein

Review of "*Mad about Physics: Braintwisters, Paradoxes and Curiosities*",
Jargodski C and Potter F, John Wiley 2000 by Peter Ford
*Physics World* July 2001 page 45

## 4.1 INTRODUCTION

We will begin with a history of the early theories of sonoluminescence up to 1990, in Sec. 4.2 to 4.6. That year, of the discovery of single bubble sonoluminescence, gradually produced an increased interest in sonoluminescence. Particularly intriguing was where does the light actually come from. This proved a fertile field for PhD students to exercise their ingenuity and knowledge of theoretical physics. Their theories, which are now largely discounted, as they cannot explain all the manifold properties of sonoluminescence, will be briefly outlined in Sec. 4.2 to 4.5. They are important in the course of attempts to explain sonoluminescence. In Sec. 4.14, we will present a modern theory by Lohse and his collaborators which is widely accepted today.

## 4.2 THE TRIBOLUMINESCENCE THEORY

Sonoluminescence was discovered by Frenzel and Schultes (1934) and many theories were put forward to explain the production of the light, for example, the triboluminescence theory.

Chambers (1936, 1937) and Flosdorf et al. (1936) measured the SL from 36 liquids at a sound frequency of 10 kHz. They found that the SL was proportional to the product of the dipole moment and viscosity of the liquid and inversely proportional to temperature. Chambers put forward this triboluminescence theory at a time when liquids were thought to have a quasi-crystalline structure. When this structure is broken as in the rupture of certain crystals, such as sugar cubes (Walton 1977) or certain varieties of mints or sweets (Highfield (1988)), light is seen.

## 4.3 THE ELECTRICAL MICRODISCHARGE THEORIES

Levshin and Rzhevkin (1937) measured the SL from water, aliphatic alcohols, benzene, xylene, carbon tetrachloride and nitrobenzene at 525 kHz. They concluded that the SL was caused by high electric potentials that "arise during the very formation of cavities as a result of the drop in pressure up to negative values, in the process of violent ultrasonic vibrations … … In the resulting rupture an electric discharge takes place by which the vapor filling the cavities is excited" and produces SL.

Harvey (1939) suggested that these electric potentials were *balloelectric* i.e. they were produced by an increase in the surface charge of liquids. When a bubble becomes smaller, its electrical capacitance decreases and the voltage increases until discharge occurs giving a weak SL. Lenard describes a similar effect on liquid droplets to give waterfall luminescence (Loeb 1958). Frenkel (1940) said that the sound field ruptures the liquid to produce lens-shaped cavities of molecular dimensions. Charges of opposite sign arise on the opposing walls of the cavity, and then give a microdischarge across the cavity as the electric field increases owing to the cavity becoming spherical. Thus, according to Frenkel, the light appears when the cavity is at its maximum radius, whereas Meyer and Kuttruff (1959) showed that the light appeared when the cavity was at its minimum radius.

Degrois and Baldo (1974) put forward an electrical hypothesis in which free atoms are absorbed on the inside of the bubble surface. Due to an asymmetric environment in the absorbed configuration, a molecule is deformed and polarized with all its kinetic energy being transformed into polarization energy. During bubble compression, this energy becomes concentrated until a discharge occurs as sonoluminescence with a free energy release of $10^5$–$10^8$ kcal mol$^{-1}$. Sehgal and Verrall (1982) and Suslick (1988) criticize this hypothesis, pointing out that the free energy released cannot exceed $10^2$ kcal mol$^{-1}$.

Golubnichii, Sytnikov and Filonenko (1987) have recently revived the electrical origin hypothesis of sonoluminescence by formulating a theory for radio frequency emission. Ben'kovskii, Golubnichii and Maslennikov (1974) had earlier observed luminescence from the oscillation of a cavity formed by a 100-kV pulse discharge in water. They calculated a bubble temperature of 9300 K.

Chincholle (1976) has also observed electrical discharges in cavitating water and oil where the local electrical potentials differ by three orders of magnitude.

Margulis (1985a,b, 1990, 2002) suggests a new electrical theory in which a bubble larger than resonant size has been made from the coalescence of smaller bubbles. This large deformed bubble has been seen and photographed when the frequency of the sound used is in the range 10 to 200 Hz. A smaller bubble is formed off the side of the large one and is joined to it by a neck. Margulis has suggested that this bubble formation produces charge separation. As the smaller bubble is pinched off, the charges from a large area of the large bubble are concentrated onto a small area of the small bubble. It is suggested that a large negative charge accumulates in the smaller bubble. A discharge, and therefore light, occurs between the negative charge on the smaller bubble and the resulting positive charge on the larger bubble. Margulis' theory has been criticized by Suslick (1990) and Lepoint-Mullie et al. (1996).

Garcia et al. (1996, 1999, 2000) proposed that single bubble sonoluminescence is caused by strong electric fields occurring in water near the surface of collapsing gas bubbles because of the *flexoelectric effect* involving polarization resulting from a pressure gradient. These large electric fields produce hot electrons which can transfer rare gas (say) atoms dissolved in water to high-energy excited states and optical transitions between these states produce the SL UV flashes in the transparency window of water.

## 4.4 THE MECHANOCHEMICAL THEORY

The mechanochemical theory of Weyl and Marboe (1949, 1951) and Marboe and Weyl (1950) supposed that ions were produced by mechanical fracture of molecules as the quasi-crystalline liquid structure is destroyed at the nascent surface of a growing bubble. The ions then recombine to radiate sonoluminescence. The simplest processes that can give rise to luminescence are

$$A + B \rightarrow AB + h\nu$$

Apart from recombination of electrons and positive ions, two-body reactions of this type might be caused by the recombination of dissociation products of diatomic or polyatomic molecules; for instance, the cold light or bioluminescence of animate matter results from the two-body reaction of luciferin with the enzyme luciferase. The spectral distribution of this light is often a continuum resembling that of a black-body radiator at 10,000 K; in that respect it resembles sonoluminescence (Harvey (1952)). Three-body reactions are more likely, but these give rise to line and band spectra only:

$$A + B + C \rightarrow AB^* + C \qquad \text{or} \qquad AB + C^*$$

where the asterisk represents an excited state (Jarman (1960)).

Saksena and Nyborg (1970), in a study of sonoluminescence from arrays of gas bubbles of uniform size, suggest that the mechanism lies in the dissociation of $H_2O$ into the OH and H during the compression phase of each cycle when the temperature is high. Photons are emitted as the radicals recombine. Calculations suggest that about $10^8$ photons $s^{-1}$ might be expected from a single 0.2-mm diameter bubble when the bubble temperature reaches 1800 K. Well over half of the detectable energy of sonoluminescence from water lies in the visible spectrum and this cannot be attributed to chemical reactions unless certain three-body collisions between gas molecules take place, with the corresponding line and band spectra broadened out into a continuum by the conditions of high pressure and density that prevail within a cavity during the final stages of collapse. However, as Jarman (1960) points out, there are more plausible hypotheses to account for the visible luminescence.

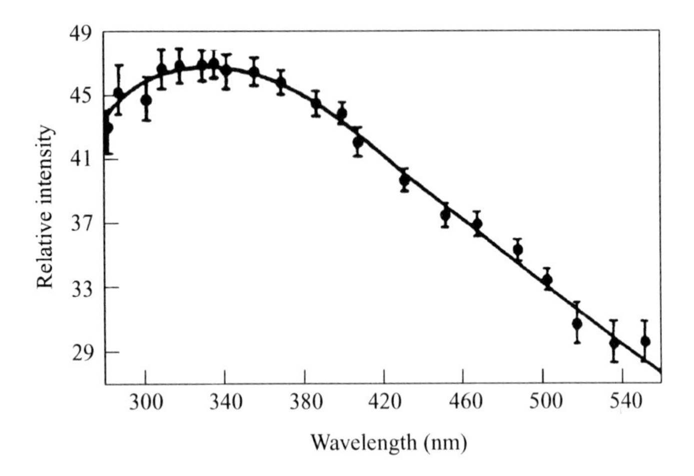


Figure 4.1 Spectral distribution for oxygen-saturated water. Solid line is the theoretical curve for black-body temperature of 8800 K. (Srinivasan and Holroyd (1961).)

## 4.5 CHEMILUMINESCENT THEORY

Griffing and Sette (1950, 1952, 1955) and Verrall and Sehgal (1987) suggested that the high temperatures caused by the cavity collapse gave rise to oxidizing agents such as $H_2O_2$ which would dissolve in the surrounding liquid, causing chemiluminescent reactions.

## 4.6 THE HOT SPOT THEORY

Figure 4.1 is a typical SL spectrum. It is generally a continuum. Srinivasan and Holroyd (1961) fitted it closely to the curve of black-body radiation at 8800 K for oxygen saturated water at 800 kHz. The absence of spectral lines also suggests that the SL bubble is very hot. This could be a hot spot from adiabatic heating. That is a Black Body where the radiation and matter are nearly in equilibrium. Or, it could be Bremsstrahlung from accelerating unbound electrons. We will discuss these ideas in Secs. 4.6.1 and 4.6.2.

### *4.6.1 The hot spot theory – Black Body*

Noltingk and Neppiras (1950) and Neppiras and Noltingk (1951) produced two famous papers suggesting that adiabatic compression of the bubble contents could result in temperatures of 5,000 to 10,000 K. They showed that cavitation is not only restricted to a finite range of frequencies $\omega$ and nuclei sizes $R_0$ but also to a fixed range of hydrostatic pressure $P_A$ and alternating pressure amplitude $P_0$. This sonoluminescence should vary in the same way as the intensity of cavitation and should decrease with $\omega$, it should only appear for nuclei of a certain range of $R_0$, show a sharp threshold at low $P_0$ (and low power) and decline above certain critical values of $P_0$ and $P_A$. These conclusions agree qualitatively with

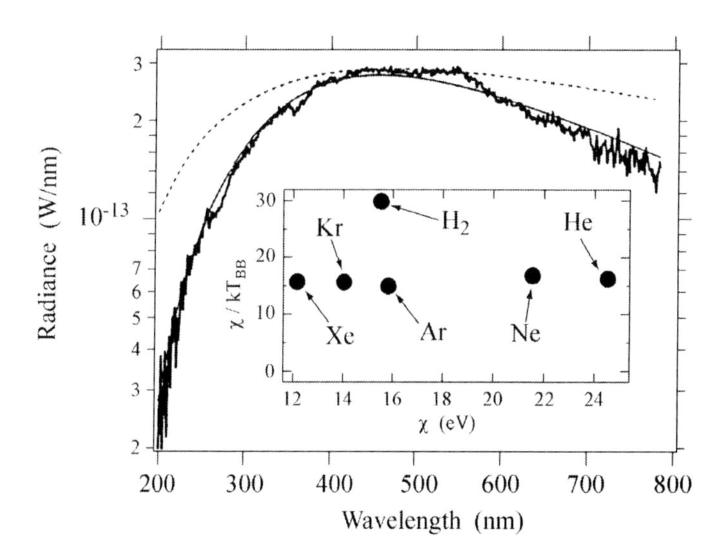


Figure 4.2 Spectrum (resolution, 24 nm FWHM) of 33-kHz SL from a bubble formed in water (23°C) into which $H_2$ was dissolved at a partial pressure of 5 torr. Solid curve, fit to a black body at 6230 K. The flash width of 110 ps requires emission from a surface with radius 0.22 μm. Dashed curve, bremsstrahlung fit with a temperature of 15000 K. Inset, plot of $\chi/T_{BB}$ as a function of $\chi$, the ionization potential of the gas used to make SL. (Vazquez et al. (2001).)

experimental observations. If vapor pressure is considered, the fall of sonoluminescence with ambient temperature can be explained, since the presence of the vapor during the collapse stage would remove energy that would otherwise be radiated. The theory cannot, however, readily explain the line spectra often observed in sonoluminescence spectra.

Many other contributions to this hydrodynamical theory of SL have been made, in particular by Löfstedt et al. (1993). In a long elegant mathematical paper, they derived the scaling laws for the maximum bubble radius and the temperature and duration of the collapse. Major contributions have been made by Neppiras (1980), Prosperetti (1984), Prosperetti et al. (1988), Prosperetti and Lezzi (1986), and Lezzi and Prosperetti (1987).

In a paper entitled "Sonoluminescence: nature's smallest black body", Vazquez et al. (2001) point out that for a mixture of hydrogen and water, the radius of a bubble surface for light emission is 0.2 μm. This is smaller than the peak of the black body spectrum, 0.5 μm from Fig. 4.2, which shows that the spectrum can match that of a black body at 6230 K which is hotter than the surface of the sun. Vazquez et al. (2001) conclude that the state of matter that would admit photon–matter equilibrium under such conditions is a mystery.

Vazquez et al. (2002) measured the spectrum of single bubble sonoluminescence from a mixture of water and hydrogen gas. They had difficulty producing the sonoluminescence from hydrogen because it is weak and has a low solubility. The author had the same difficulty with hydrogen for multibubble sonoluminescence (Young (1976)). Figure 4.3 shows the spectrum. It is well matched by an ideal 6,000 K black body radiating from a surface with a radius of less than $\frac{1}{4}$ μm, according to Planck's black body formula. The fit is remarkably close. Vazquez et al. (2002) discuss whether sonoluminescence is black body radiation and make a number of points.

Using the hot spot theory of bubble dynamics, Sochard et al. (1997) calculated the production of free radicals in the collapse of an air bubble in water. They used driving

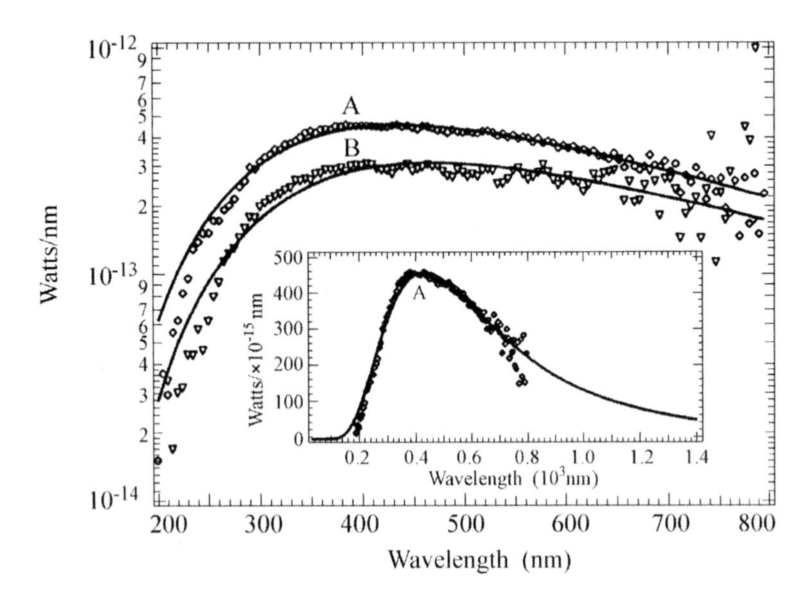


Figure 4.3 Spectrum of a sonoluminescing bubble formed from a mixture of hydrogen and water at 0°C (A) and 20°C (B). The solid lines are fits to Planck's formula for 6644 and 6200 K. The radius of the surface of emission is also fit to 0.25 μm at 20°C and 0.2 μm at 0°C. The uncertainty in $T_{BB}$ of 200 K is dominated by scatter in the spectral data. The uncertainty in $R_e$ of 0.02 μm is dominated by imprecise determination of the solid angle subtended by the photodetector. The inset displays a linear scale. (Vazquez et al. (2002).)

frequencies of 20 kHz and 500 kHz. Other papers on free radicals in cavitation are Henglein (1987, 1993), Suslick (1990), and von Sonntag et al. (1999).

### *4.6.2 The hot spot theory – Bremsstrahlung*

A free electron traveling through the electric field of an ion in an ionized gas can either emit a photon without losing all its kinetic energy and remain free, or it can absorb a photon and acquire additional kinetic energy. These free–free transitions are usually called *Bremsstrahlung* (from the German: *Bremse* brake and *Strahlung* radiation), since the electron is slowed down in the field of the ion and loses a part of its energy in the radiation process. There are various cases:

(1) *Electron ion bremsstrahlung* is the light emission from an electron slowed down in the Coulomb field of a *positive ion* and has a radiative power emitted per unit volume given by Wu and Roberts (1993, 1994), Rybicki and Lightman (1979) and Yasui (1999a), of

$$P = 1.57 \times 10^{-40} \; q^2 N^2 T^{1/2} \; \mathrm{W\,m^{-2}}$$

where $q$ is the degree of ionization, $N$ is the number density of atoms and $T$ is the temperature.

(2) *Electron atom bremsstrahlung*. Bremsstrahlung can also occur when an electron passes through the field of a *neutral* atom. Compared with the field of an ion, the field of a

neutral atom decreases rapidly with distance, and therefore the electron must pass very close to the atom to ensure the emission or absorption of light. Therefore, the probability of bremsstrahlung with the participation of a neutral atom is much smaller than the probability of this process with an ion. Taylor and Caledonia (1969) and Yasui (1999a, 2003) crudely estimate the radiative power per unit volume for electron atom bremsstrahlung to be

$$P = 4.6 \times 10^{-44}\ q^2 N^2 T\ \ \mathrm{W\ m^{-2}}$$

(3) *Polarization bremsstrahlung* is the case of electron–atom bremsstrahlung where the *neutral* atom is polarized in the Coulomb field of the electron in the fly-by encounter; the resulting time-varying dipole moment emits light. However, Hammer and Frommhold (2001b, 2002) conclude that under sonoluminescence conditions, polarization bremsstrahlung amounts to less than 10% of the electron-atom bremsstrahlung contributions.

Yasui (1999a) shows that sonoluminescence is mainly by radiative recombination of electrons and ions, and electron-atom bremsstrahlung.

Bremsstrahlung was considered as the sonoluminescence light emission process as long ago as 1970 by Saksena and Nyborg. They concluded that bremsstrahlung was negligible under the conditions discussed of stable cavitation.

I can do no better than quote Cheeke (1997): The shock wave model of Wu and Roberts (1993, 1994) integrates a mechanical and a thermodynamic model to describe the collapsing bubble. With the assumptions used, very high temperatures at the center are predicted. They then propose that the air in the bubble is fully ionized by shock compression, which gives rise to bremsstrahlung. The mechanism is based firmly on standard bremsstrahlung theory. They calculate the degree of ionization $q$, assuming that air has atomic mass 14.4 and is singly ionized with ionization potential 14.5 eV, that thermodynamic equilibrium is attained and that the Saha formula (Yasui (1999a), Rybicki and Lightman (1979)) applies. They then use the standard beam power emission formula to obtain the detailed spectrum of bremsstrahlung, which is

$$\mathrm{d}P = 1.57 \times 10^{-40}\ q^2 N^2 A\, T^{-1/2} \lambda^{-2} \exp\frac{-A}{\lambda T} d\lambda \quad \mathrm{W\ m^{-3}}$$

where $A = 1.44 \times 10^{-2}$ mK and $N \sim 4.16 \times 10^{25}$ pm$^{-3}$ is the number density of atoms.

The bremsstrahlung model does indeed predict several of the observed properties. It predicts a continuous spectrum with the right order of magnitude intensity (of the order of $10^{-12}$ J per flash) and the correct time scale, which comes naturally out of the shock-wave model, Fig. 4.4a (Putterman (1995)). The latter also appears to provide the correct functional dependence for $P_A$, sound frequency and atomic mass.

However, the bremsstrahlung model has several shortcomings. As presented, it is a simplistic adaptation of a known model, and no attempt is made to examine the microscopic processes associated with the particular environment inside the bubble. Eberlein (1996a) criticizes the model in that the intensity predicted below 180 nm would be so high as to not have observable macroscopic (presumably radiative) consequences in the liquid surrounding the bubble. The temperatures required are an order of magnitude higher than those that have been measured inside bubbles, and there simply does not exist any concrete experimental evidence for the existence of such high temperatures. Finally, the theory is

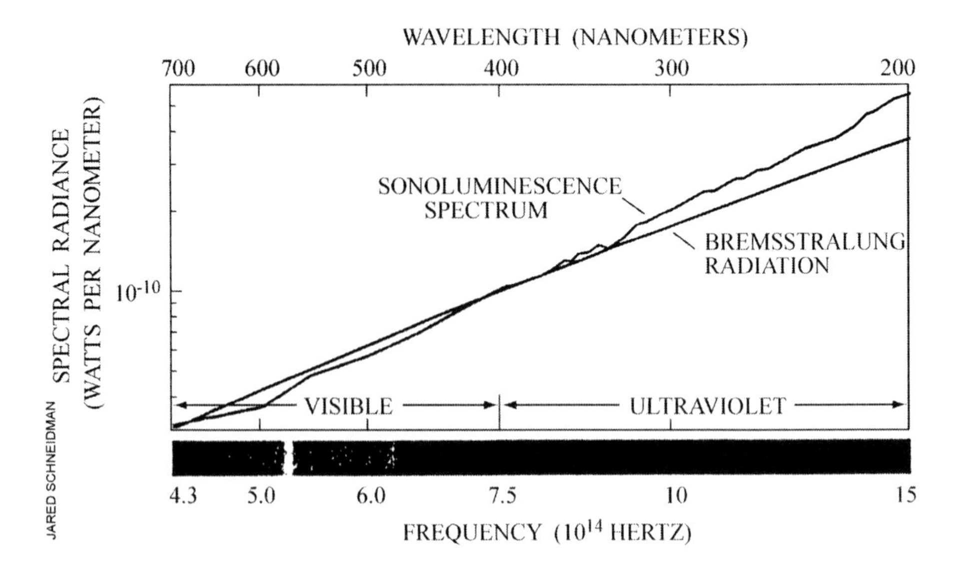


Figure 4.4a Spectrum of sonoluminescence shows that most of the emitted light is ultraviolet. As pointed out by Paul H. Roberts and Cheng-Chin Wu of the University of California at Los Angeles, the signal compares closely with bremsstrahlung radiation – that is, light emitted by a plasma at 100,000 K. (Putterman (1995).)

incomplete in that the van der Waals equation of state is certainly not applicable at such high temperatures and pressures, and radiative, thermal, and mass transport at the interface have not been included.

The bremsstrahlung model cannot be ruled out or confirmed at this stage; the precise theoretical predictions must be quantified and more relevant experimental evidence must be available before a definite conclusion can be reached.

Kondić et al. (1995) extend the analysis to include various energy loss mechanisms, especially due to radiation. They consider a thermal approach based on the coupling of the radiative transfer equation with the gas dynamics equations. They also consider a bremsstrahlung approach and stipulate the formation of a plasma in the center of the bubble. Figure 4.4b shows Kondić's (1995) results. Moss et al. (1997) describe a collapsing bubble as a partially ionized plasma in which the emitted light is neither a pure Planckian nor a pure bremsstrahlung spectrum, but a convolution of the two.

Kwak's group at Chung-Ang University, Seoul, Korea, have made a careful study of bremsstrahlung as the mechanism for the emission of SL (Kwak and Na (1996, 1997), Jeon et al. (2000)). They argue that in the ultra high temperature just before bubble collapse, the gas molecules may dissociate and/or ionize, so that elastic collisions can occur between ions and electrons. These collisions produce light. They conclude that the emission mechanism is the same as black body radiation with finite absorption which confirms that the principal mechanism of SBSL is bremsstrahlung with slight black body emission. The spectrum was measured (Jeon et al. (2000)). Experimental and calculated values yield common spectral behavior in the visible region. Gompf et al. (1997) and Hiller et al. (1998) used time-correlated single-photon counting to find that the width of sonoluminescence flashes ranged from under 40 to over 350 ps for mixtures of various gases in water. Flash widths and emission times are independent of wavelength from the ultraviolet to the

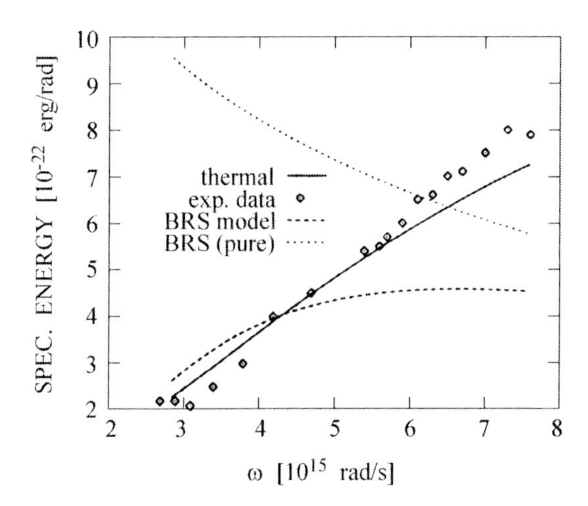


Figure 4.4b Spectrum of emitted SL radiation resulting from the thermal and bremsstrahlung approach. The "pure" bremsstrahlung results are obtained without corrections due to the presence of a plasma and absorption. The acoustic pressure amplitude is 1.325 atm for both the theoretical and experimental results. The attenuation of radiation in the water and the container walls was not taken into account. The theoretical results were scaled down to the experimental ones. (Kondić et al. (1995).)

infrared. The measurements suggest that SL originates as a shock wave–plasma–bremsstrahlung model (Wu and Roberts (1993,1994)). See Sec. 3.9.

Frommhold (1998) suggests that electron–neutral-atom bremsstrahlung may be a principal mechanism for the SL from rare gas bubbles if a weakly ionized environment is assumed. He calculated the spectra for temperatures from 5,000 to 40,000 K, and found them to agree with experiment. For the heavier rare gases, computed intensities and spectral line shapes compare favorably with the measurement of electron "temperatures" of roughly 20,000 K. Figure 4.4c shows the calculated number of photons emitted per sec, per $cm^3$, per $amagat^2$, by bremsstrahlung collisional electron–rare gas atom pairs, as a function of wavelength, in the spectral window of liquid water. To compare this calculated and observed photon yields, Frommhold multiplies the integrals by four factors and arrives at, for argon bubbles, for temperatures around 20,000 K, about $0.5 \times 10^6$ photons per flash. This compares with the about $10^6$ photons per flash measured by Barber and Putterman (1991).

Hammer and Frommhold (2000a,b, 2001a,b, 2002a) have made steady contributions, and in Hammer and Frommhold (2002b) they conclude that in conventional single bubble sonoluminescence, *electron ion bremsstrahlung* amounts to but a small percentage of the overall emission, and that *electron neutral bremsstrahlung* is likely to be the principal radiative mechanism. Figure 4.4d shows a spectrum computed from Hammer and Frommold's 2002b model for argon bubbles in freezing water. The predominance of *electron neutral bremsstrahlung* is shown.

Xu et al. (1998, 1999, 2000) consider bremsstrahlung to be the cause of SL. They consider an argon bubble undergoing electron collisional ionization, recombination, and radiative energy loss. They found that near minimum radius the atoms inside a very thin layer around the origin of the bubble are strongly ionized, and the SL occurs nearly simultaneously. They conclude that multiple ionization and recombination, which occur mainly in the thin layer of plasma, play a dramatically important role. The processes of radiation

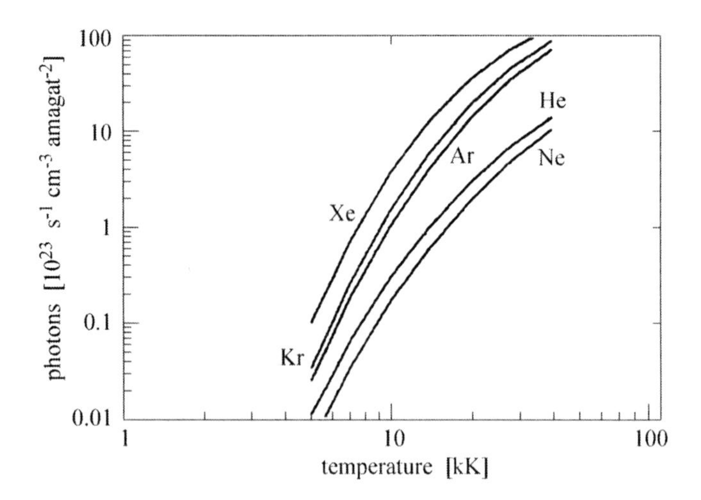


Figure 4.4c Number of photons emitted per second, per cm³, per amagat², by collisional electron–rare-gas atom pairs, as a function of wavelength, for temperatures from 5,000 to 40,000 K, in the spectral window of liquid water. (Frommhold (1998).)

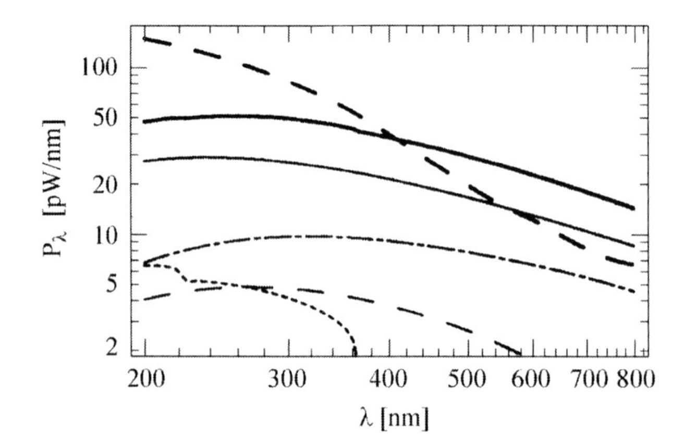


Figure 4.4d Single bubble sonoluminescence (SBSL) spectrum and its composition for argon bubbles in water at freezing temperature. An SBSL bubble of ambient radius 5.5 μm, driven at an ultrasound frequency of 33.4 kHz with an amplitude of 1.48 bar, was assumed. Shown are the total spectrum (thick solid line) and contributions due to electron-neutral argon bremsstrahlung (thin solid line), $O^-$ radiation (dotted line), $H^-$ radiation (dashed line), and $e$-$Ar^+$ bremsstrahlung (dash-dotted line). For comparison, the measured spectrum is indicated by the thick dashed line. (Hammer and Frommhold (2002b).)

must be extraordinarily complex due to high density and pressure. Shock is still the cause of heating. Figure 4.5 shows the light power, with a wavelength from 180 to 750 nm, calculated by assuming bremsstrahlung as the mechanism of the radiation. The maximum power is about 19 mW, and the full width at half maximum of the pulse is about 33 ps. The highest temperature they get is about 20 eV (200,000 K) which is rather higher than experimental values. Xu et al. (2000) claim that the magnitude of the electric field within the shock wave of a collapsing sonoluminescing bubble approaches the magnitude of the

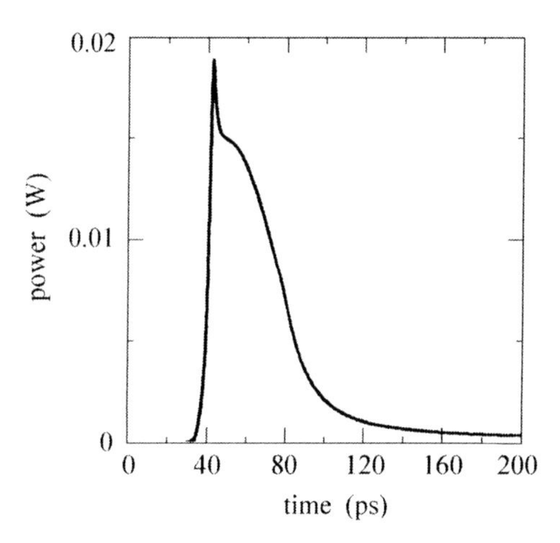


Figure 4.5 The emission power of the sonoluminescing bubble in the final 200 ps starting from $t$ = 22.4289 μs. The wavelength range of the light calculated here is from 180 to 750 nm. (Xu et al. (1999).)

electric field at the Bohr radius of the hydrogen atom ($\approx 5 \times 10^{11}$ V/m). Moss (2000) doubts this.

1999 heralds an upsurge in interest in bremsstrahlung as an emittive mechanism. After odd papers each year, as I write this page on the 23rd March 2001, I have seven papers from 1999 and five papers from 2000 considering bremsstrahlung.

In 1995, Yasui of Japan arrived on the scene and has written 16 thought provoking theoretical studies of SL (1995, 1996, 1997a,b, 1998a,b,c, 1999a,b,c,d, 2000, 2001a,b,c, 2002, 2003). In his 1999a paper, Yasui tackles the mechanism of single bubble sonoluminescence. He postulates a *quasiadiabatic model* (Yasui (1999a)), in which no shock wave develops in the bubble and the whole bubble is heated up by the quasiadiabatic compression ("quasi" means that appreciable thermal conduction takes place between the bubble and the surrounding liquid). The model allows for thermal conduction both inside and outside the bubble, for non-equilibrium evaporation and condensation of water vapor at the bubble wall, and for chemical reactions inside the bubble. Figure 4.6 shows the light intensity around the minimum bubble radius due to radiative recombinations (dash-dotted line), by election–atom bremsstrahlung (dotted line), and by electron-ion bremsstrahlung (dashed line). The light is mainly due to radiative recombinations. Yasui (1999a) also calculates the spectra of a mixture of 2% Xe and 98% N in water, Fig. 4.7. The open circles are experiment data from Hiller (1994), the line is the calculated black body spectrum with the effective temperature of 10,000 K, and the dash-dotted line is that of thermal bremsstrahlung with the effective temperature of 32,000 K. The question arises, why does the experimental spectrum fit the black body formula remarkably well despite the fact that the computer simulation indicates that SL is not black body radiation? Yasui explains this using Kirchhoff's law. Yasui applies the quasiadiabatic model to a hydrogen bubble (1999c) and an argon bubble (1999d), and to argon, xenon and helium bubbles (2001).

With Kozuka et al. (2000) and Hatanaka et al. (2000, 2001) Yasui describes the observation of an SL bubble using a stroboscope, the difference in threshold between SL and sonochemical luminescence, and the quenching mechanism of multibubble SL at excessive sound pressure, respectively.

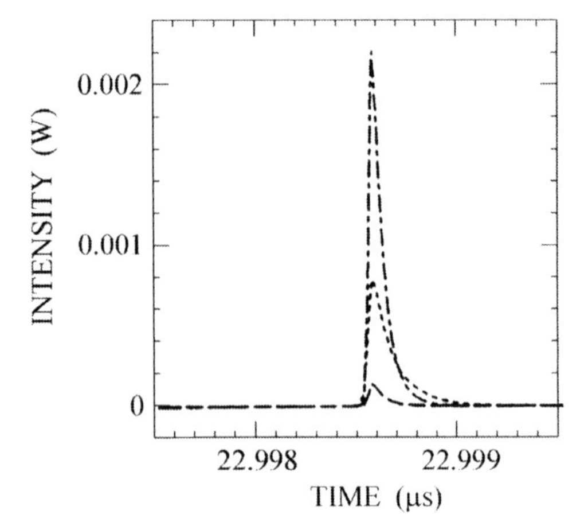


Figure 4.6 Calculated results at around the minimum bubble radius as functions of time for 2000 ps (0.002 μs). The intensity of the light emitted by radiative recombinations (dash-dotted line), that by electron atom bremsstrahlung (dotted line), and that by electron ion bremsstrahlung (dashed line). (Yasui (1999a).)

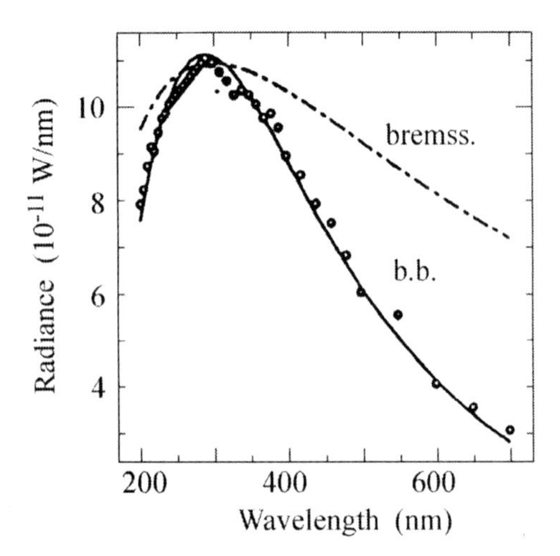


Figure 4.7 Spectrum of SBSL from a bubble of the mixture of 2% Xe and 98% $N_2$ in water. The open circles are the experimental data [14], the line is the calculated black body spectrum with the effective temperature of 10,000 K, and the dash-dotted line is that of thermal bremsstrahlung with the effective temperature of 32,000 K. It should be noted that the absolute values of the calculated spectra are arbitrarily determined just to see the shape of the spectra. (Yasui (1999a).)

Suslick and his group have amassed considerable evidence in favor of thermal molecular emission luminescence in multibubble SL. From the presence of clearly identifiable molecular bands and the absence of other lines associated with discharges, Suslick (1990) deduces the thermal origin of the emission. Suslick et al. (1986) used comparative rate thermometry to evaluate the temperature from the intensity of the different spectrum lines. The temperatures obtained are very consistent and are confirmed by different methods of spectral analysis by Flint and Suslick (1991). We will take as an

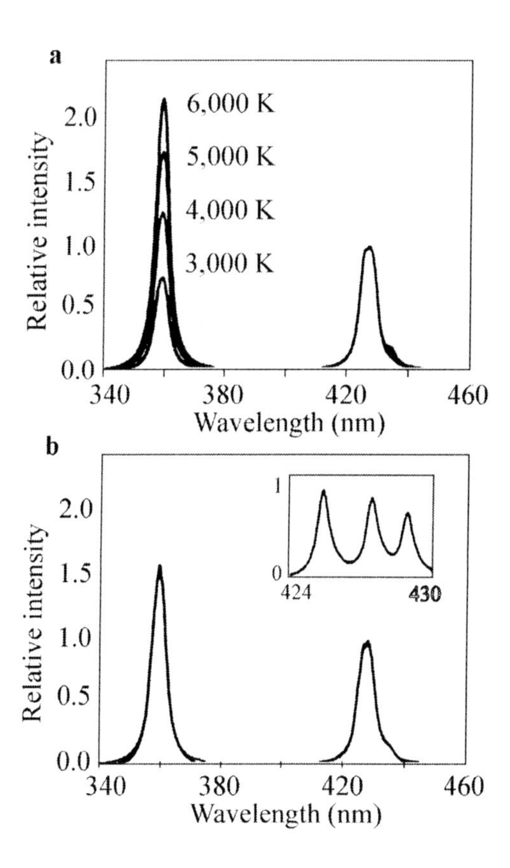


Figure 4.8 Multi-bubble sonoluminescence (MBSL) emission from excited states of Cr atoms. ***a***, Calculated spectra as a function of temperature. ***b***, The observed emission spectrum compared to the best-fit calculated spectrum (4700 K). Inset, observed spectrum at higher resolution, which resolves the individual atomic emission lines. Sonoluminescence was generated from 2.5 mM solutions of metal carbonyls (0.25 mM for $Mo(CO)_6$ due to solution absorbance) in poly(dimethylsiloxane) (Dow 200 silicone oil, 100 centistokes viscosity) irradiated at ~90 W $cm^{-2}$ at 20 kHz (Heat Systems Mysonix 375, with 0.5 inch Ti horn), and spectra collected with an 0.5 m spectrograph (Acton Research 505F with a Princeton Instruments IRY 512N diode array). The solutions were continuously sparged with the appropriate gas during sonication, and each spectrum is the average of at least 15 10-second collections. All spectra were corrected for absorbance by the solution and for the response of the optical system. Imaging of the MBSL shows that the emission comes from a dense cloud of emitting bubbles rather than from a few exceptional ones. (Reprinted with permission from *Nature*, McNamara III et al. **401**,772. Copyright (1999) Macmillan Magazines Limited.)

example the careful work of McNamara III et al. (1999). They studied the multibubble SL from the excited states of chromium atoms in a solution of $Cr(CO)_6$. The improved spectral resolution using Ar at 20 kHz are shown in Fig. 4.8. The calculated spectra as a function of temperature are shown in the top diagram a, as a function of temperature. The observed spectra are shown in the lower diagram b. The relative integrated intensities ($I_1/I_2$) of the two atomic lines emitted from the different excited states of the same metal atom are given by

$$\frac{I_1}{I_2} = \frac{g_1 A_1 \lambda_2}{g_2 A_2 \lambda_1} \exp[(E_1 - E_2)/kT_e]$$

where $g$ is the degeneracy of the electronic state, $A$ is the Einstein transition probability for the electronic transition, $\lambda$ is the wavelength of the emitted line, $E$ is the energy of the electronic state, and $T_e$ is the electronic temperature. The result for Cr atom emission under Ar at 20 kHz and 90 W $cm^{-2}$ was

$$\boxed{4700 \pm 300 \text{ K}}$$

## 4.7 SHOCK WAVE THEORY

Section 3.5.6, which contains an account of *experiments* on shock waves from bubbles, should be read in conjunction with Sec. 4.7.

### *4.7.1 Early theories*

As long ago as 1952, Trilling computed the velocity and pressure field in a slightly compressible liquid resulting from the collapse of a spherical bubble as a function of the pressure at the bubble wall. Trilling supposed the bubble to be filled with inviscid perfect non-conducting gas. His investigation of the gas motion involved a series of shock waves.

Jarman (1960) and Taylor and Jarman (1970) suggested that SL may result from the high temperatures attained by spherical microshocks propagating within the imploding bubble as they converge towards the center of the bubble and are reflected.

Vaughan and Leeman (1986, 1987) discuss shock waves as an alternative to the hot spot theory and adiabatic compression for producing SL.

All the above is for multibubble sonoluminescence. Now we come to single bubble sonoluminescence, discovered in 1989.

In 1992, Barber and Putterman, in a classic paper, used light scattering to measure the dynamics of the repetitive collapse of a sonoluminescing bubble of air trapped in water. It was found that the surface of the bubble was collapsing with a supersonic velocity at about the time of light emission which in turn precedes the minimum bubble radius by about 0.03% of the period of the acoustic drive. These observations suggested that the shedding of an imploding shock mediates between the bubble collapse and light emission.

Greenspan and Nadim (1993), in a pioneer paper, used the governing equations of gas dynamics inside the bubble to show that a shock forms during the bubble collapse. This converging spherical shock grows in strength as it propagates towards the center. Once the shock has strengthened to the extent that it can be classified as a "strong shock", the Chisnell–Whitham characteristic rule, as expounded in Whitham (1974), can be used thereafter to calculate all quantities of interest in the collapse (e.g. Mach number, temperature and pressure). For a converging spherical shock, one finds the shock Mach number varies with its instantaneous radius according to

$$M \propto r^{-0.394} \quad \text{for } \gamma = 1.4 \quad \text{[Whitham (1974) pp. 270–274]}$$

Here the Mach number, $M$, is defined as the ratio of the shock speed to the sound speed $c_0$, in the undisturbed liquid ahead of it. Therefore, as the radius of the converging shock tends to zero, its Mach number, and thus the temperature and pressure behind it, all tend to infinity. If the bubble is small enough, the converging shock can give rise to sufficiently high temperature near the bubble center to produce a flash of light.

Suslick and Crum (1997), in an excellent review article entitled "Sonochemistry and Sonoluminescence" for the Encyclopedia of Acoustics, discuss the hot spot theory and the possibility of an imploding shock wave being created in the gas bubble during the final stages of collapse.

### 4.7.2 *Wu and Roberts*

Wu and Roberts (1993, 1994, 1996) and Roberts and Wu (1996) advanced the subject in a series of papers. They set up a model for an air bubble in water based on the Rayleigh–Plesset equation

$$R\ddot{R} + \frac{3}{2}\dot{R}^2 = \frac{1}{\rho_W}[p(R,t) - P_a(t) - P_0] + \frac{R}{\rho_W c_W}\frac{d}{dt}[p(R,t) - P_0(t)] - 4\nu\frac{\dot{R}}{R} \quad (4.1)$$

where $\rho_W$ = water density, $R(t)$ = bubble radius,
$p$ = air pressure, $P_0$ = ambient pressure,
$P_A$ = driving pressure = $-P_A \sin \omega t$,
$\nu$ = kinematic viscosity of the water and
$c_W$ is the speed of sound in water.

Because of the enormous compressions in the bubble, they modeled the trapped air by a hard-core van der Waals equation of state, for which the pressure $p$, specific volume $V$, temperature $T$, internal energy $e$, and entropy $S$ are related by

$$p = \frac{RT}{V-b}, \qquad e = c_V T = \frac{V-b}{\gamma - 1}p, \qquad S = c_V \ln[p(V-b)^\gamma] + \text{const} \quad (4.2)$$

where $R$ is the gas constant, $c_V = R/(\gamma - 1)$ is the specific heat at constant volume, $\gamma = 1.4$ is the ratio of specific heats, and $b = {}^1/\rho_m$ is the van der Waals excluded volume.

Roberts and Wu (2003) have provided an excellent fundamental account of sonoluminescence and the shock wave theory of its production. They warn that no experimental fact has as yet ruled out the shock wave theory of sonoluminescence, though none has positively supported it. Roberts and Wu (2003) also say that the most popular rival to the spherical shock wave theory, at the time of writing, supposes that the shape instability of the bubble launches a high speed, possibly supersonic, *jet* of fluid across the gas cavity (Chanine (1994), Longuet-Higgins (1996), and Prosperetti (1996, 1997)). Other possibilities have been studied (Patel and Rao (1996)).

### 4.7.3 *Moss et al.*

Moss et al. (1994, 1996, 1997, 1999a,b) and Moss (1997) in a series of refined papers extended the work of Wu and Roberts by including vibrational contributions, dissociation,

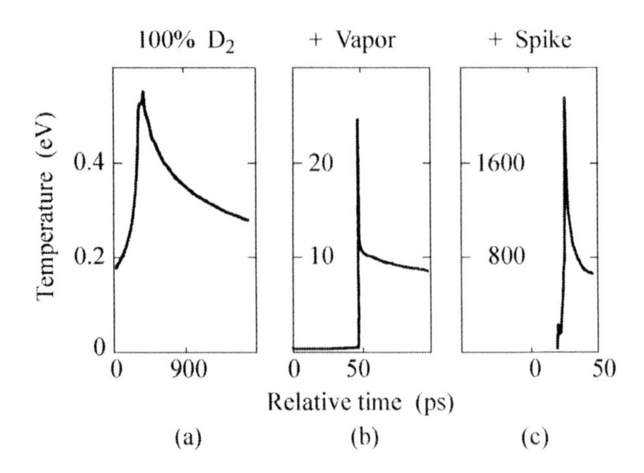


Figure 4.9 Temperature at the center of the bubble, as a function of time (relative to an offset origin) for (*a*) pure deuterium ($t_0$ = 38.2956 μs), (*b*) 85% molar deuterium with 15% molar "vapor", approximated using air ($t_0$ = 38.29578 μs), and (*c*) the mixture in (*b*) with a spike added to the sinusoidal driving pressure ($t_0$ = 37.6854 μs). The spike has a large effect on the temperature at the center of the bubble. (Moss et al. (1996). Credit is given to the University of California, Lawrence Livermore National Laboratory, and the Department of Energy under whose auspices the work was performed.)

translational energy of the atoms and electrons, and ionization energy in the energy equation of the equation of state. They note that extremely high temperatures occur, unless dissociation and ionization of the air are included in the equation of state. They also note that narrow pulse widths do not occur when the air is described by an ideal gas and they show that temperatures approaching one kilowatt ($10^7$ K) might be achieved, leading to the possibility of microthermonuclear fusion. Moss et al. (1996) showed that by adding a pressure spike to the driving pressure, a small but measurable number of thermonuclear neutrons may be generated. Moss et al. (1996), in Fig. 4.9, showed the calculated temperature at the center of the bubble as a function of time for (a) pure deuterium, (b) pure deuterium with 15% molar water vapor and (c) the mixture as in (b) with a spike added to the driving pressure. Noting the different temperature scales, the spike has a large effect on the temperature at the center of the bubble.

In Moss et al. (1997a), a SL bubble was modeled as a thermally conducting, partially ionized two-component *plasma*. The model showed that shock waves generated during the collapse of the bubble and electron conduction and the rapidly changing opacity of the plasma produce the short pulse widths and the inferred high temperatures. The calculations show that the emitted light is described by neither a pure Planckian nor a pure bremsstrahlung spectrum, but a convolution of the two. During the collapse of the bubble, a shock is generated that compresses and heats its contents. More heating occurs at the center of the bubble than at its boundary because the shock strength increases as it approaches the bubble's center. The hotter regions begin to ionize and create a two-component plasma of ions and electrons. The hot matter emits light by an energy cascade from the ions, to the electrons, to the photons. In the final step of the cascade, the electrons lose energy to the photon field to produce the SL flash. The electron–photon coupling can be due to bremsstrahlung, Compton scattering, photo-ionization, and line transitions. Moss et al. (1997a) believe bremsstrahlung to be the dominant mechanism. The calculations show that light is emitted from regions that are optically thick and thin. If the emitting region is optically thick, then the coupling mechanism is irrelevant. The emitted optical energy flux comes only from the surface and can be calculated from the black body expression $\sigma T^4$ where $\sigma$ is the Stefan–Boltzmann constant. However, if the region is optically thin, then the coupling mechanism is important, and the emitted optical power is a volume integral of the emissivity. Typical SL spectra (Hiller et al. (1992, 1994)) can be integrated to show that *the optical energy per flash is at most one millionth of the thermal energy of the compressed bubble.*

### 4.7.4 Kondić et al.

Kondić et al. (1995), in a detailed classic paper, calculated the SL from a collapsing SL bubble. The gas dynamic equations coupled to bubble wall dynamics found shock waves and high temperatures. They considered (1) heat flow between liquid, (2) thermal conduction in the gas, (3) inclusion of vibrational degrees of freedom in the equation of state, (4) dissociation of the gas, (5) ionization of the gas, (6) radiation losses – bremsstrahlung. They started with the Rayleigh–Plesset equation

$$R\ddot{R} + \frac{3}{2}\dot{R}^2 = \frac{1}{\rho_1}[P_{\rm g}(t) - P_{\rm A}(t) - P_0] + \frac{R}{\rho_1 c_1}\frac{\rm d}{{\rm d}t}[P_{\rm g}(R) - P_{\rm A}(t)] - 4\nu_1\frac{\dot{R}}{R} \tag{4.3}$$

where $P_{\rm g}(t)$ is the pressure in the gas next to the bubble wall, $P_{\rm A}(t)$ is the acoustic pressure given by $P_{\rm A}(t) = P_{\rm A}\sin(\omega_a t)$, $P_0$ is the hydrostatic pressure, $\rho_1$ is the density of the liquid, $c_1$ is the speed of sound in the liquid, and $\nu_1$ is its kinematic viscosity.

Assuming an adiabatic compression of the gas in the bubble down to the van der Waals hard core radius $a$ (for air $R_0/a \sim 8.5$) where $R_0$ is the equilibrium radius of the bubble, we have

$$P_{\rm g} = P_0\frac{(R_0^3 - a^3)^\gamma}{(R^3 - a^3)^\gamma} \tag{4.4}$$

where $\gamma$ is the ratio of the specific heats. Equations (4.3) and (4.4) were solved numerically. They referred to this solution as an adiabatic solution. However, on this approach it did not seem possible to explain the production of a shock wave in the bubble, as it was not clear if the speed of the bubble wall becomes supersonic with respect to the speed of sound in the gas. Also, the emission of visible light requires a much higher temperature than that which can be expected on the adiabatic assumption. To completely solve the problem, especially close to the point of maximum contraction, it was necessary to solve Eq. (4.3) coupled with the gas dynamics equations.

### 4.7.5 Kwak and Yang

Kwak and Yang (1995) solved the conservation equations for the gas inside the bubble analytically. They considered heat transfer in the liquid layer adjacent to the bubble wall,

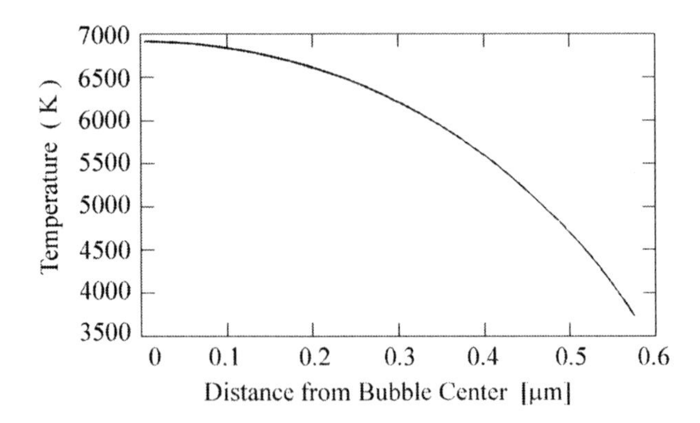


Figure 4.10 Temperature distribution inside an air bubble in water. (Kwak and Yang (1995).)

and identified the launch condition and the Hugoniot curve for the shock propagation. The calculated shock duration of 2.7 to 17 ps, which is comparable to experimental values, was obtained with the use of a similarity solution for a converging spherical shock. The air bubble temperature after shock focusing was found to be 7,000–44,000 K, depending on the equilibrium bubble radius and the acoustic pressure. Kwak and Yang (1995) solved the conservation equations analytically to obtain the density, pressure and temperature *distribution* of the gas in the bubble. The heat transfer in the thermal boundary layer adjacent to the bubble wall was treated by solving the liquid energy equation in this layer. And the time dependent radius and wall velocity of the oscillating bubble under sound was found by using the equation derived from the Keller and Miksis (1980) formulation. The temperature distribution was derived by assuming that the thermal conductivity of the gas in the bubble is linearly dependent on the gas temperature by

$$k_g = AT + B$$

For air, $A = 5.528 \times 10^{-5}$ W/mK$^2$ and $B = 1.165 \times 10^{-2}$ W/mK from Prosperetti et al. (1988).

For this approximation and Fourier law, Kwak and Yang (1995) obtained the following temperature profile inside the bubble by solving the energy equation for the heat flux, with uniform temperature approximation

$$T_b(r) = \frac{B}{A}\left(-1 + \sqrt{\left(1 + \frac{A}{B}T_{b0}\right)^2 - 2\eta\left(\frac{A}{B}\right)(T_{bl} - T_\infty)\left(\frac{r}{R_b}\right)^2}\right)$$

where $T_{b0}$ = temperature of center of bubble

$T_{bl}$ = temperature of bubble wall

$R_b$ = radius of bubble wall

$\eta = \dfrac{R_b k_l}{\delta B}$, where $k_l$ is the thermal conductivity of the liquid and $\delta$ is the bubble wall thickness.

Figure 4.10 shows the temperature distribution inside an air bubble in water at the collapse point with $R_0 = 4.5$ μm, $P_A = 1.35$ atm and $f = 26.5$ kHz.

Lee et al. (1997) continued the above treatment using the Kirkwood–Bethe hypothesis for the outgoing wave from the SL collapse. The rise time and the magnitude of the shock pulse were in good agreement with the observed values, and may provide an approximate value of the gas pressure near the collapse point of the SL bubble.

However, in 1999, Kwak et al. took up the fundamental question of whether a shock wave or a pressure wave is formed inside a SL bubble. They found that a *pressure wave* rather than a shock wave is launched into the bubble center just prior to the bubble collapse and that the wave is reflected near the center. *The temperature rise associated with the pressure wave developed inside the bubble turned out to be insufficient for emitting light.*

### *4.7.6 Evans*

Evans (1996) studied the nearly spherical converging shock wave in a van der Waals gas with regard to SL. Calculations showed that the small deformations of a nearly spherical converging shock wave increased as the shock converged. This suggested that at some critical radius, the spherical shape would be completely lost and the convergence disrupted. In this sense, a converging shock wave would be unstable. Perturbations to the spherical shape would grow slowly, as an inverse power of the radius. In SL this weakness would be vitally important because it would allow shock waves to focus energy in a very small region at the center of a bubble, even when slight deviations from spherical symmetry exist. Evans estimated that the amount of asymmetry in the bubble surface needed to disrupt SL would be about one part in twenty.

### *4.7.7 Vuong and Szeri*

Vuong and Szeri (1996) concisely presented a new model for the gas dynamics in a bubble at conditions that lead to SL. They solved the spherically symmetric Navier–Stokes equations with variable properties together with momentum and energy equations in water. Calculations were done for bubbles of Ar, He and Xe. The first main result was that in contrast to previous models of air bubbles in water, there are *no sharp shocks* focusing at the origin of the bubble. Vuong and Szeri (1996) included diffusive transport in the model and showed that this dramatically slowed the steepening of a disturbance into a sharp shock in the rare gas bubbles. Because the bubble is so small at its minimum radius and because the disturbance travels so fast, there is insufficient "fetch" for the development of a steep shock that focuses on the origin. The second main result concerns an observed correlation between SL and the thermal conductivity of the gas (Young (1976)), which suggests that heat transfer plays a dominant role in the focusing of acoustic energy. Vuong and Szeri (1996) showed instead that mechanical effects associated with the molecular mass of the gas figured prominently in determining the peak temperature and pressures in the bubble, when the bubble is forced strongly enough to engender wavy disturbances that focus on the bubble center. Figure 4.11 shows their calculated temperatures at the origin of the bubble versus time. Incidentally, Vuong and Szeri (1996) pointed out the importance of the ratio of the specific heats of the gas $\gamma$ by a simple example showing that due to the higher $\gamma$, the rare gases become much hotter upon rapid compression. Suppose a 50 μm bubble at 300 K is compressed (without shocks) to 0.5 μm. If the bubble is of a rare gas, $\gamma = 1.67$, the temperature changes to 3,000,000 K; if air with $\gamma = 1.4$, the final temperature is only 75,000 K.

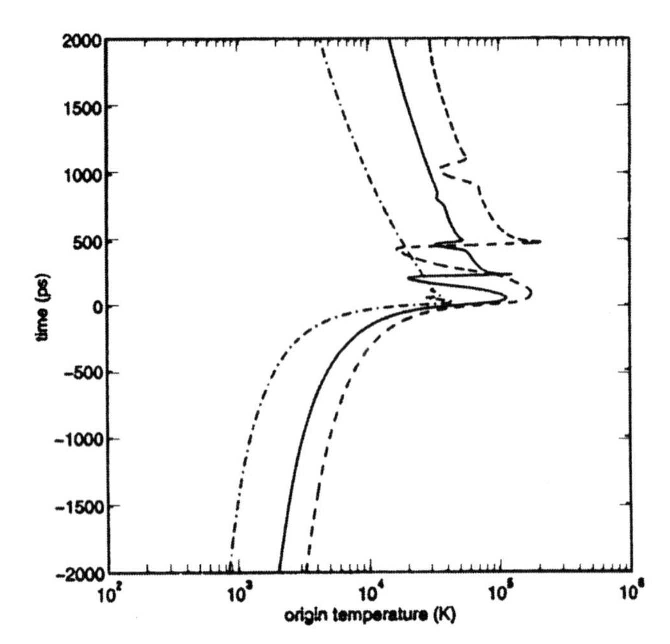


Figure 4.11 Temperature at the origin versus time for argon (solid curve), helium (dot-dashed curve) and xenon (dashed curve), near the end of the first collapse of the bubble. $P_A = 1.3$ atm. $R_0 = 4.5$ μm. (Reprinted with permission from Vuong and Szeri 1996 *Physics of Fluids* **8**, 2354. American Institute of Physics.)

Vuong et al. (1999) analyze the main collapse of the bubble wall. They state that although the wall does indeed launch an inwardly-traveling compression wave, and although the subsequent reflection of the wave at the bubble center produces a very rapid temperature peak, the wave is prevented from steepening into a sharp shock by an adverse gradient in the sound speed caused by heat transfer. In a later paper, Lin and Szeri (2001) discuss the fact that the sound speed in the gas increases towards the center of the bubble preventing the compression wave from steepening into a sharp shock before it reaches the center of the bubble. This adverse gradient in sound speed forms because: (i) the bubble collapse to one millionth the maximum volume causes enormous compression heating of the contents, and (ii) rapid heat conduction of the bubble reduces the temperature near the wall.

### *4.7.8 Chu*

Chu (1996) presented a homologous solution to the spherical hydrodynamic equations governing the oscillation of an SL bubble. The dynamics of the collapse had an interesting and close resemblance to that of a pre-supernova stella core collapse (Goldreich and Weber (1980)). Chu (1996) verified in a numerical simulation that the solution dominated the contraction phase of the bubble until the shock waves were generated. The stability of the homologous solution ensured that the spherical symmetry of the bubble was preserved during the scaling contraction. This may explain partially the regularity and stability of an SL bubble. The stability analysis and numerical simulation both break down when shock waves are generated.

### 4.7.9 Barber et al.

Barber et al. (1997a) concluded that SL is accompanied by the supersonic collapse of the gas bubble. The strongly supersonic motion suggested that the bubble launched an imploding shock wave which focuses energy by many orders of magnitude. However, they point out that shock wave model has a number of limitations. (1) It does not explain the range of sound field intensities for which SL can be observed (Barber et al. (1994)). (2) It does not explain the observed bubble radii which in many cases (e.g. air in water) cannot be accounted for by diffusive equilibrium (Barber et al. (1995), Löfstedt et al. (1993, 1995)). (3) It does not explain the role of rare gas dopants. (4) It does not explain the effects of cooling the liquid. (5) It does not explain why water is the friendliest fluid for SL and finally (6) it does not explain why the spectra of bubbles in water and commercially available (Isotech) heavy water (Hiller and Putterman (1995)) can be dramatically different.

### 4.7.10 MacIntyre

MacIntyre (1997) mentioned that black body emission and bremsstrahlung are too slow to explain the short pulses observed by Barber et al. (1992). MacIntyre stated that this leaves quantum vacuum radiation (Eberlein (1996a,b)), and collision-induced emission (Frommhold and Atchley (1994)). MacIntyre states that these two approaches are not mutually exclusive, but differ diametrically over the fundamental question of whether the bubble gas is optically active, or exerts its 750-fold influence on SL intensity indirectly through bubble dynamics and interface crossing. MacIntyre (1997) goes on to state that the UCLA approach (Barber (1992), Löfstedt (1995) and the work of Wu and Roberts (1993)) is consistent with an optically inactive gas, but also ignores the complexities which would then make gas composition important.

MacIntyre (1997) performed a thermodynamic analysis of Wu and Robert's case (1993). Figure 4.12 is an idealized cycle in $P-V$ space. We will simplify MacIntyre's painstaking treatment and start with an ambient bubble at $R_0$. We have 4 legs to the cycle.

Leg 1. The bubble expands relatively slowly and isothermally to $R_{max}$, the maximum radius.

Leg 2. The rapid collapse from $R_{max}$ to $R_{\wedge}$ is adiabatic. During this leg, the bubble absorbs work from the acoustic field, taking us to some 62,000 K (5.34 eV) – about one third of the 15 eV (174,000 K) ionization potential of air.

Leg 3. The very interesting Leg 3 is isochoric i.e. constant volume. It represents the passage of the shock wave out of the bubble, with the order-of-magnitude decrease in $P$ and $T$ reflecting conditions before and after the hydraulic jump of the shock. Mechanical coupling at the wall, normally weak because of the mismatch in acoustic impedance $\rho c$ between gas and liquid, is enhanced here. The impedance $\rho c$ (water) $\approx 1.5\times 10^6$ kgm$^{-2}$s$^{-1}$ should be matched at some point along Leg 3, where – in the event that the usual gas equation held for the velocity of sound in the bubble interior under these extreme conditions – $\rho c$ (gas) would drop from $2.5\times 10^6$ at $R_{\wedge}$ to $0.84\times 10^6$ at $R_{min}$. MacIntyre shows from Table 2 in his paper, that the bubble is 90% efficient in converting acoustic to shock energy; remarkably high for a mechanical process.

Leg 4 which returns from $R_{min}$ to $R_0$ is adiabatic.

Figure 4.12 is a near Carnot cycle. Thermodynamically the cycle is a heat pump, tracing its $P-V$ loop counterclockwise and pumping energy uphill from the sound field to the electromagnetic field.

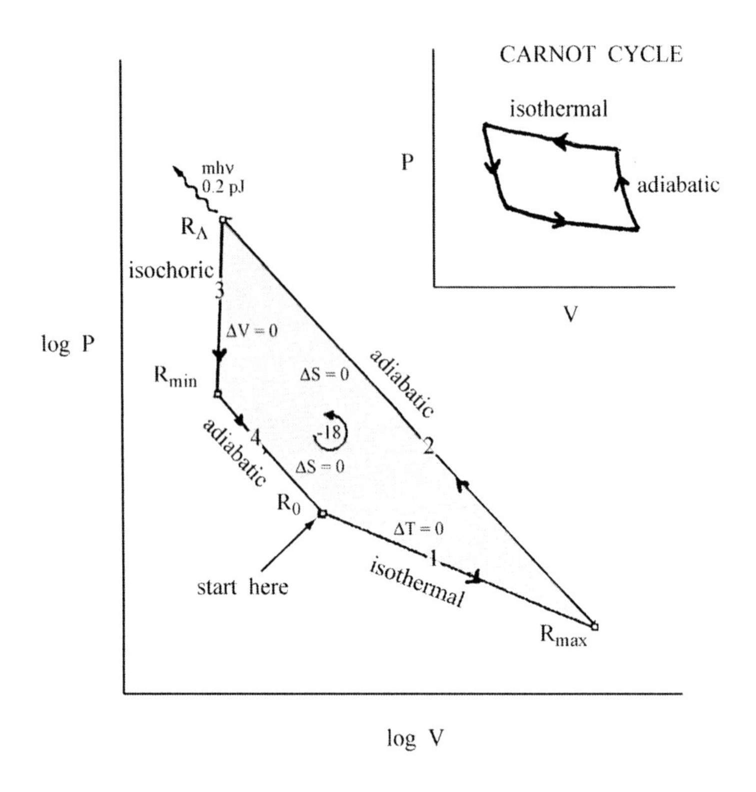


Figure 4.12 Idealized SL cycle.
Start at $R_0$; expand isothermally to $R_{max}$, collapse adiabatically to $R_\wedge$ at the emission SL temperature. $R_\wedge$ to $R_{min}$ is the emission of the shock wave. $R_{min}$ to $R_0$ is the adiabatic return. $S$ – entropy. (MacIntyre (1997).)

I am writing this in May 2001. In the four years since MacIntyre published in 1997, I have not seen one citation of his paper. Are sonoluminescent researchers frightened of thermodynamics?

### 4.7.11 *Yuan et al.*

Yuan et al. (1998) developed three Rayleigh–Plesset (RP) equations:

(1) RP1 is for an incompressible flow and modified by Keller and Kolodner (1956) to include acoustic radiation.

$$R\ddot{R} + \frac{3}{2}\dot{R}^2 = \frac{1}{\rho_{1_\alpha}}\left[P_g(R,t) - \frac{2\sigma}{R} - \frac{4\eta\dot{R}}{R} - P_0 - P\right] + \frac{t_R}{\rho_{1_\alpha}}\frac{d}{dt}[P_g(R,t) - P]$$

where $\rho_{1_\alpha}$ is the ambient liquid density, $P_g(R, t)$ the gas pressure, $P_0$ the ambient pressure, $P = -P_A \sin(\omega t)$ the pressure of the sound field of frequency $\omega$ and amplitude $P_A$,

$t_R \equiv R/c_{1_\alpha}$, $c_{1_\alpha}$ is the speed of sound in the liquid, $\sigma$ the surface tension, and $\eta$ the dynamic viscosity of the liquid.

(2) RP2 follows from the Keller–Miksis formulation (1980) which includes the effect of liquid compressibility:

$$(1-M)R\ddot{R}+\frac{3}{2}\left(1-\frac{M}{3}\right)\dot{R}^2 = \frac{1}{\rho_{1_\alpha}}(1+M)[P_{\mathrm{L}}-P_0-P(t+t_R)]+\frac{t_R}{\rho_{1_\alpha}}\frac{\mathrm{d}P_{\mathrm{L}}}{\mathrm{d}t}$$

This equation contains terms that depend on the bubble-wall Mach number $M \equiv \dot{R}/c_{1_\alpha}$, which characterizes the liquid compressibility. $P_{\mathrm{b}}(t)$ is the pressure on the liquid side of the bubble wall. Yuan (1988) shows how RP2 can reduce to RP1.

(3) RP3 is found by a more accurate model of liquid compressibility. The speed of sound is made to depend on the equation of state of the liquid as done in the original Gilmore (1952) form. The modified Keller equation then gives RP3 as from Kamath et al. (1987).

$$(1-M)R\ddot{R}+\frac{3}{2}\left(1-\frac{M}{3}\right)\dot{R}^2 = (1+M)\left[H_{\mathrm{b}}-\frac{1}{\rho_1}P(t+t_R)\right]+t_R\frac{\mathrm{d}}{\mathrm{d}t}H_{\mathrm{b}}$$

where $\rho_1$, $c_1$ and $H_{\mathrm{b}}$ are the density, speed of sound, and enthalpy of the liquid. Yuan et al. (1988) shows how RP3 can reduce to RP2.

Yuan et al. (1998) deal in detail with the conservation equations for the gas in the bubble, the energy equation in the liquid, liquid compressibility, heat conduction, surface tension, and the existence of a shock-free picosecond pulse. They show many graphs of velocity, density, pressure, temperature against time, for the three equations. Just two interesting graphs will be shown here, Figs. 4.13 and 4.14 for the temperature and average power for an air bubble. Figure 4.13 has $P_{\mathrm{A}} = 1.275$ atm and Fig. 4.14 has $P_{\mathrm{A}} = 1.35$ atm. Note the different temperature and power scales on the two figures. The temperature above 12,000 K lasts about 120 ps in (a) in agreement with previous numerical results of Moss et al. (1994) and Vuong and Szeri (1996), and more than 200 ps in (b). The wiggle in the temperature in (b) is related to the reflection of compressional waves at the center.

Yuan et al. (1998) conclude that even though the effect of liquid compressibility is important only during a very short duration of bubble collapse, it can damp out shock waves and make the solution shock-free. Surface tension has the same effect as liquid compressibility to reduce the violence of the bubble motion. On the other hand, heat conduction between liquid and gas is important during the slow expansion phase, as it leads to a larger expansion ratio $R_{\mathrm{max}}/R_0$, enhancing the violence of the bubble collapse.

### *4.7.12 Cheng et al.*

In a further paper, Cheng et al. (1998) using the material in Yuan et al. (1998) conclude that shock waves are absent in the stable argon single bubble SL regime, and their formation in air bubbles is very sensitive to the various physical parameters and equations of state used.

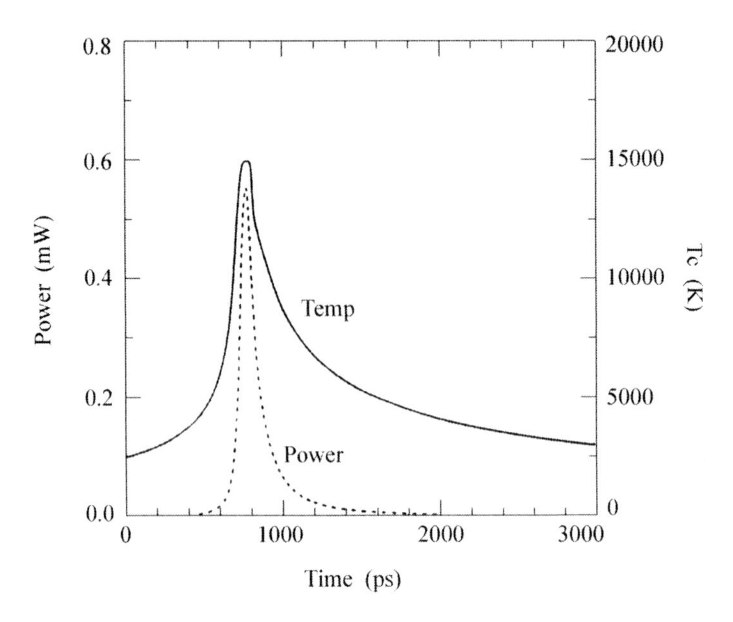


Figure 4.13 Temperature at the bubble center (solid line) and radiation power (dashed line) vs time, for an *air* bubble with $R_0 = 4.5$ μm, $P_A = 1.275$ atm, $f = 26.4$ kHz, $\sigma = 0.0725$ kg s$^{-2}$. (Yuan et al. (1998).)

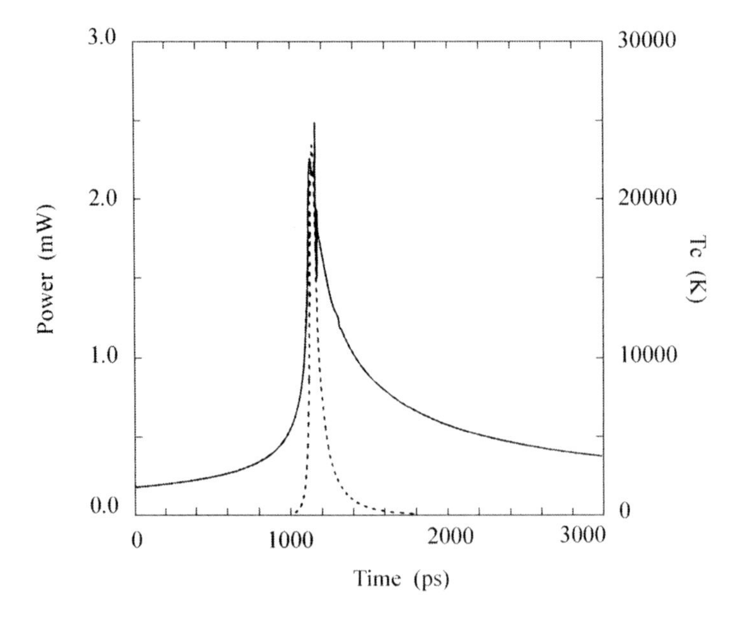


Figure 4.14 Same as Fig. 4.13 but with $P_A = 1.35$ atm. (Yuan et al. (1998).)

For both air and argon bubbles, smooth compressional waves are much more robust and are already adequate to explain the radiation power and pulse length of SL.

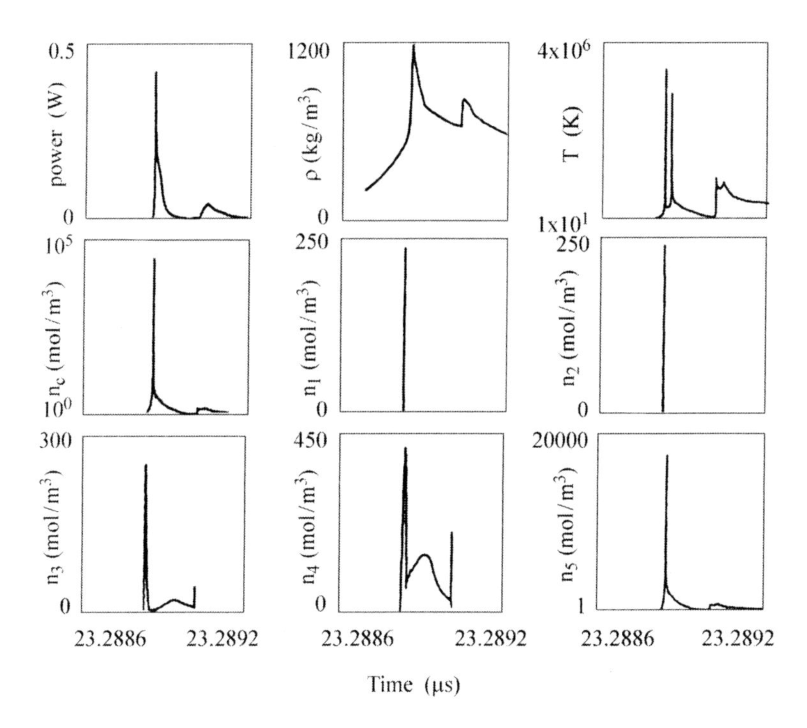


Figure 4.15 The time evolution for the case that the bubble is driven by 1.5 atm acoustic pressure. The duration is about 0.6 ns starting from 23.2886 μs. (Xu et al. (1998).)

### *4.7.13 Xu et al.*

Xu et al. (1998, 1999) calculate a hydrodynamic simulation of an argon bubble including electron-collisional ionization, recombination and radiative energy. They found that near the moment of bubble minimum radius, the atoms inside a very thin layer around the center of the bubble are strongly ionized, and the light emission occurs nearly simultaneously. For the case of 1.5 atm acoustic pressure, Fig. 4.15 shows two distinct thermal spikes. They conclude that near the minimum bubble radius the gases undergo strong multiple ionization and sharp recombination. This state of plasma leads to the SL.

### *4.7.14 Putterman and Weninger*

If the shock wave theory is tenable, the shock should be detectable. The outgoing shock wave has recently been photographed by Weninger et al. (2000a). Figure 4.16 is a shadowgraph obtained 5–10 ns after the moment of collapse of a 100 torr 1% Xe in oxygen bubble at 40 kHz. The 3 ns flash of light shows a spherical bubble surrounded by the emitted outgoing shock wave. Also visible round the bubble are three particles of dirt which appeared to be trapped to the bubble.

See also Sec. 3.5.6.

Putterman and Weninger (2000) summarized the exploding shock wave/plasma/ bremsstrahlung model succinctly. “The trapped bubble collapses according to Rayleigh’s

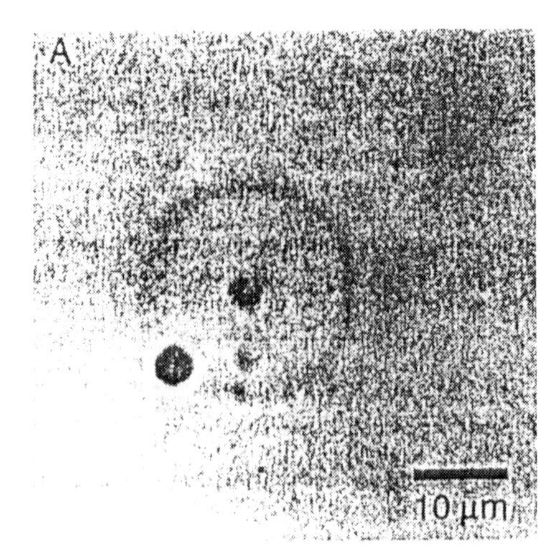


Figure 4.16 Emission of an outgoing shock wave from a collapsing bubble (100 torr 1% Xe in oxygen at 40 kHz). Two-dimensional (extinction) shadowgraph photo obtained 5–10 ns after the moment of collapse with a 3 ns flash of light showing a spherical bubble surrounded by the emitted outgoing shockwave. Also visible around the bubble are three particles of dirt which appeared to be trapped to the bubble. (Weninger et al. (2000a).)

equation. As the velocity of collapse becomes supersonic, there is a handover to an imploding shock wave which further concentrates the energy. As the shock reaches its minimum radius, there is a sudden and dramatic heating that ionizes the contents of the bubble. The ionization quenches as the shock expands, and the system cools back down through the ionization temperature. Light comes out only while the plasma exists, which accounts for the flash width being independent of color. The means of light emission is thermal bremsstrahlung from the accelerating free electrons. These electrons will accelerate and radiate light as they collide with the ions".

### 4.7.15 *Water vapor theories*

To study the effect of water vapor on the collapse, Storey and Szeri (2000) considered a bubble initially consisting of argon, with $R_0 = 4.5$ μm and driven at $P_A = 1.2$ atm and $f =$ 26.5 kHz. Including water vapor, the maximum temperature is reduced from 20900 K (cf. Vuong and Szeri (1996)) to 9700 K, due to the lower polytropic exponent $\Gamma$. *No shock waves* are observed under these conditions. During the intense volume oscillations of the bubble, water is constantly evaporating and condensing, driven by dis-equilibrium between the partial pressure of the water vapor inside the bubble and the saturation vapor pressure at the interface. The amount of water in the bubble is not constant at all, as can be seen from Fig. 4.17. A large amount of water evaporates into the bubble during the main expansion when the pressure is low. At the bubble maximum, about 90% of the bubble content is water. Water vapor condenses at the bubble wall at the collapse, but not completely since the time scale of the collapse $\tau_{dyn} = R/|\dot{R}|$ becomes much smaller than the time scale for the transport of water vapor out of the bubble. The water vapor transport is a two-step process, consisting of diffusion to the wall and condensation, so that it involves two time scales, one for vapor diffusion in the bubble,

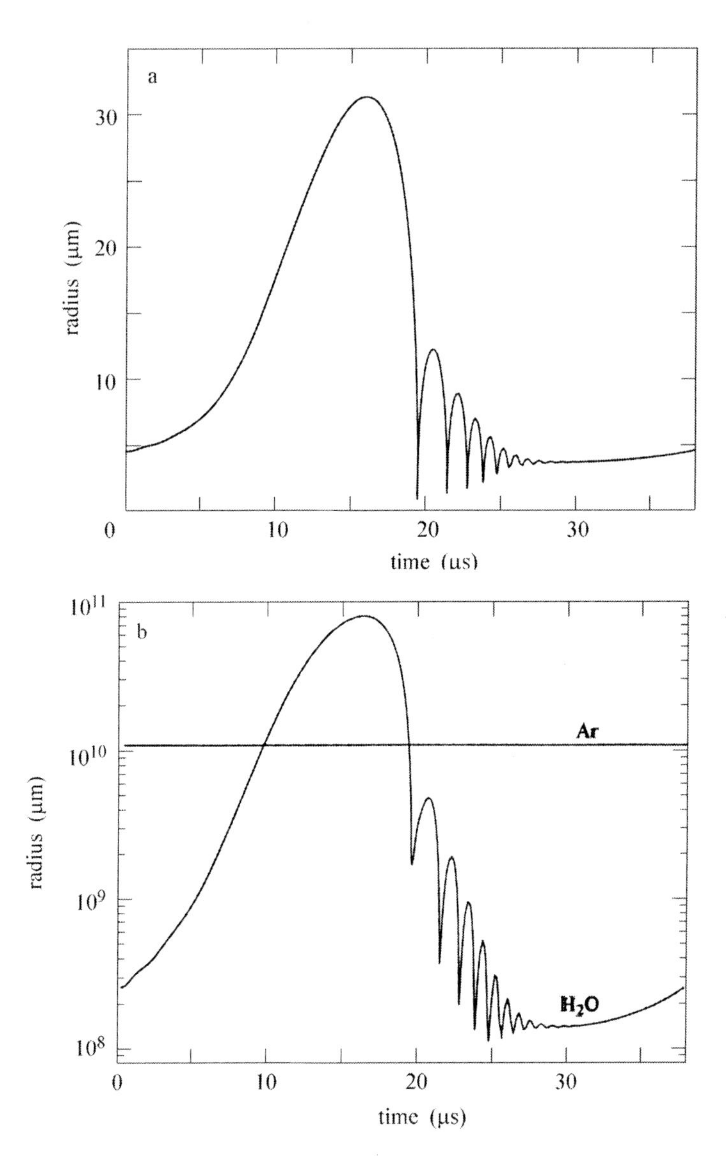


Figure 4.17 Figures taken from Storey and Szeri (2000). Bubble dynamics (*a*) and number of water molecules (*b*) as a function of time for an (initially) $R_0 = 4.5$ μm argon bubble driven at $P_a$ = 1.2 atm and $f$ = 26.5 kHz. Note the comparison in (*b*) to the constant number of argon atoms.

$$\tau_{\text{dif}} = \frac{R^2}{D_{H_2O}(R,T)} \approx \frac{1}{D_{H_2O}(R_0,T_0)} \frac{R_0^3 T_0^{1/2}}{RT^{1/2}}$$

and one for condensation at the wall,

$$\tau_{\text{con}} = \frac{R}{c_g} \sqrt{\frac{2\pi \Gamma M_{H_2O} T_0}{\sigma_a^2 q M_0 T_{\text{int}}}}$$

where $D_{H_2O}$ is the diffusion coefficient of water vapor in the gas mixture.
$M_{H_2O}$ is the molecular mass of the bubble contents at time $T_{int}$.
$M_0$ is the initial mass of the bubble content.
$\Gamma$ is a correction for bulk motion to the interface and remains close to one for mass transfer.

$c_g = \sqrt{Tp_{g0}/\rho_{g0}}$ is the sound velocity of the gas in the initial state.
$\sigma_a$ is the accommodation coefficient (sticking probability).

Researchers have upscaled sonoluminescence to find the limits of energy focusing and light emission by inducing more violent bubble collapses. Examples are by applying non-sinusoidal driving pressures (Holzfuss et al. (1998)), or by reduced driving frequencies (Hilgenfeldt and Lohse (1999) and Toegel et al. (2000)). But the larger expansion ratios achieved with these techniques lead to a larger water vapor content of the bubble, which again limits the heating at collapse. A theoretical study by Toegel et al. (2000) shows that these two effects roughly cancel each other, essentially to the same temperature and the same amount of light at lower driving frequencies for otherwise identical bubble parameters. A residual upscaling effect may still be observed (as in Barber and Putterman (1991)) due to the possibility of stabilizing bubbles with larger $R_0$ at lower driving frequencies.

### 4.7.16 *Conclusions*

In spite of the tremendous amount of work on the relevance of shock waves to single bubble sonoluminescence there is no convincing evidence, experimentally or theoretically, that shock waves occur *inside* a single sonoluminescing bubble. Thus shock waves cannot be the source of sonoluminescence.

### 4.7.17 *Experimental studies of shock waves from a sonoluminescing bubble*

See §3.5.6.

## 4.8 THE COLLISION-INDUCED EMISSION (CIE)

Frommhold and Atchley (1994) considered CIE arising from dipoles induced by *intermolecular* interactions or collisions. They say that if the temperature of a gas is insufficient for excitation of electronic states, molecules with inversion symmetry, e.g. $N_2$, Ar, etc., do not absorb or emit radiation in the visible or infrared region. However, interacting *pairs* of molecules, e.g. $N_2$–$N_2$, $N_2$–Ar, etc., possess interaction-induced dipoles which typically absorb radiation at frequencies ranging from the far infrared to the visible region. At high temperatures, if sufficient vibrational excitation exists, these dipoles *emit* radiation in the various vibrational bands, possibly even in high overtone bands in the visible region. Three mechanisms are known which generate dipoles in binary interactions: overlap-induced dipoles, multipole-induced dipoles, dispersion-induced dipoles. The known facts concerning induced dipole moments have been collected by Frommhold (1993).

For an understanding of SL, they considered that overlap-induced dipole components would be the most significant dipole component at the temperatures of interest (i.e. many

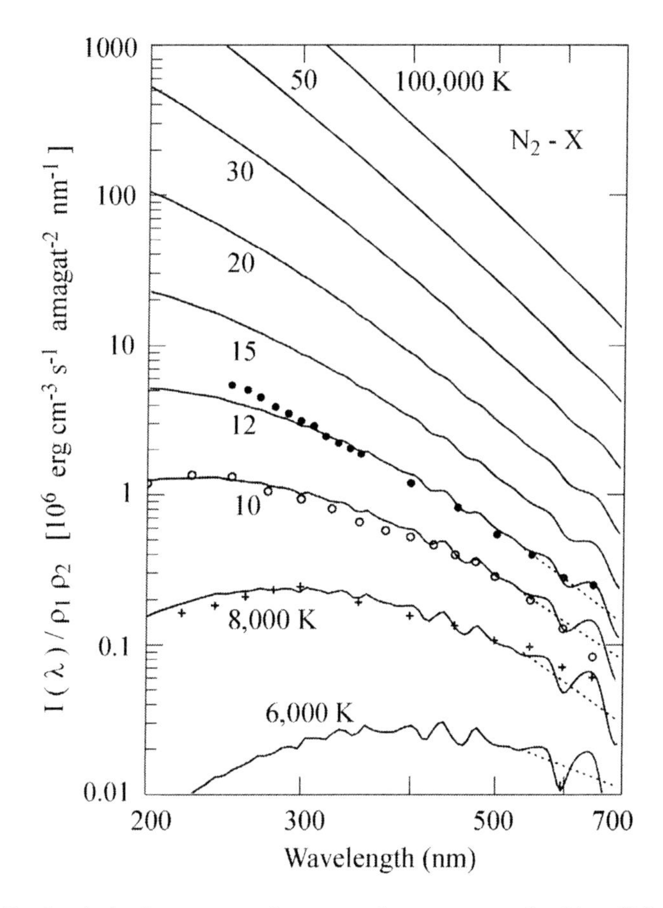


Figure 4.18 Total emission in watt per unit source volume, over gas densities of $N_2$ and $X$, per 1 nm wavelength, for 11 temperatures from 6000 to 100,000 K; the numbers 10–50 are short for 10000–50000 K; continuous wave emission is assumed. Also shown is a measurement (circles). The structures discernible at low temperature are artifacts. (Revised figure sent privately from Frommhold correcting Fig. 1 from Frommhold and Atchley (1994).)

thousands of Kelvin). After considering the modulation of the dipole moment by the vibrational and rotational motions of the interacting molecules, Frommhold and Atchley (1994) and Hammer and Frommhold (2001a) plotted the emitted intensities per unit wavelength $\lambda$, per volume, and per amagat squared (1 amagat is the molar volume $V$ of a real gas at $p$ = 101325 $P_A$ Pascals where $P_A$ is in atmospheres, $T$ = 273.15 K and is approximately equal to $22.4 \times 10^{-3}$ m$^3$/mol) in Fig. 4.18. Results were shown for interacting pairs of $N_2$ and X, for 11 temperatures from 6,000 to 100,000 K. The circles in Fig. 4.18 represent the measured spectrum from air bubbles in water from Carlson et al. (1993). The spectrum was scaled arbitrarily to fit on the graph. The trends of the two spectra are similar.

In the visible and near UV, the emission spectra are interaction-induced high overtone and "hot" bands, which are more or less familiar from interaction-induced absorption spectra at low temperatures. Electronic excitation and bremsstrahlung are not needed to explain the emitted intensities at high gas densities and temperatures such as the ones shown above; bremsstrahlung requires the presence of free electrons (i.e. substantial ionization) and much higher temperatures, in the $10^6$ or $10^7$ K range.

Frommhold and Atchley note that several interaction-induced overtone spectra, up to $0 \rightarrow 3$, were studied in detail, mostly of $H_2$, by Reddy (1985). Similarly, in planetary atmospheres, overtones up to $0 \rightarrow 4$ were reported by Herzberg (1952) and Spinrad (1964).

Motyka and Sadzikowski (1999) tackle the case of inelastic collisions between *atoms* of different kinds as a potential source of photons in SL. They calculated the number of photons in one flash as about $5 \times 10^5$ which agrees with experiment. However the resulting photon spectrum is somewhat too steep. The spectral density is featureless and universal. This model requires the gas temperature to be approximately 30,000 K. The bubble should contain atoms of two different gases: a rare gas and a gas with a smaller ionization e.g. oxygen; however one of these gases may appear at a much smaller concentration than the other.

## 4.9 THE QUANTUM RADIATION THEORY

Quantum theory has been invoked to explain sonoluminescence by Schwinger (1992a,b, 1993a,b,c,d, 1994), Eberlein (1992a,b, 1993, 1996a,b,c), Barton (1999) and Barton and Eberlein (1993). Schwinger used the Casimir (1948) effect which says that parallel conducting plates attract each other due to zero-point fluctuations in the fields.

The formidable team of Liberati, Visser, Belgiorno have written five long papers on SL and the Quantum Electrodynamics (Liberati et al. (1999a,b, 2000a,b) and Belgiorno (2000)). Jensen and Brevik (2000), Stepanovsky and Sergeeva (2000), Milton (1980, 1995, 1996), Milton and Ng (1996), Carlson et al. (1997a,b), Molina-Paris and Visser (1997), Brevik, Marachevsky and Milton (1999), Chodos and Groff (1999) and Chen et al. (2000) all contribute lengthily to the detailed discussion. Their work must have required a lot of time, energy and application.

In particular, Eberlein (1996a,b) argued that SL is closely related to the Unruh effect, the dynamic generalization of the Casimir effect. However, Kwak and Na (1997) point out that the Unruh temperature $T_{\text{Unruh}} = \hbar a/(2\pi kc)$, where $a$ is the acceleration, $k$ is Boltzmann's constant and $c$ is the velocity of light, is close to absolute zero with the maximum achievable bubble wall acceleration, $a$, of $10^{13}$ m/s$^2$ (Hiller et al. (1992), Kwak and Na (1996)). Thus the radiation due to the Casimir effect is nothing but a quantum fluctuation at zero temperature.

## 4.10 THE COOPERATIVE MANY BODY MODEL

Mohanty and Khare (1998) and Mohanty (2000a,b) proposed a new mechanism to explain SL as a cooperative interaction of the matter in the bubble with a radiation field. As the bubble contracts and reaches its minimum size, gas in the bubble undergoes an order of magnitude increase in its pressure and temperature which makes the gas highly excited or ionized. This results in a population inversion of such excited atoms or molecules. This completes the first step. Atoms in the excited states may return to their ground states by the spontaneous emission of radiation.

## 4.11 THE SUPER-RADIATION MODEL

Trentalange and Pandey (1996) used Bose–Einstein correlations to measure the spatial and temporal characteristics of the light emitting region. Bose–Einstein correlations have been used for many years in astronomy and nuclear physics to extend the accessible range of time and length scales. In these fields, the usefulness of this tool has been limited by the low brightness and nonreproducible nature of the sources, as well as the final state interactions and unwanted correlations inherent in the particle production process. These limitations are absent in SL. Trentalange and Pandey (1996) described how the size of the light-emitting region could be measured by the Hanbury Brown–Twiss (1958) effect.

## 4.12 THE TWO COMPONENT MODEL

Tsiklauri (1997) suggested that SL originates from two sources. The first source is an isotropic black body or a bremsstrahlung emitting core. The second source is a dipole radiation-emitting shell of accelerated electrons driven by the liquid–bubble interface. Tsiklauri (1997) and Weninger et al. (1996) argued that the light from the black body, or bremsstrahlung emitted from the air after it has been ionized by shock compression, is refracted from the nonspherical liquid–bubble interface, which results in a dipole angular distribution of the light.

## 4.13 THE PROTON-TUNNELING MODEL

This section closely follows the paper by Willison (1998) who suggested that SL does not come from a hot plasma inside the bubble, but is the radiation from a large number of current impulses (proton tunneling events) which occur as the water around the bubble undergoes a phase transition. This phase transition is initiated by the abrupt pressure transient that is coincident with the bubble reaching minimum size.

Water is a very polar molecule that forms structures and has unusual physical properties due to the strong dipole–dipole interaction (hydrogen bonding) between molecules (Jeffrey (1997)). The structure of water depends on its temperature and pressure, as detailed in its phase diagram (Pauling (1970)). During phase transitions water molecules move to new positions and change orientations. Classically, water molecules rotate to these new orientations. However, the protons are light enough, the distances that they need to move are small enough, and the potential barriers are low enough that protons can tunnel to their new positions (Pauling (1970)).

The bubble collapse is very abrupt, with its diameter changing by a factor of 2 in the final nanosecond before reaching minimum size, Fig. 4.19 from Barber et al. (1997b). The pressure at the surface of the bubble at this time has been deduced from needle hydrophone measurement, Fig. 4.20 from Barber et al. (1997b), and computation (Hickling (1994)), to be over 6,000 atm. Liquid water undergoes a phase transition to ice V1 (with a density of about 1.37) at 6200 atm (Pauling (1970)). During the phase transition the orientation of the

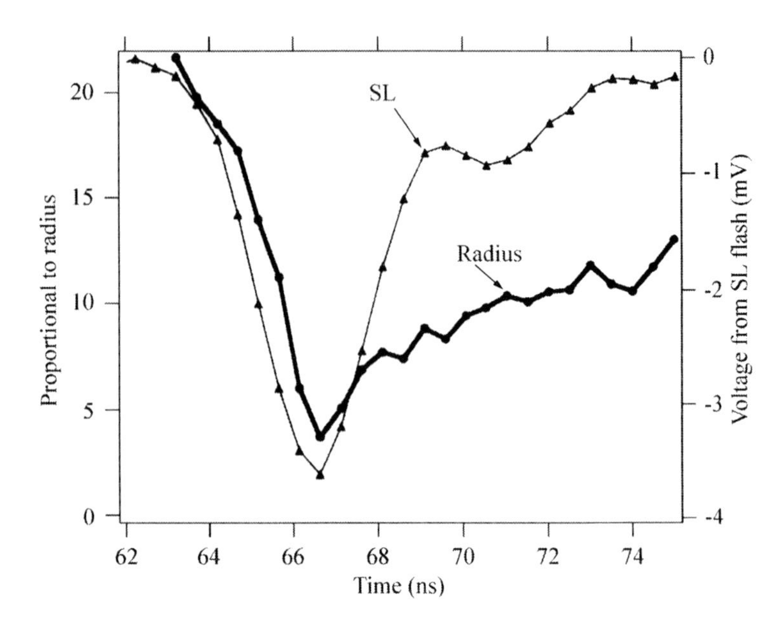


Figure 4.19 Flash of SL and light scattering as resolved by the same photomultiplier tube. (Reprinted from *Physics Reports* **281**, 65, Barber et al. 1997b, with permission from Elsevier Science.)

water molecules is known to change on a picosecond time scale (Marconcelli (1988)). Willison (1998) says that the agility of water molecules, together with the nanosecond bubble dynamics, is responsible for the time scale of SL.

One component of the reorientation of water molecules during a phase transition involves the tunneling of protons by about 0.75 Å along the line between two oxygen atoms as shown in Fig. 4.21. This tunneling event exchanges the covalent and hydrogen bonds, and flips the electric dipole moments of both molecules (Pauling (1970)). When the proton tunnels to the right, the electron distributions of both oxygen atoms will move to the left, enhancing the apparent current impulse. This proton-tunneling motion and corresponding electron current are thought to be the most important current components for the observed emissions.

## 4.14 THE HYDRODYNAMIC THEORY

This section is taken from Hilgenfeldt, Grossman and Lohse (1999b).

### *4.14.1 Calculation of the gas temperature*

The bubble stability (Brenner et al. (1995) and Hilgenfeldt et al. (1996)) and the dissociation and outward diffusion of gases under the extreme conditions induced by the collapse (Lohse et al. (1997) and Lohse and Hilgenfeldt (1997)) are important precepts of the theory.

All theories start with the Rayleigh–Plesset equation as in Eqs. (4.1) and (4.3), and Hilgenfeldt, Grossman and Lohse (1999a,b) use it again to first calculate the gas

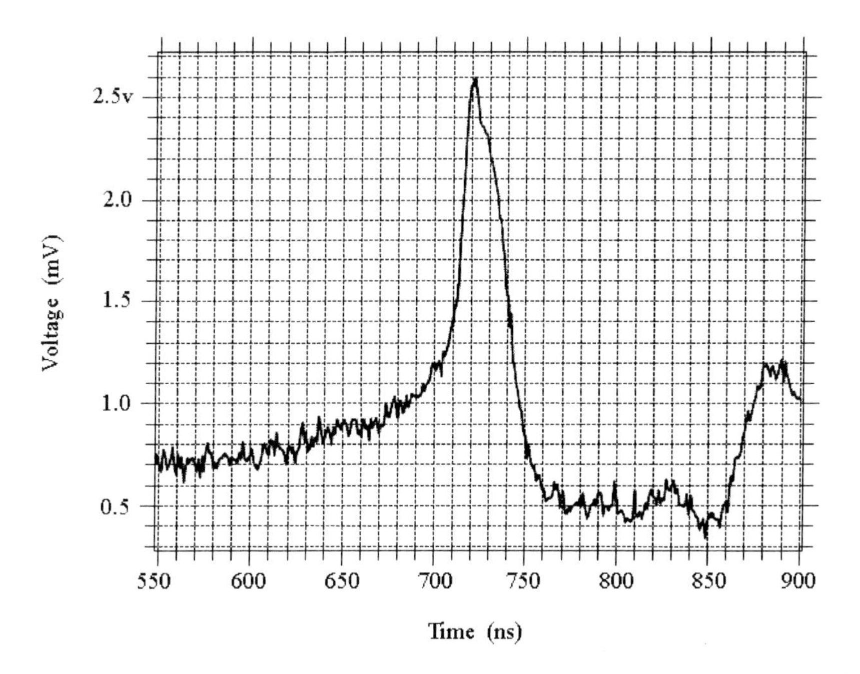


Figure 4.20 Response of a PVDF needle hydrophone to the short pulse of sound emitted by the collapsing bubble. This is the spike shown on trace (*b*) of Fig. 3.5 with an expanded timescale. The rise time of 10 ns is instrument limited and indicates the presence of frequencies greater than 30 MHz. (Hallaj et al. (1996) report the measurement of a 5 ns risetime that is still instrument limited.) According to the calibration of the hydrophone, the signal at 1 mm from the bubble is about 3 atm. Correcting for geometric dispersion the amplitude at the point of emission (0.5 μm) is over 6000 atm. The attenuation of sound is also considerable for high frequency pulses. For instance a 300 MHz pulse is reduced in amplitude by a factor of 10000 after having traveled 1 mm. Inclusion of this effect would substantially increase the estimate of the launching amplitude of the outgoing pulse. (Barber et al. (1997b).)

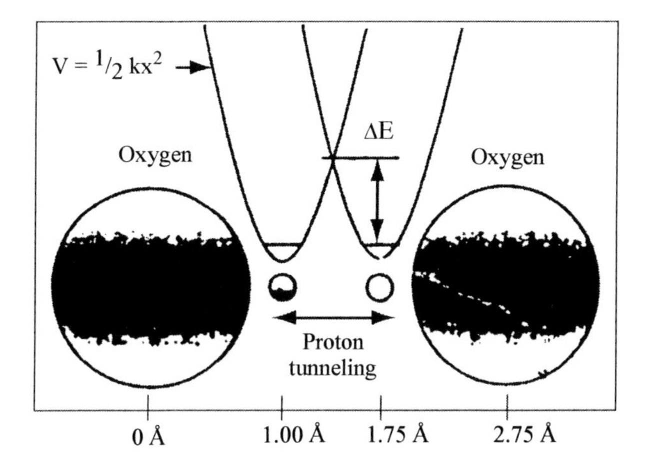


Figure 4.21 A sketch of the position, binding potentials, energy levels, and tunneling barrier ($\Delta E$) for a single proton between two oxygen atoms. (Willison (1998).)

temperature from the classical Rayleigh–Plesset equation of bubble dynamics for the radius $R(t)$ of the spherical bubble of a rare gas in water (Rayleigh (1917), Plesset (1949), Brenner (1945) and Löfstedt et al. (1993)):

$$p_1\left(R\ddot{R} + \frac{3}{2}\dot{R}^2\right) = p_{\text{gas}}(R(t)) + p_{\text{vap}} - P(t) - P_0 + \frac{R}{c_1}\frac{\mathrm{d}}{\mathrm{d}t}p_{\text{gas}}(R(t)) - 4\eta_1\frac{\dot{R}}{R} - \frac{2\sigma}{R} \tag{4.5}$$

where $\eta$ is the liquid viscosity, $\rho$ is the liquid density, $c$ is the speed of sound in the liquid and $\sigma$ is the surface tension. Although the RP equation is not accurate at the bubble collapse, when the bubble wall velocity ($\dot{R}$) is greater than the speed of sound in water, over the oscillation cycle as a whole it is remarkably robust as shown by Hilgenfeldt et al. (1999a).

The gas pressure $p_{\text{gas}}$ can be represented by a van der Waals-type process equation

$$\dot{p}_{\text{gas}}(R, t) = \frac{\mathrm{d}}{\mathrm{d}t}p_{\text{gas}}(R(t)) = -\gamma(R, \dot{R}, T)\frac{3R^2\dot{R}}{R^3 - h^3}p_{\text{gas}} \tag{4.6}$$

where $h$ is the van der Waals hard core radius. An effective polytropic exponent $\gamma$ is used to describe the thermodynamics of the bubble motion in every time step.

The bubble behaves isothermally when $\gamma = 1$ and adiabatically when $\gamma \to 5/3$ (for rare gases). The exponent depends on time via the dynamic variables $R$, $\dot{R}$ and $T$. Prosperetti (1977) has performed a rigorous *analytical* computation of the effective polytropic exponent for the limiting case of linear bubble oscillations. In this case, $\gamma$ depends only on the Péclet number $P_e = R(t)|\dot{R}(t)|/\chi(t)$, where $\chi$ is the thermal diffusivity. Because $\chi$ is, in general, a function of density and temperature, it also depends on time. Hilgenfeldt et al. (1999b) use Enskog theory, in Hirschfelder et al. (1954), to obtain the function $\chi(t) = \chi(R(t), T(t))$ and use Prosperetti's (1997) values for $\gamma(P_e)$ to determine the polytropic exponent.

While the assumption of linear oscillations is clearly violated for sonoluminescing bubbles, Prosperetti's approach still yields physically reasonable values, viz., $\gamma \to 1$ (isothermal) for low $P_e$ and $\gamma \to 5/3$ (adiabatic) for $P_e \to \infty$. The latter case of course occurs only in the immediate vicinity of the bubble collapse. During the rest of the oscillation cycle (i.e. almost all the time) the bubble behaves isothermally.

We now use the van der Waals equation of state to translate Eq. (4.6) into an ODE for $T(t)$,

$$\dot{T} = -[\gamma(P_e) - 1]\frac{3R^2\dot{R}}{R^3 - h^3}T \tag{4.7}$$

Solving the coupled system of ODEs (4.5) and (4.7) yields the time series of temperatures used to evaluate the photon absorption coefficients and the resulting light emission. The maximum temperatures achieved usually lie between 20,000 K and 30,000 K and show only moderate variations over the parameter range of sonoluminescing bubbles.

### 4.14.2 Photon absorption coefficients in a weakly ionized gas

The first reliable measurements of the pulse width of SBSL, and its virtual independence on wavelength (Gompf et al. (1997), Pecha et al. (1998), Moran and Sweider (1998) and

Hiller et al. (1998)) showed that simple Planck emission from a black body is not sufficient to explain the experimental observations, because it would result in much longer pulses at long wavelengths than in the short wavelength regime. Moss and collaborators (1997a) realized that the temperature-dependent photon absorption of the gas has to be taken into account to explain the deviation from black body radiation. *They found that the bubble is not an ideal absorber (that is, not a surface emitter like an ideal black body)*, but because of its tiny size most photons can escape without reabsorption and therefore the bubble is transparent for its own radiation most of the time, that is, it is a *volume emitter*. This means that the detected radiation does not come just from a thin surface layer, but from the whole volume, as the photons are only weakly absorbed inside the bubble. The photons carry information about the light production processes, and the radiation differs from black body emission in both intensity and spectral shape.

To calculate the light emission, we need the absorption coefficients in the gas.

#### 4.14.2.1 Introduction

All absorption processes described here occur only when at least a small number of ions and/or electrons are present (all absorption coefficients are linearly, or quadratically proportional to the degree of ionization $\alpha$). The Saha equation gives $\alpha$ explicitly as

$$\alpha[T] = \left(\frac{2\pi m_e k_B T}{h^2}\right)^{3/4} \left(\frac{2u_+}{nu_0}\right)^{1/2} \exp\left(-\frac{E_{\text{ion}}}{2k_B T}\right)$$

with the electron mass $m_e$, Boltzmann and Planck constants $k_B$ and $h$, and statistical weights $u_+$, $u_0$ for the ground states of the ion and the neutral gas atom, respectively. We follow the hydrogen-like atom model in Zel'dovich and Raizer (1966) and set $2u_+/u_0 = 1$. Considering the large ionization energies $E_{\text{ion}} = 15.8$ eV or 12.1 eV for Xe, it is apparent from the exponential factor $\exp(-E_{\text{ion}}/2k_B T)$ in the equation above that at maximum temperatures of $\approx$2–3 eV (20,000–30,000 K) ionization is rare.

When evaluating different contributions to photon absorption from Zel'dovich and Raizer (1966), we find three terms to be of vital importance in at least part of the parameter range of sonoluminescence explored today in experiment. They are described in the following subsections.

#### 4.14.2.2 Free–free transitions of electrons near ions

At high or moderate degrees of ionization, photons will predominately be absorbed by free electrons, where an ion acts as a third partner to satisfy conservation of both energy and momentum. Zel'dovich and Raizer (1966) give the absorption coefficient for this process as

$$\kappa_\lambda^{(i)}[T] = \frac{4}{3}\left(\frac{2\pi}{3m_e k_B T}\right)^{1/2} \frac{Z^2 e^6 \lambda^3}{(4\pi\varepsilon_0)^3 h c^4 m_e} (\alpha[T])^2 n^2$$

with an effective charge $Z$ of ions (here, $Z = 1$), the elementary charge $e$, the speed of light $c$, and the vacuum permeability $\varepsilon_0$. The number density of electrons (and therefore of ions) is denoted by $n$. The inverse (emission) process corresponding to this absorption is the well-known *bremsstrahlung* of electrons in the field of ions.

#### 4.14.2.3 Free–free transitions of electrons near neutral atoms

As the degree of ionization $\alpha$ is small for typical SBSL situations, there are far more neutral atoms than ions present in the bubble. Therefore, collisions of electrons with neutral atoms are important despite their much smaller interaction cross section (compared to electron–ion collisions). This *transport scattering cross section*, $\sigma_{tr}$, from Ginzburg (1962), depends on the energy of the electron; for argon we have with good accuracy the linear relation from Brown (1966),

$$\sigma_{tr}(E_e) \approx c_{tr} E_e + d_{tr}$$

for the relevant thermal electron energies $E_e$ (a few eV) here, with constants $c_{tr} \approx 1.6\times10^{-20}$ m$^2$/eV and $d_{tr} \approx -0.6\times10^{-20}$. Using this expression and averaging the photon absorption coefficient for this process over the Maxwell-distributed electron energies as by Zel'dovich and Raizer (1966), we obtain

$$\kappa_\lambda^{(ii)}[T] = \frac{e^2}{\pi\varepsilon_0 h^{3/2} c^3 m_e^{3/4} \pi^{3/4}} (2k_B T)^{9/4} n^{3/2} \lambda^2 \left(c_{tr} + \frac{d_{tr}}{3k_B T}\right) \exp\left(-\frac{E_{ion}}{2k_B T}\right)$$

Here, the inverse process is *neutral bremsstrahlung*, i.e. photon emission from electrons colliding with neutral atoms. When computed numerically, the size of this contribution is indeed comparable to $\kappa_\lambda^{(i)}$ in a wide range of parameters in the SBSL range.

#### 4.14.2.4 Bound–free transitions of electrons

Finally, we consider ionization of electrons through absorption of photons. In the optical wavelength regime we are interested in, an ionization from the ground state $j = i$ of a rare gas is impossible. However, the gap between the ground state and the first excited state $j = 2$ is so wide that at least for a range of $\lambda$, all levels $j \geq 2$ can participate in the process. More precisely, for every $\lambda$ there exists a $j = j^*$ which is the smallest $j$ for which the $j$-th energy level $E_j = E_{ion}/j^2$ of a hydrogen-like atom is smaller than $hc/\lambda$. Electrons on all levels with $j \geq j^*$ can then be ionized by a photon of wavelength $\lambda$ and their contribution must be summed up to yield the absorption coefficient

$$\kappa_\lambda^{(iii)}[T] = \frac{64\pi^4}{3\sqrt{3}} \frac{e^{10} n}{(4\pi\varepsilon_0)^5 h^6 c^4} \lambda^3 \sum_{j=j^*}^{\infty} \frac{1}{j^3} \exp\left(-\frac{E_{ion} - E_j}{k_B T}\right)$$

Above the first excited state, the level spacing becomes so narrow, as in Bacher and Goudsmit (1932), that a continuum model is appropriate as in Zel'dovich and Raizer (1966). This model can be simplified and added to $\kappa_\lambda^{(i)}$ to obtain a corporate absorption coefficient for both processes involving ions,

$$\kappa_\lambda^{(i)+(iii)}[T] = \frac{16\pi^2}{3\sqrt{3}} \frac{e^6 k_B T n}{(4\pi\varepsilon_0)^3 h^4 c^4} \lambda^3 \exp\left(-\frac{E_{ion} - hc/\max\{\lambda, \lambda_2\}}{k_B T}\right)$$

As $\lambda$ decreases, more and more electron energy levels become available for ionization, so that $\kappa_\lambda^{ion}$ increases. But for photons with energies higher than that of the first excited

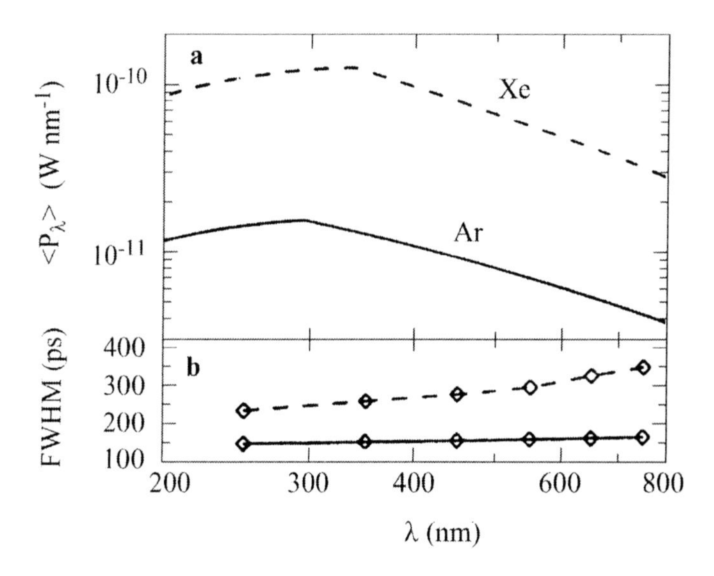


Figure 4.22 Spectral variation of intensity and pulse width of sonoluminescence pulses from xenon and argon bubbles. ***a***, Spectral radiance from an argon (solid line) and a xenon (dashed) bubble, for $P_a = 1.3$ atm, $R_0 = 5.0$ μm. The intensity and shape of the experimentally reported spectra is reproduced very well, although the measured spectra decay slightly faster towards the red. ***b***, The variation of pulse width with $\lambda$, computed for the total power contained in intervals of $\Delta\lambda = 100$ nm. This corresponds to experiments using filters of ~100 nm bandwidth. (Reprinted with permission from *Nature*, Hilgenfeldt et al. **398**, 402. Copyright (1999) Macmillan Magazines Limited.)

level $E_2 \equiv hc/\lambda_2$, there are no additional energy levels available (the ground state lying too low to be reached). Therefore, the ionic absorption at wavelength $\lambda_2$ will not increase for $\lambda < \lambda_2$ but instead will decrease with decreasing $\lambda$, as does the neutral absorption (Hirschfelder et al. (1954). The resulting maximum in absorptivity is reflected by an emissivity maximum in the observed spectrum, whose location is very close to $\lambda_2$ (see Fig. 4.22).

As photon absorption and emission must balance in local thermodynamic equilibrium (Kirchhoff's law), every contribution to the absorption corresponds to a light emission process. Sonoluminescence is therefore generated by thermal bremsstrahlung (electron ion bremsstrahlung, electron atom bremsstrahlung, see Sec. 4.6.2) and recombination radiation (inverse of this Sec. 4.14.2.4).

Adding the contributions $\kappa_\lambda^{(i)+(iii)}$ and $\kappa_\lambda^{(ii)}$ we arrive at the quantitative formula for the *total* absorption coefficient $\kappa_\lambda[T(t)]$.

### 4.14.3 Light emission intensity

Emitted radiation that has traveled a distance $s$ through a bubble of radius $R$ at a spatially uniform temperature $T$ (and therefore uniform $\kappa_\lambda[T(t)]$) shows a spectral intensity given by Zel'dovich and Raizer (1966) and Siegel and Howell (1972)

$$I_\lambda(s,t) = I_\lambda^{\mathrm{Pl}}[T(t)](1-\exp(-\kappa_\lambda[T(t)]s)), \qquad 0 < s < 2R$$

This quantity represents energy per unit time, wavelength interval, solid angle, and projected surface area. It becomes the Planck intensity $I_\lambda^{\mathrm{Pl}}(t)$ for $\kappa_\lambda R \to \infty$, but in the case of sonoluminescing bubbles this product is small compared to one. Integrating over the projected surface of the bubble and all solid angles, we arrive at the total emitted power from the bubble per wavelength interval at wavelength $\lambda$,

$$P_\lambda(t)\mathrm{d}\lambda = 4\pi^2 R^2 I_\lambda^{\mathrm{Pl}}[T(t)]\left(1 + \frac{\exp(-2\kappa_\lambda R)}{\kappa_\lambda R} + \frac{\exp(-2\kappa_\lambda R) - 1}{2\kappa_\lambda^2 R^2}\right)\mathrm{d}\lambda \tag{4.8}$$

For $\kappa_\lambda R \gg 1$, this becomes the power emitted from an ideal black body (Planck emitter), whereas for $\kappa_\lambda R \ll 1$ we find a volume emitter of the form

$$P_\lambda(t)\mathrm{d}\lambda = 4\pi\kappa_\lambda I_\lambda^{\mathrm{Pl}}[T(t)]\frac{4\pi R^3}{3}\mathrm{d}\lambda = \frac{4}{3}\kappa_\lambda R P_\lambda^{\mathrm{Pl}}(t)\mathrm{d}\lambda$$

The Planck emission strength is thus reduced in proportion to value of $\kappa_\lambda R$. As this limit is usually valid in typical SBSL parameter regimes, it explains the smaller photon yields as compared to an ideal black body calculation. The sensitive dependence of all contributions to the absorption coefficient $\kappa_\lambda$ on $T$ via the degree of ionization accounts for the shortness and wavelength dependence of sonoluminescence pulses.

### 4.14.4 Discussion

Figure 4.22 shows examples of the spectral radiance from Eq. (4.6) for strongly driven argon and xenon bubbles in water. They are in good agreement with experiment (not shown: compare Barber et al. (1997b), and the stronger light emission for xenon (mainly due to its lower ionization energy) at equal parameters is reproduced. For both gases, a spectral maximum is predicted. It results from an absorption edge which reflects the detailed energy level structure of the rare gases (Hilgenfeldt et al. (1999a)). In reality, these maxima should lie at somewhat smaller $\lambda$ as the level structure becomes modified at the high pressures inside the collapsed bubble. For xenon, the spectral maximum has been observed at about 300 nm by Barber et al. (1997b); for argon, the experiments are not conclusive because of the light absorption of water which modifies the spectrum below $\lambda = 250$ nm. Diagram b in Fig. 4.22 confirms that the full width at half maximum for the argon bubble shows practically no dependence on $\lambda$ (as in experiments by Gompf et al. (1997), Pecha et al. (1998) and Hiller et al. (1998)), whereas the present model predicts a quite pronounced variation of the largest and brightest xenon bubbles (though much smaller than in the black body case). This variation should be detectable and awaits experimental verification.

A significant advantage of the present approach is the possibility to scan the whole parameter space of SBSL, because the calculations are very simple, and thus to reproduce experimentally observed parameter dependences of SBSL light emission. In an experiment, the driving frequency $f$, the forcing pressure $P_A$, the water temperature $T_w$, and the rare gas concentration $c_\infty$ in the liquid far away from the bubble are the crucial adjustable parameters. To calculate $R(t)$, we need the bubble's ambient radius $R_a$ for a given $f$, $P_A$, $T_w$ and $c_\infty$. It has been shown by Hilgenfeldt et al. (1996) and Fyrillas and Szeri (1994) that $R_0(f, P_A, T_w, c_\infty)$ is given by the condition of diffusive stability, i.e. the dynamical equilibrium of gas exchange between the bubble and liquid.

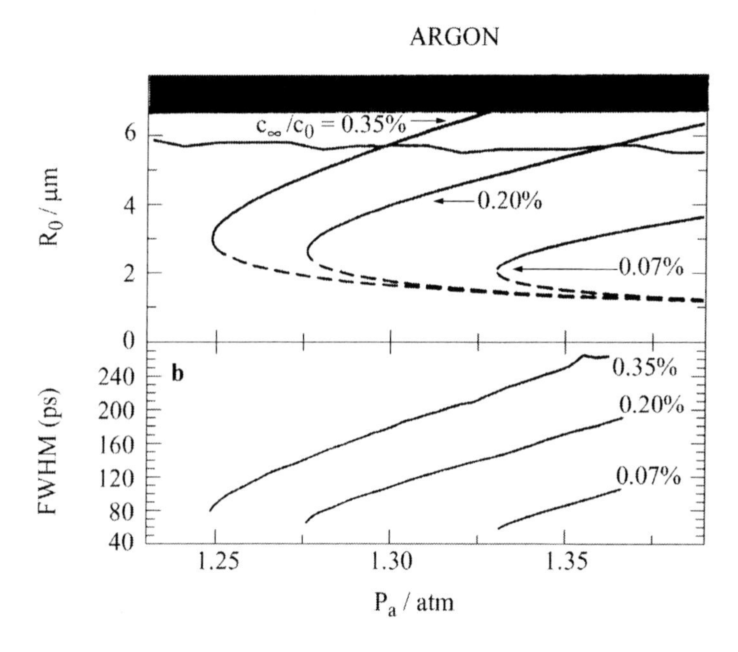


Figure 4.23 Size and light emission of diffusively stable argon bubbles. ***a***, Curves $R_0(P_a)$ of stable (solid thick lines) and unstable (dashed) diffusive equilibria for $f$ = 20 kHz at dissolved gas concentration $c_\infty/c_0$ = 0.07%,0.20% and 0.35% (ref. 13), all given relative to the gas stability $c_0$. The parametric surface instability restricts $R_0$ to ~5.5 μm (calculated, thin solid line) or ~ 7 μm (experimental, hatched region). ***b***, Calculated pulse widths for the stable combinations of $P_a$ and $R_0$ from ***a***. The pulse widths and their dependence on $P_a$ are in good agreement with experiments. (Reprinted with permission from *Nature*, Hilgenfeldt et al., **398**, 402. Copyright (1999) Macmillan Magazines Limited.)

Figure 4.23, from Hilgenfeldt et al. (1996b), shows the curves of diffusive equilibria in $P-R$ space (the branches with positive slope are stable) for several gas concentrations $c_\infty$ used in experiments by Gompf et al. (1997) and Pecha et al. (1998). As the experimenter changes $P_A$, the corresponding curve determines the response of the ambient size R0 of the SBSL bubble, and calculating the light emission for these specific pairs ($P_A$, $R_0$) should therefore reproduce the measured photon numbers, pulse widths and . so on. The upper limits for $P_A$ and $R_0$ are imposed by the onset of shape instabilities (compare Fig. 4.23a and Hilgenfeldt et al. (1996), Holt and Gaitan (1996) and Prosperetti and Hao (1999)). Figure 4.23b shows the results for the widths (full-width at half maximum (PWHM)) of the emitted SBSL pulses, as a function of $P_A$. The range of pulse widths is in excellent agreement with experiment. The shift in the $P_A$ axis (~0.10–0.15 atm) with respect to the data in Pecha et al. (1998) could be due to the simplified modeling, but also to experimental inaccuracies of at least 0.1 atm in measuring $P$ (B. Gompf, personal communication).

A more stringent test of the theory presented here can be made by comparing its predictions with the extensive experimental data of Hiller et al. (1998) on spectral pulse widths and intensities, both of which are accurately measurable quantities. In Hiller et al. (1998) the SBSL intensity values are given relative to the maximum intensity of an air bubble dissolved in water at 150 mm Hg partial pressure. This normalizing intensity was determined with the present model (the relevant partial pressure being only that of argon i.e. 1.5 mm Hg, because of molecular dissociation), and used as the intensity unit for all

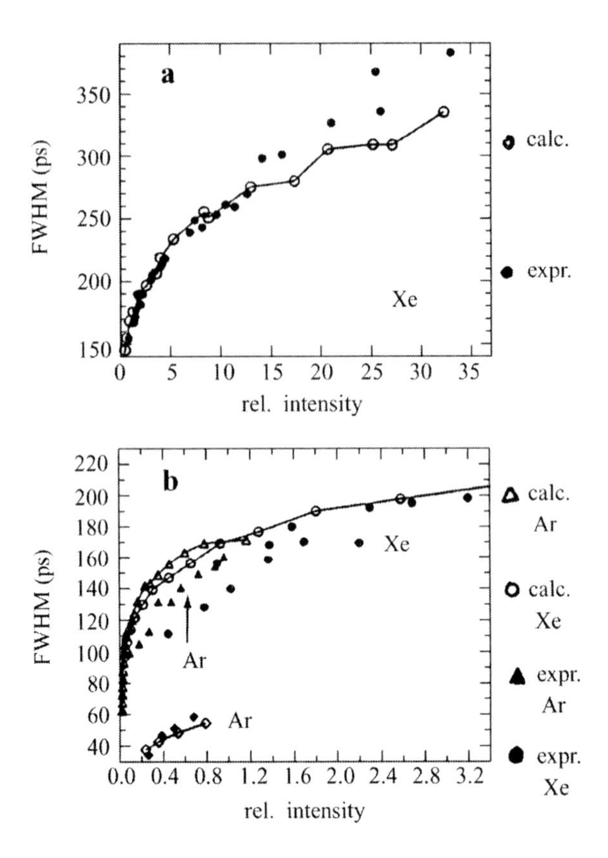


Figure 4.24 ($a$) Calculated pulse widths (open circles) for diffusively stable xenon bubbles at $f = 34$ kHz and $c_\infty/c_0 = 0.4\%$ (as reported in Hiller et al. (1998)) as a function of pulse intensity. The filled symbols are the experimental values of Hiller et al. (1998). The intensities are given relative to the maximum intensity for 2 atm (150 torr) air (argon concentration $c_\infty/c_0 \approx 0.2\%$). In the experiment the brightest bubbles were obtained at lower water temperatures, where stronger forcing and larger bubble radii are possible (Hilgenfeldt et al. (1998b)). ($b$) A comparison of the pulse widths for diffusively stable Ar (triangles) and Xe (circles) bubbles for the same parameters as in ($a$), now in the range of smaller intensities and widths. Solid symbols denote experimental data, open symbols theoretical results. Diamonds represent values for argon at $c_\infty/c_0 = 0.3\%$, corresponding to the 0.026 atm (20 torr) air experiment reported in Hiller et al. (1998).

theoretical results in Fig. 4.24 from Hilgenfeldt et al. (1999a). In Fig. 4.24a we compare for xenon bubbles, the experimentally found dependence of FWHM on intensity (filled symbols) to our calculations (open symbols). The agreement is, despite the approximations involved in the model, excellent. Not only is the qualitative shape of the curve reproduced, but also a quantitative comparison shows that the pulse widths differ from the experimental values by no more than ~5% for moderate intensities, and by no more than ~15% for high intensities. Other features from experiment are reproduced in Fig. 4.24 for argon and xenon bubbles in the range of smaller intensities. Note first that both argon and xenon bubbles follow almost exactly the same line in this diagram if the same rare gas concentration is applied (here $c_\infty/c_0 = 0.4\%$), although the values of $P$ and $R$ are quite different at a given intensity. Second, Fig. 4.24b, bottom left, shows results obtained with strongly degassed air, which results in the tiny pulse widths and intensities (diamonds in Fig. 4.24b). If the dissociation hypothesis is correct, these bubbles should not be air bubbles at $c_\infty/c_0 = 3\%$, but *argon* bubbles at $c_\infty/c_0 = 0.03\%$. The corresponding calculation indeed reproduces the experimental results (Brenner et al. (2002))

Hilgenfeldt et al. (1999b) conclude by saying that the Rayleigh–Plesset equation, shape stability, diffusive stability, chemical processes in the bubble, and, finally, light emission as thermal emission from an optically thin body, fit together and give a consistent picture of SBSL in its whole parameter range. There are no free parameters in the calculations; no functions or constants are chosen arbitrarily. Due to the simplicity of the approach, the complete parameter space of SL is accessible to these calculations. Compared with many of the theories described earlier in this chapter, this hydrodynamic theory, cleverly and painstakingly worked out alongside experimental data by Lohse and his collaborators is a testament to their scholarship. The basic ideas of the theory are simple physics, and as Occam's razor tells us, we should not choose a more complicated model, if a simple one will do.

Putterman et al. (2001) says that the above Hydrodynamic theory of Lohse and his collaborators (1999a,b), Sec. 4.14 is too simple, on the grounds that it cannot account for some well-established observations and that it involves the application of the hydrodynamic (Rayleigh–Plesset) equation outside its range of validity. Putterman et al. (2001) advance three comments which will be listed below with the replies from Hilgenfeldt et al. (2001).

1. *Putterman et al.* The Rayleigh–Plesset equation is used in a limit where it is not valid.
   *Hilgenfeldt et al.* It is well known that the assumptions used to derive the Rayleigh–Plesset equation indeed break down at bubble collapse, and that there are many ways to extend the equation to higher orders in $\dot{R}/c$ (Prosperetti and Lezzi (1986)). The Rayleigh–Plesset equation has provided useful results when applied over the whole oscillation cycle of the bubble, as evidenced in the pioneering work of Gaitan (1990, 1992) and the later studies of the Putterman group in Barber et al. (1997b), in which the criticized Rayleigh–Plesset equation was used to fit various parameters to experimental data on $R(t)$.
2. *Putterman et al.* Although He and Xe have very different physical properties, the similarity of their SL constitutes a litmus test for theories. When dissolved in water at 3 torr partial pressure, these gases can form light-emitting bubbles with $R_0 = 4$ μm and a maximum radius of $R_{max} = 30$ μm at 40 kHz driving frequency. The RP equation yields a collapse temperature, $T_c$ of 17500 K when the bubble has been compressed to a van der Waals hard core radius $a$. According to standard formulae of plasma physics, the photon–matter interaction length is large compared to $a$ (Zel'dovich and Raizer (1966)). Therefore the proposed spectrum (Hilgenfeldt et al. (1999b) and Wu and Roberts (1993) is not that of a black body, but bremsstrahlung from a thermally ionized plasma. At 17500 K, the rare gases are only weakly ionized, and the radiation is approximately proportional to the square of the degree of ionization, $e^{-c/kT}$, where $\chi$ is the ionization potential (25 eV for He and 12 eV for Xe). Thus radiation from a He bubble should be less than that from a Xe bubble by more than four orders of magnitude for electron ion bremsstrahlung, and by two orders of magnitude for electron-neutral bremsstrahlung. But in fact, the observed emission from He is less than that from Xe by only a factor of 3 – see Fig. 3.39 – a discrepancy between theory (Hilgenfeldt et al. (1999b)) and experiment of one to three orders of magnitude. Interestingly enough, Young (1976), for

multibubble sonoluminescence, measured the light emission from He to be two orders of magnitude greater than that from Xe.

*Hilgenfeldt et al.* As He has a huge ionization energy of 25 eV, the light-emission process is dominated by water and its reaction products (H and O). Thus it is possible that with He, it is rather the water vapor and the radicals H and O derived from the water vapor that produce light emission as their ionization energies are lower (in the range of $\approx$14 eV for both H and O. Hilgenfeldt et al. have now included water vapor in their model (Toegel et al. (2000, 2002)).

3. *Putterman et al.* The adiabatic equations of state for the gas temperature $T_g$ and pressure $P_g$ inside a compressed bubble is used:

$$P_g(R) = \frac{P_0 R_0^{3\gamma}}{(R^3 - a^3)^\gamma}; \qquad T_g(R) = \frac{T_0 R_0^{3(\gamma-1)}}{(R^3 - a^3)^{\gamma-1}}$$

where $\gamma$ is the ratio of heat capacities $C_p/C_V$ and $a$ is the van der Waals hard core radius.

*Hilgenfeldt et al.* Contrary to what Putterman et al. (2001) state, the equations above were never used in their model. Rather, they changed $\gamma$ dynamically, following Prosperetti (1977, 1991), which allows for a more realistic heating of the bubble interior.

Hilgenfeldt and Lohse (2000) provide a very readable account of sonoluminescence in which the above hydrodynamic theory is succinctly summarized in 132 words.

## 4.15 KWAK'S CONTRIBUTION

Kwak et al. (2001) point out that the light emission of SBSL occurs during the 0.5 ns before collapse (Kwak and Na (1996)). Hence the bubble dynamics theory in the last nanosecond before collapse is crucial to understanding SBSL. Since the Rayleigh–Plesset equation breaks down if the speed of the bubble wall is greater than the sound speed in the liquid, a proper numerical procedure for the bubble wall motion at the moment of collapse is needed. Kwak et al. (2001) overcame this problem by using two time scales, one for the driving force and one for the bubble motion. Radius-time curves are calculated for various parameters and one of these is shown in Fig. 4.25, which is for two air bubbles of equilibrium radii $R_0 = 4.5$ μm and $R_0 = 6.5$ μm with $P_A = 1.35$ atm and $f = 26.5$ kHz. This nine-page paper is well worth studying.

## 4.16 MOLECULAR DYNAMICS AS A METHOD OF MODELING THE BUBBLE INTERIOR

A typical single sonoluminescing bubble driven at 30 kHz has an ambient radius of 6 μm and contains about $10^{10}$ particles. At the extreme, single sonoluminescence bubbles have been observed at 1 MHz having a measured ambient radius of 0.3–0.6 μm (Weninger et al. (2000b)). These bubbles will only contain several million particles.

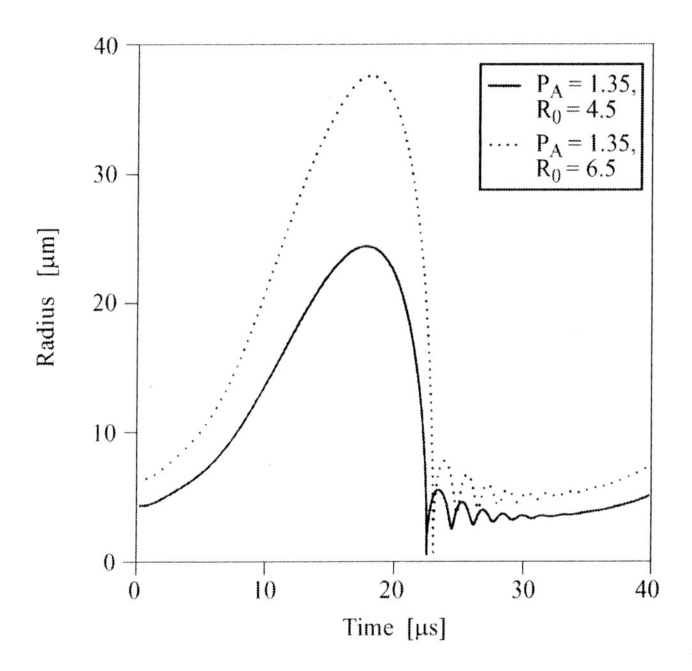


Figure 4.25 Theoretical radius–time curves under the ultrasound of $P_A$ = 1.35 atm and $f$ = 26.5 kHz for air bubble of $R_0$ = 4.5 (—) and $R_0$ = 6.5 (- - -). (Kwak et al. (2001).)

This number compares well with molecular dynamics simulations (Rapaport (1998)) where systems of about $10^6$ particles can be studied.

Matsumoto et al. (2000) made a start by studying a collapsing bubble of 62,500 particles as a three-dimensional Lennard–Jones (12–6) fluid.

By using a sophisticated event-driven algorithm (Rapaport (1980, 1997)) and up-to-date computer power, Metten and Lauterborn (2000), Metten (2001), Kurz et al. (2001), Kurz et al. (2002), Lauterborn (2002) and Lauterborn et al. (2002), were able to increase the number of collective model particles to several millions. All the particles in the bubble were modeled as hard spheres. The bubble could contain a mixture of rare gases, water vapor and its dissociation products. Diffusion, heat conduction in the gas and through the walls, evaporation and condensation, viscosity, rotational degrees of freedom of the water, ionization, ion transport, and generation of bremsstrahlung were taken into account. The bubble's radial dynamics was calculated by an extended Rayleigh–Plesset equation that included liquid viscosity and damping, viz. Eq. (3.3). The macroscopic quantities (density, pressure, velocity, temperature) were obtained by averaging over a grid of cells of fixed volume (concentric cells) over a certain time interval. To calculate the sonoluminescence, a simple bremsstrahlung model was assumed, where the degree of ionization was derived from the temperature of the medium by means of the Saha equation (Zel'dovich and Raizer (1966) pages 194 and 444). Figure 4.26, from Metten (2001), shows the internal temperature distribution inside a collapsing sonoluminescing bubble containing 1 million atoms of *argon*, with near adiabatic conditions (reflection of the atoms at the inner wall of the bubble).

Ruuth et al. (2002) also modeled dynamically the interior of a collapsing rare gas bubble controlled by the Rayleigh–Plesset equation. Energy losses due to ionization of the

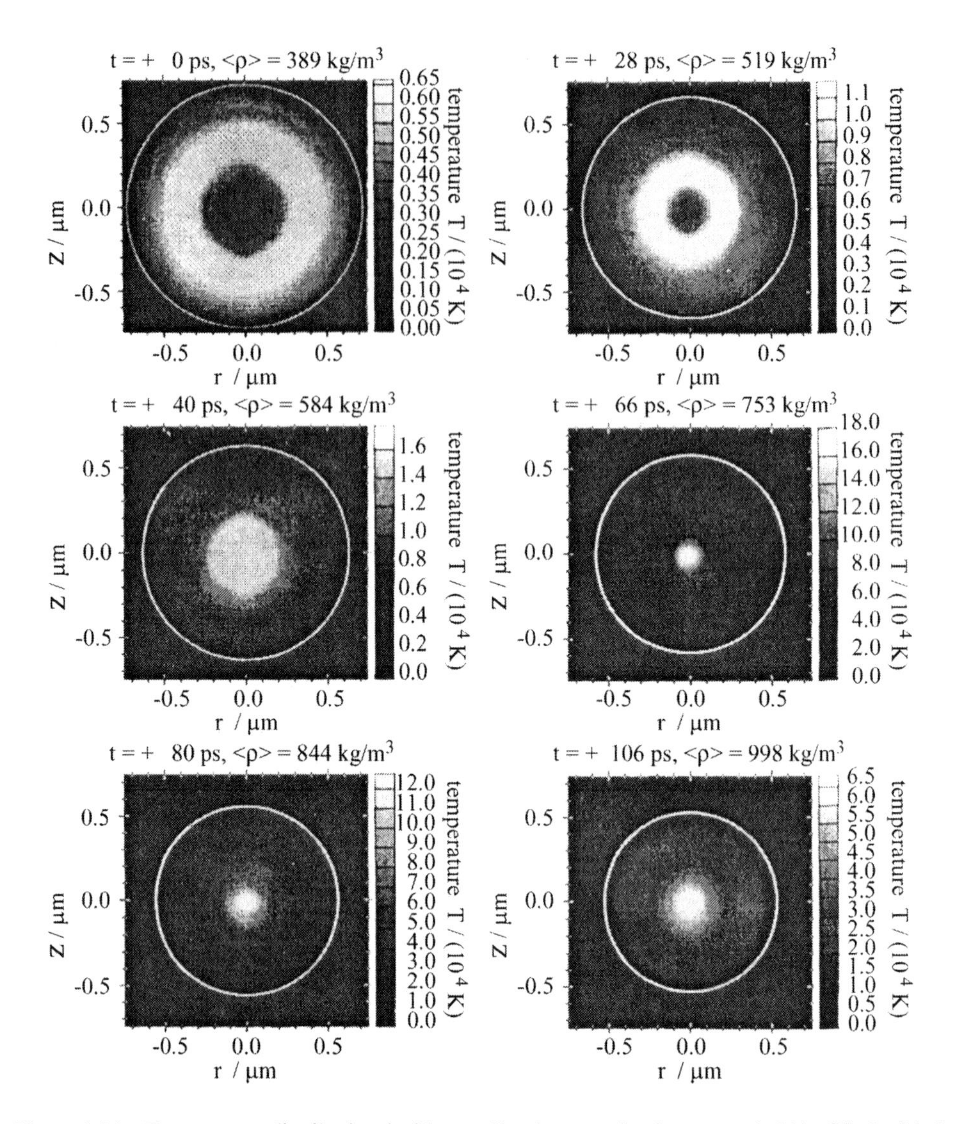


Figure 4.26 Temperature distribution inside a collapsing sonoluminescence bubble filled with 1 million particles of argon, radius of the bubble at rest 4.5 μm, sound pressure amplitude 130 kPa, sound field frequency 26.5 kHz, total time covered 106 ps. (Metten (2001).)

gas at the high temperatures were included. Water vapor was neglected. By using fast, tree-based algorithms, they were able to follow the dynamics of 1 million particle systems during the collapse. Peak temperatures ranged from 40,000 K for He to 500,000 K for Xe. One example of Ruuth's et al. (2002) computations is given in Fig. 4.27 for an *argon* bubble of 1 million particles with specular boundary collisions (where particles reflect from the boundary with a speed equal to the incident speed), constant diameter particles, and ionization.

It will be seen that Metten's (2001) and Ruuth's et al. (2002) final collapse temperatures for an *argon* bubble are 160,000 K and 90,000 K respectively, agreeing to an order of magnitude, although the parameters may be different.

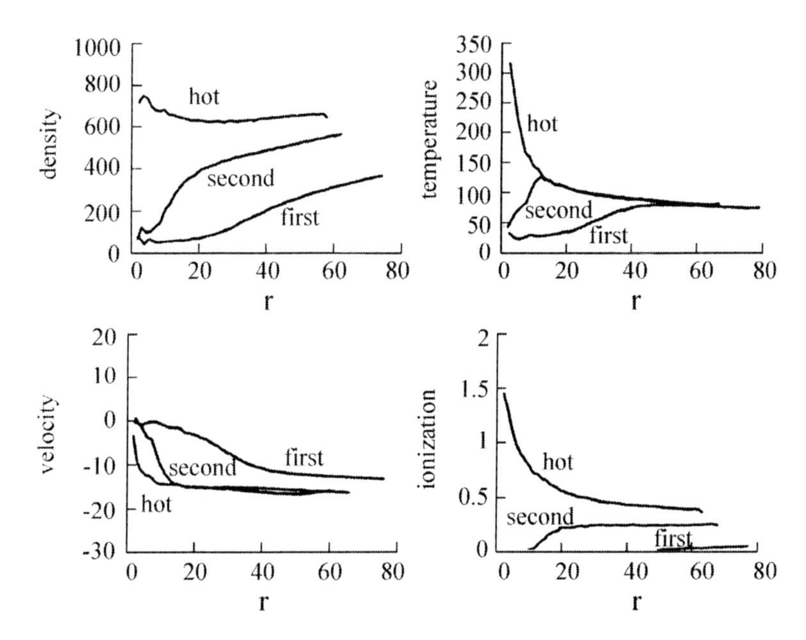


Figure 4.27 The argon bubble with specular BCs, constant diameter particles, and ionization. “hot” gives properties at the peak temperature. In this simulation, the peak temperature occurs just before the minimum radius (of 58.2) is attained and corresponds to a radius of 60.3. “first” and “second” give properties before the peak temperature is attained. Here, $R_{first}$ = 77.6 and $R_{second}$ = 64.9. (The coordinates are dimensionless e.g. the dimensionless temperature is the temperature divided by the ambient temperature $T_0$ (Ruuth et al. (2002).)

Xiao et al. (2002) used a different approach. They let an empty bubble collapse and considered water molecules going inside from the surrounding water to fill it. Obviously without chemical reactions. They considered simulations of a Lennard-Jones liquid. First, the “hottest” molecules from the high kinetic energy tail in the Maxwell–Boltzmann distribution diffused into the empty cavity. This was followed by a gradual filling in of the cavity until the density in the center was a little lower than that of the bulk liquid. The bubble filled in an oscillatory manner, by partly filling in, and then partially emptying, and so on, with ever decreasing amplitude towards the final uniform liquid state. The maximum temperature occurs typically at the end of the initial filling-in stage and is about 6,000 K for water, enough for sonoluminescence.

Okumura and Ito (2003) used *molecular dynamics* to investigate the case where a liquid is locally heated to a temperature higher than its boiling point. A phase transition then occurs from the liquid to gas, in other words, *a bubble* is created. After the heating is stopped, the surrounding liquid compresses the bubble and annihilates it. Constant pressure molecular dynamics simulations were performed by the Anderson (1980) method with the number of particles $N$ = 16,875 in the cubic unit cell. 80 atoms near the center of the cell were instantaneously heated from $T^*$ = 1.0 to $T^*$ = 11.0 where $T^*$ is the reduced temperature $k_B T/\varepsilon$ where $k_B$ = Boltzmann’s constant and $\varepsilon$ is the energy scale of the Lennard-Jones pair potential of two particles (see Xiao et al. (2002)). The atomic movement was then observed over 3,000 steps. Figure 4.28 shows snapshots of the bubble. These snapshots show the atoms that are in a 10% slab of the simulation slab in thickness. Figure 4.28 shows that the heated atoms scatter the neighboring nonheated atoms and create a bubble. Then the bubble

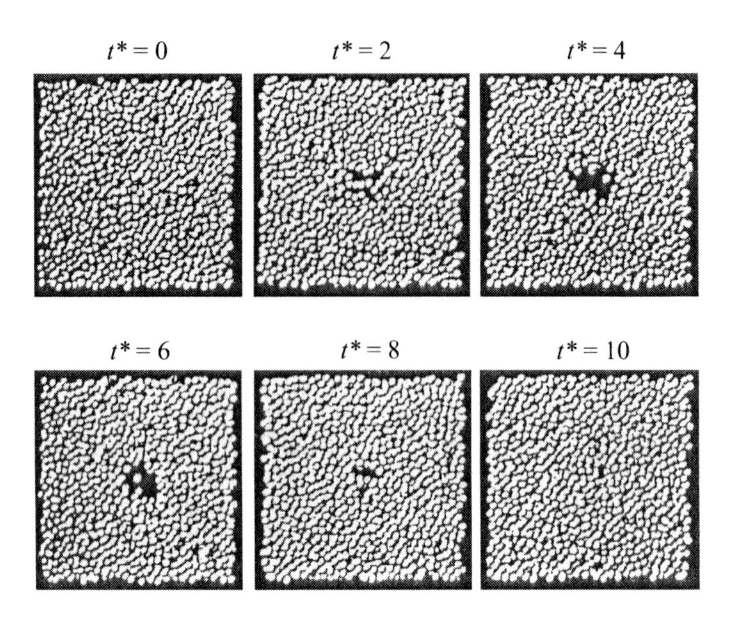


Figure 4.28 Snapshots of the bubble. The heated atoms scatter the surrounding nonheated atoms and make a bubble, and then the bubble is compressed by the surrounding liquid. $t^*$ = the reduced time = $t\sqrt{\varepsilon/m\sigma^2}$, where $m$ = mass of atom, and $\sigma$ and $\varepsilon$ are the distance and energy of the two particles in a Lennard-Jones pair potential. (See Xiao et al. (2002).) (Okumara and Ito (2003).)

is cooled and compressed by the surrounding liquid. Okumura and Ito (2003) compared this bubble dynamics to the Rayleigh–Plesset equation which describes the expansion and contraction of a bubble in terms of macroscopic hydrodynamics. Good agreement was obtained by choosing the appropriate parameters.

Rickayzen and Powles (2002) have presented the first non-equilibrium statistical mechanical theory for the mechanical and thermal behavior of the collapse of a microscopically small bubble in a liquid. First the number density and temperature space–time profiles for the special case of weakly interacting particles, the perfect gas model, are obtained. This is then generalized to a model in which the motion of the molecules is characterized by a single finite diffusion constant. The results for the collapse of a small bubble in a typical fluid are compared with those recently obtained through computer simulation. The agreement with the simulation is remarkably good for the perfect gas model; very high temperatures, sufficient for sonoluminescence, appear in a simple and natural way. An unexpected conclusion is that the perfect gas model agrees better with computer simulation than the model characterized by a single bulk diffusion constant. This may be because the collapse of the bubble is controlled by the leading shell of the fluid where the fluid density is low.

Useful books for studying this subject are Pryde (1966), Egelstaff (1992) and Heyes (1997).

## REFERENCES


Anderson HC 1980 Molecular dynamic simulations at constant pressure and/or temperature, *J Chem Phys* **72**, 2384.

Atkins PW and Friedman RS 1997 *Molecular Quantum Mechanics*, Oxford University Press, 3rd ed.

Bacher RF and Goudsmit S 1932 *Atomic energy states*, McGraw-Hill, New York.

Barber BP, Hiller A, Löfstedt R, Putterman SJ and Weninger KR 1997b Defining the unknowns of sonoluminescence, *Phys Rep* **281**, 65.

Barber BP and Putterman SJ 1991 Observation of synchronous picosecond sonoluminescence, *Nature* **352**, 318.

Barber BP and Putterman SJ 1992 Light scattering measurements of the repetitive supersonic implosion of a sonoluminescing bubble, *Phys Rev Lett* **69**, 3839.

Barber BP, Weninger KR, Löfstedt R and Putterman SJ 1995 Observation of a new phase of sonoluminescence at low partial pressures, *Phys Rev Lett* **74**, 5276.

Barber BP, Weninger KR and Putterman SJ 1997a Sonoluminescence, *Phil Trans Roy Soc London A* 355, 641.

Barber BP, Wu CC, Löfstedt R, Roberts PH and Putterman SJ 1994 Sensitivity of sonoluminescence to experimental parameters, *Phys Rev Lett* **72**,1380.

Barton G 1999 Perturbative check on the Casimir energies of nondispersive dielectric spheres, *J Phys A* **32**, 525.

Barton G and Eberlein C 1993 On quantum radiation from a moving body with finite refractive index, *Ann Phys* **227**, 222.

Belgiorno F, Liberati S, Visser M and Sciama DW 2000 Sonoluminescence: Two-photon correlations as a test of thermality, *Phys Lett A* **271**, 308 (quant-ph/9904018).

Ben'kovskii VG, Golubinchii PI and Maslennikov SI 1974 Pulsed electrohydrodynamic sonoluminescence accompanying a high-voltage discharge in water, *Sov Phys Acoust* 20, 14.

Bernstein LS and Zakin MR 1995 Confined electron model for single-bubble sonoluminescence, *J Phys Chem* **99**,14619.

Birrell ND and Davies PCW 1982 *Quantum Fields in Curved Space*, Cambridge University Press p 46.

Brennen CE 1995 *Cavitation and Bubble Dynamics*, Oxford University Press.

Brenner MP, Hilgenfeldt S, Lohse D and Rosales RR 1996 Acoustic energy storage in single bubble sonoluminescence, *Phys Rev Lett* 77, 3467.

Brenner MP, Hilgenfeldt S and Lohse D 2002 *Rev Mod Phys* **74**, 425.

Brenner MP, Lohse D and Dupont TF 1995 Bubble shape oscillations and the onset of sonoluminescence, *Phys Rev Lett* **75**, 954.

Brevik I, Marachevsky VN and Milton KA 1999 Identity of the van der Waals force and the Casimir effect and the irrelevance of these phenomena to sonoluminescence, *Phys Rev Lett* **82**, 3948.

Brown SC 1966 *Basic Data of Plasma Physics*, MIT Press, Cambridge US.

Carlson CE, Molina-Paris C, Perez-Mercader J and Visser M 1997a Schwinger's dynamical Casimir effect: bulk energy contribution, *Phys Lett B* **395**, 76.

Carlson CE, Molina-Paris C, Perez-Mercader J and Visser M 1997b Casimir effect in dielectrics: Bulk energy contribution, *Phys Rev D* **56**, 1262.

Carlson JT, Lewia SD, Atchley AA, Gaitan DF and Maruyama XF 1993 Spectra of Picosecond Sonoluminescence, Advances in Nonlinear Acoustics, Proc 13th ISNA Bergen ed Hobaek H, World Scientific.

Casimir HBG 1948 On the attraction between two perfectly conducting plates, *Proc K Ned Akad Wet* **51**, 793.

Chahine GL 1994 *Strong interactions bubble/bubble and bubble/flow, Bubble and Interface Phenomena*, ed Blake JR et al. Kluwer, Dordrecht, 195.
Chambers LA 1936 The emission of visible light from pure liquids during acoustic excitation, *Phys Rev* **49**, 881 (Abstract).
Chambers LA 1937 The emission of visible light from cavitated liquids, *J Chem Phys* **5**, 290.
Cheeke JDN 1997 Single-bubble sonoluminescence: "bubble, bubble, toil and trouble", *Can J Phys* **75**, 77.
Chen TW, Leung PT and Chu M-C 2000 Optical emissions in a sonoluminescing bubble, *Phys Rev E* **62**, 6584.
Cheng HY, Chu M-C, Leung PT and Yuan L 1998 How important are shock waves to single-bubble sonoluminescence? *Phys Rev E* **58**, R2705.
Chincholle L 1976 Effets physiques des ultrasons et conséquences éventuelles dans le domaine biomedical, *Onde Electrique* **56**, 28.
Chodos A and Groff S 1999 Modelling sonoluminescence, *Phys Rev E* **59**, 3001.
Ghu M-C 1996 The homologous contraction of a sonoluminescing bubble, *Phys Rev Lett* **76**, 4632.
Crum LA 1994 Sonoluminescence, *Phys Today* **47** Sept, 22.
Degrois M and Baldo P 1974 A new electrical hypothesis explaining sonoluminescence, chemical actions and other effects produced in gaseous cavitation, *Ultrasonics* **12**, 25.
Eberlein C 1992a Fluctuations of Casimir forces on finite objects: I. Spheres and hemispheres, *J Phys A* **25**, 3015.
Eberlein C 1992b Fluctuations of Casimir forces on finite objects: II. Flat circular disk, *J Phys A* **25**, 3039.
Eberlein C 1993 Radiation – reaction force on a moving mirror, *J Phys I France* **3**, 2151.
Eberlein C 1996a Theory of quantum radiation observed as sonoluminescence, *Phys Rev A* **53**, 2772.
Eberlein C 1996b Sonoluminescence as quantum vacuum radiation, *Phys Rev Lett* **76**, 3842.
Eberlein C 1996c Reply to Comment on "Sonoluminescence as quantum vacuum radiation" by Unnikrishnan and Mukhopadhyay, *Phys Rev Lett* **77**, 4691.
Egelstaff PA 1992 *An Introduction to the Liquid State*, 2nd edition, paperback, Academic Press.
Evans AK 1996 Instability of converging shock waves and sonoluminescence, *Phys Rev E* **54**, 5004.
Flint EB and Suslick KS 1991 Sonoluminescence from alkali-metal salt solutions, *J Phys Chem* **95**, 1484.
Flosdorf EW, Chambers LA and Malisoff W 1936 Sonic activation in chemical systems: Oxidations at audible frequencies, *J Am Chem Soc* **58**, 1069.
Frenkel J 1940 On electrical phenomena associated with cavitation due to ultrasonic vibrations in liquids, *Acta Physicochimica* **12**, 317.
Frenzel H and Schultes H 1934 Luminescenz im ultraschallbeschickten Wasser. Kurze Mitteilung, *Zeit für Phys Chem* **327**, 421.
Frommhold L 1993 *Collision-Induced Absorption in Gases*, Cambridge University Press.
Frommhold L 1998 Electron–atom bremsstrahlung and the sonoluminescence of rare gas bubbles, *Phys Rev E* **58**, 1899.
Frommhold L and Atchley AA 1994 Is sonoluminescence due to collision-induced emission? *Phys Rev Lett* **73**, 2883.
Fyrillas MM and Szeri AJ 1994 Dissolution or growth of soluble spherical oscillating bubbles, *J Fluid Mech* **277**, 381.
Gaitan DF 1990 An experimental investigation of acoustic cavitation in gaseous liquids, PhD thesis, University of Mississippi.

Gaitan DF, Crum LA, Church CC and Roy RA 1992 Sonoluminescence and bubble dynamics for a single, stable cavitation bubble, *J Acoust Soc Am* **91**, 3166.
Garcia N and Levanyuk AP 1996 Sonoluminescence: A new electrical breakdown hypothesis, *JETP Lett* **64**, 907.
Garcia N, Levanyuk AP and Osipov VV 1999 Scenario of the electric breakdown and UV radiation spectra in single-bubble sonoluminescence, *JETP Lett* **70**, 431.
Garcia N, Levanyuk AP and Osipov VV 2000 Nature of sonoluminescence: Noble gas radiation excited by hot electrons in cold water, *Phys Rev E* **62**, 2168.
Gilmore FR 1952 The Growth or Collapse of a Spherical Bubble in a Viscous Compressible Liquid, California Institute of Technology Hydrodynamics Laboratory, Report 26-4.
Ginzburg VL 1962 *Propagation of electromagnetic waves in plasma*, Gordon and Breach.
Godunov SK 1959 *Math Sbornik* **47**, 271 (in Russian).
Goldreich P and Weber SV 1980 Homologously collapsing stellar cores, *Astrophys J* **238**, 991.
Golubnichii PI, Sytnikov AM and Filoneko AD 1987 Electrical nature of ultrasonic luminescence and possibility of testing the hypothesis experimentally, *Sov Phys Acoust* **33**, 265.
Gompf B, Günther R, Nick G, Pecha R and Eisenmenger W 1997 Resolving sonoluminescence pulse width with time-correlated single photon counting, *Phys Rev Lett* **79**, 1405.
Greenland PT 1999 Sonoluminescence, *Contemporary Physics* **40**, 11.
Greenspan HP and Nadim A 1993 On sonoluminescence of an oscillating gas bubble, *Phys Fluids A* **5**,1065.
Griffing V 1950 Theoretical explanation of the chemical effects of ultrasonics, *J Chem Phys* **18**, 997.
Griffing V and Sette D 1952 The chemical effects of ultrasonics, *J Chem Phys* **20**, 939.
Griffing V and Sette D 1955 Luminescence produced as a result of intense ultrasonic waves, *J Chem Phys* **23**, 503.
Guderley G 1942 Starke kugelige und zylindrische Verdichtungsstöße in der Nähe des Kugelmittelpunktes bzw. der Zylinderachse, *Luftfahrtforschung* **19**, 302.
Hallaj IM, Matula TJ, Roy RA and Crum LA 1996 Measurements of the acoustic emission from glowing bubbles, *J Acoust Soc Am* **100**, 2717 (Abstract).
Hammer D and Frommhold L 2000a Spectra of sonoluminescent rare-gas bubbles, *Phys Rev Lett* **85**, 1326.
Hammer D and Frommhold L 2000b Comment on Hilgenfeldt S, Grossmann S and Lohse D 1999 *Phys Fluids* **11**, 1318, *Phys Fluids* **12**, 472.
Hammer D and Frommhold L 2001a Topical review. Sonoluminescence: how bubbles glow, *J Mod Opt* **48**, 239.
Hammer D and Frommhold L 2001b Polarization bremsstrahlung spectra of electron-rare-gas atom collisions from 5 to 40 kK, *Phys Rev A* **64**, 024705.
Hammer D and Frommhold L 2002a Light emission of sonoluminescent bubbles containing a rare gas and water vapour, *Phys Rev E* **65**, 046309.
Hammer D and Frommhold L 2002b Electron-ion bremsstrahlung spectra calculations for sonoluminescence, *Phys Rev E* **66**, 056303.
Hanbury Brown R and Twiss RQ 1958 Interferometry of the intensity fluctuations in light, II. An experimental test of the theory for partially coherent light, *Proc Roy Soc Ser A* **243**, 291.
Harvey EN 1939 Sonoluminescence and sonic chemiluminescence, *J Am Chem Soc* **61**, 2392.
Harvey EN 1952 *Bioluminescence*, Academic Press.

Hatanaka S, Yasui K, Tuziuto T and Mitome H 2000 Difference in threshold between sono- and sonochemical luminescence, *Jpn J Appl Phys* **39**, 2962.
Hatanaka S, Yasui K, Tuziuti T, Kozuka T and Mitome H 2001 Quenching mechanism of multibubble sonoluminescence at excessive sound pressure, *Jpn J Appl Phys* **40**, 3856.
Henglein A 1987 Sonochemistry: historical development and modern aspects, *Ultrasonics* **25**, 6.
Henglein A 1993 Contributions to various aspects of cavitation chemistry, *Advances in Sonochemistry* Vol 3, 17, JAI Press.
Herzberg G 1952 Spectroscopic evidence of molecular hydrogen in the atmospheres of Uranus and Neptune, *Astrophys J* **115**, 337.
Heyes DM 1997 *The Liquid State: Application of Molecular Simulations*, Wiley, Chichester, UK.
Hickling R 1994 Transient, high-pressure solidification associated with cavitation in water, *Phys Rev Lett* **73**, 2853.
Highfield R 6 July 1988 Mints may be flashes of inspiration, *Daily Telegraph*, London p 2.
Hilgenfeldt S, Grossmann S and Lohse D 1999a Sonoluminescence light emission, *Phys Fluids* **11**, 1318; see also Comment by Hammer D and Frommhold L 2000 *Phys Fluids* **12**, 472; and Response by Hilgenfeldt et al. 2000 *Phys Fluids* **12**, 474.
Hilgenfeldt S, Grossmann S and Lohse D 1999b A simple explanation of light emission in sonoluminescence, *Nature* **398**, 402.
Hilgenfeldt S and Lohse D 1999 Predictions for upscaling sonoluminescence, *Phys Rev Lett* **82**, 1036.
Hilgenfeldt S and Lohse D 2000 Sonoluminescence: When bubbles glow, *Current Science* **78**, 238.
Hilgenfeldt S, Lohse D and Brenner MP 1996 Phase diagrams for sonoluminescing bubbles, *Phys Fluids* **8**, 2808.
Hiller RA and Putterman SJ 1995 Observation of isotope effects in sonoluminescence, *Phys Rev Lett* **75**, 3549.
Hiller RA, Putterman SJ and Barber BP 1992 spectrum of synchronous picosecond sonoluminescence, *Phys Rev Lett* **69**, 1182.
Hiller RA, Putterman SJ and Weninger KR 1998 Time-resolved spectra of sonoluminescence, *Phys Rev Lett* **80**, 1090.
Hiller RA, Weninger KR, Putterman SJ and Barber BP 1994 Effect of noble gas doping in single-bubble sonoluminescence, *Science* **266**, 248.
Hirschfelder JO, Curtiss CF and Bird RB 1954 *Molecular Theory of Gases and Liquids*, Wiley, New York.
Holt M 1977 *Numerical Methods in Fluid Dynamics*, Springer-Verlag, New York.
Holt RG and Gaitan DF 1996 Observation of stability boundaries in the parameter space of single bubble sonoluminescence, *Phys Rev Lett* **77**, 3791.
Holzfuss J, Rüggeberg M and Mettin R 1998 Boosting sonoluminescence, *Phys Rev Lett* **81**, 1961.
Jarman P 1960 Sonoluminescence: A discussion, *J Acoust Soc Am* **32**, 1459.
Jeffrey GA 1947 *An Introduction to Hydrogen Bonding*, Oxford University Press p 100.
Jensen B and Brevik I 2000 Transition radiation and the origin of sonoluminescence, *Phys Rev E* **61**, 6639.
Jeon J, Yang I, Na J and Kwak H 2000 Radiation mechanism for a single bubble sonoluminescence, *J Phys Soc Japan* **69**,112.
Keller JB and Kolodner II 1956 Damping of underwater explosion bubble oscillations, *J Appl Phys* **27**, 1152.
Keller JB and Miksis M 1980 Bubble oscillations of large amplitude, *J Acoust Soc Am* **68**, 628.

Kling L and Hammitt PG 1972 Study of spark-induced cavitation bubble collapse, *Trans Am Soc Mech Eng Series D J Basic Engineering* **94**, 825.
Kondić L, Gersten JI and Yuan C 1995 Theoretical studies of sonoluminescence radiation: Radiative transfer and parametric dependence, *Phys Rev E* **52**, 4976.
Kozuka T, Hatanaka S, Tuziuti T, Yasui K and Mitome H 2000 Observation of a sonoluminescing bubble using a stroboscope, *Jpn J Appl Phys* **39**, 2967.
Kozuka T, Hatanaka S, Yasui K, Tuziuti T and Mitome H 2002 Simultaneous observation of motion and size of a sonoluminescing bubble, *Jpn J Appl Phys* **41**, 3248.
Kurz T, Lauterborn W and Metten B 2001 Molecular dynamics approach to bubble dynamics and sonoluminescence, Proc 17th Int Congress on Acoustics, Rome, CD Vol. 1.
Kurz T, Metten B, Schanz D and Lauterborn W 2002 Molecular dynamics of bubble collapse at sonoluminescence conditions, Forum Acusticum, Seville, PACS REFERENCE: 43.35.HI.
Kwak H, Lee YP and Karng SW 1999 Pressure wave propagation inside a sonoluminescing gas bubble, *J Phys Soc Japan* **68**, 705.
Kwak H, Lee J and Karng SW 2001 Bubble dynamics for single bubble sonoluminescence, *J Phys Soc Japan* **70**, 2909.
Kwak H and Na JH 1996 Hydrodynamic solutions for a sonoluminescing gas bubble, *Phys Rev Lett* 77, 4454.
Kwak H and Na JH 1997 Physical processes for single bubble sonoluminescence, *J Phys Soc Japan* **66**, 3074.
Kwak H and Yang H 1995 An aspect of sonoluminescence from hydrodynamic theory, *J Phys Soc Japan* **64**, 1980.
Landau LD and Lifshitz EM 1975 *The Classical Theory of Fields*, Pergamon Press.
Landau LD and Lifshitz EM 1987 *Fluid Mechanics*, Pergamon Press.
Lauterborn W 2002 Nonlinear Acoustics and Acoustic Chaos, Lecture Notes in Physics, *Sound-Flow Interactions*, Aurégan Y, Maurel A, Pagneux V and Pinton J-F (eds), Springer, page 265.
Lauterborn W and Bolle H 1975 Experimental investigations of cavitation-bubble collapse in the neighbourhood of a solid boundary, *J Fluid Mech* **72**, 391.
Lauterborn W, Kurz T, Mettin R, Metten B, Krefting D, Koch P, Appel J and Schanz D 2002 Acoustic Cavitation and Bubble Dynamics, Proc Int Symp on Innovative Materials Processing by Controlling Chemical Reaction Field (IMP2002), Miyagi, Japan, page 109.
Lee Y, Karng SW, Jeon J and Kwak H 1997 Shock pulse from a sonoluminescing gas bubble, *J Phys Soc Japan* **66**, 2537.
Lentz WJ, Atchley AA and Gaitan DF 1995 Mie scattering from a sonoluminescing air bubble in water, *Appl Oct* **34**, 2648.
Lepoint-Mullie F, Pauw DD and Lepoint T 1996 Analysis of the "new electrical model" of sonoluminescence, *Ultrasonics Sonochem* **3**, 73.
Levshin VL and Rzhevkin SN 1937 *Doklady Akad Nauk SSSR* **16**, 407.
Lezzi A and Prosperetti A 1987 Bubble dynamics in a compressible liquid. Part 2. Second-order theory, *J Fluid Mech* **185**, 269.
Liberati S, Visser M, Belgiorno F and Sciama DW 1999a Sonoluminescence and the QED vacuum, quant-ph/9904008.
Liberati S, Visser M, Belgiorno F and Sciama DW 1999b Sonoluminescence as a QED vacuum effect: Probing Schwinger's proposal, quant-ph/9805031.
Liberati S, Visser M, Belgiorno F and Sciama DW 2000a Sonoluminescence as a QED vacuum effect. I. The physical scenario, *Phys Rev D* **61**, 085023.
Liberati S, Visser M, Belgiorno F and Sciama DW 2000b Sonoluminescence as a QED vacuum effect. II. Finite volume effects, *Phys Rev D* **61**, 085024.

Lin H and Szeri AJ 2001 Shock formation in the presence of entropy gradients, *J Fluid Mech* **431**, 161.
Loeb LB 1958 *Static Electrification*, Springer, Berlin.
Löfstedt R, Barber BP and Putterman SJ 1993 Toward a hydrodynamic theory of sonoluminescence, *Phys Fluids A* **5**, 2911.
Löfstedt R, Weninger KR, Putterman S and Barber BP 1995 Sonoluminescing bubbles and mass diffusion, *Phys Rev E* **51**, 4400.
Lohse D, Brenner MP, Dupont TF, Hilgenfeldt S and Johnston B 1997 Sonoluminescing air bubbles rectify argon, *Phys Rev Lett* **78**, 1359.
Lohse D and Hilgenfeldt S 1997 Inert gas accumulation in sonoluminescing bubbles, *J Chem Phys* **107**, 6986.
Longuet-Higgins MS 1996 Shedding of vortex rings by collapsing cavities, with application to single-bubble sonoluminescence, *J Acoust Soc Am* **100**, 2678 (Abstract).
Louisell WH 1973 *Quantum Statistical Properties of Radiation*, John Wiley.
MacIntyre P 1997 Some comments on the thermodynamics of sonoluminescence – a personal view, *Ultrasonics Sonochem* **4**, 85.
Marboe EC and Weyl WA 1950 Mechano chemistry of the dispersion of mercury in liquids in an ultrasonic field, *J Appl Phys* **21**, 937.
Margulis MA 1985a Study of electrical phenomena associated with cavitation. II. Theory of the development of sonoluminescence in acoustochemical reactions, *Russ J Phys Chem* **59**, 882.
Margulis MA 1985b Sonoluminescence and sonochemical reactions in cavitation fields. A review, *Ultrasonics* **23**, 157.
Margulis MA 1990 *The Nature of Sonochemical Reactions and Sonoluminescence*, Advances in Sonochemistry Vol 1 JAI Press page 39.
Margulis MA and Margulis IM 2002 Contemporary review on nature of sonoluminescence and sonochemical reactions, *Ultrasonics Sonochem* **9**, 1.
Marinesco N and Trillat JJ 1933 Actions des ultrasons sur les plaques photographiques, *Comptes Rendus Acad Sci* **196**, 858.
Maroncelli M 1988 Computer simulation of the dynamics of aqueous solvation, *J Chem Phys* **89**, 5044.
Matsumoto M, Miyamoto K, Ohguchi K and Kinjo T 2000 Molecular dynamics simulation of a collapsing bubble, *Prog Theor Phys Suppl* **138**, 728.
McNamara III WB, Didenko YT and Suslick KS 1999 Sonoluminescence temperatures during multi-bubble cavitation, *Nature* **401**, 772.
Mehra J and Milton K 2000 *Climbing the Mountain. The Scientific Biography of Julian Schwinger*, Oxford University Press.
Metten B 2001 *Molekulardynamik – Simulationen zur Sonolumineszenz*, Der Andere Verlag, Osnabrück, Germany.
Metten B and Lauterborn W 2000 Molecular Dynamics Approach to Single-Bubble Sonoluminescence, *in* Nonlinear Acoustics at the Turn of the Millennium ed by W Lauterborn and T Kurz, American Institute of Physics Conf Proc No 524 (American Institute of Physics, Melville, NY) page 429.
Meyer E and Kuttfuff H 1959 Zur Phasenbeziehung zwischen Sonolumineszenz und Kavitationvorgang bei periodischer Anregung, *Zeit angew Phys* **9**, 325.
Milton KA 1980 Semiclassical electron models: Casimir self-stress in dielectric and conducting balls, *Ann Phys* **127**, 49.
Milton KA 1995 Casimir energy for a sphericavity in a dielectric: Toward a model for sonoluminescence? hep-th/9510091.
Milton KA 1996 Casimir energy for a spherical cavity in a dielectric: toward a model for sonoluminescence, *in* Quantum field theory under the influence of external conditions, ed. Bordag M, Tuebner Verlagsgesellschaft Stuttgart, pages 13–23. See also hep-th/9607186.

Milton KA and Ng YJ 1996 Casimir energy for a spherical cavity in a dielectric: Applications to sonoluminescence, hep-th/9607186.
Milton KA and Ng YJ 1998 Observability of the bulk Casimir effect: Can the dynamical Casimir effect be relevant to sonoluminescence? *Phys Rev E* **57**, 5504.
Mohanty P 2000a An analytic description of light emission in sonoluminescence, cond-mat/9912271.
Mohanty P 2000b Electromagnetic field correlation inside a sonoluminescing bubble, cond-mat/0005233.
Mohanty P and Khare SV 1998 Sonoluminescence as a cooperative many body phenomenon, *Phys Rev Lett* **80**, 189.
Molina-Paris C and Visser M 1997 Casimir effect in dielectrics: Surface area contribution, *Phys Rev D* **56**, 6629.
Moran MJ, Haigh RE, Lowry ME, Sweider DR, Abel GR, Carlson JT, Lewia SD, Atchley AA, Gaitan DF and Maruyama XF 1995 Direct observations of single sonoluminescence pulses, *Nucl Instrum Methods Phys Res Sect B* **96**, 651.
Moran MJ and Sweider D 1998 Measurements of sonoluminescence temporal pulse shape, *Phys Rev Lett* **80**, 4987.
Moss WC 1997 Understanding the periodic driving pressure in the Rayleigh–Plesset equation, *J Acoust Soc Am* **101**, 1187.
Moss WC 2000 Comment on "Extreme electrostatic phenomena in a single sonoluminescing bubble", *Phys Rev Lett* **85**, 4837.
Moss WC, Clarke DB, White JW and Young DA 1994 Hydrodynamic simulations of bubble collapse and picosecond sonoluminescence, *Phys Fluids* **6**, 2979.
Moss WC, Clarke DB, White JW and Young DA 1996 Sonoluminescence and the prospects for table-top micro-thermonuclear fusion, *Phys Lett A* **211**, 69.
Moss WC, Clarke DB and Young DA 1997 Calculated pulse widths and spectra of a single sonoluminescing bubble, *Science* **276**, 1398.
Moss WC, Clarke DB and Young DA 1999 *Star in a Jar, Sonochemistry and Sonoluminescence* Crum LA et al. (eds) Kluwer, Netherlands, p. 159.
Moss WC, Levatin JL and Szeri AJ 2000 A new damping mechanism in strongly collapsing bubbles, *Proc Roy Soc Lond A* **456**, 2983.
Moss WC, Young DA, Harte JA, Levatin JL, Rozsnyai BF, Zimmerman GB and Zimmerman IH 1999a Computed optical emissions from a sonoluminescing bubble, *Phys Rev E* **59**, 2986.
Motyka L and Sadzikowski M 1999 Atomic collisions and sonoluminescence, arXiv:physics/9912013.
Neppiras EA 1980 Acoustic cavitation, *Phys Rep* **61**, 159.
Neppiras EA and Noltingk BE 1951 Cavitation produced by ultrasonics: Theoretical conditions for the onset of cavitation, *Proc Phys Soc B (Lond)* **64B**, 1032.
Noltingk BE and Neppiras EA 1950 Cavitation produced by ultrasonics, *Proc Phys Soc B (Lond)* **63B**, 674.
Okumura H and Ito N 2003 Nonequilibrium molecular dynamics simulations of a bubble, *Phys Rev E* **67**, 045301 (R).
Patel NH and Ranga Rao MP 1996 Imploding shocks in a non-ideal medium, *J Eng Math* **30**, 683.
Pauling L 1970 *General Chemistry*, WH Freeman, 3rd ed, Chap 12.
Pecha R, Gompf B, Nick G, Wang ZQ and Eisenmerger W 1998 Resolving the sonoluminescence pulse shape with a streak camera, *Phys Rev Lett* **81**, 717.
Perisco F and Power EA 1988 Vacuum in non-relativistic matter–radiation systems, *Phys Scripta* T21.
Plesset MS 1949 The dynamics of cavitation bubbles, *J Appl Mech* **16**, 277.
Prosperetti A 1977 Thermal effects and damping mechanisms in the forced radial oscillations of gas bubbles in liquids, *J Acoust Soc Am* **61**, 17.

Prosperetti A 1984 Bubble phenomena in sound fields: part one, *Ultrasonics* **22**, 69.
Prosperetti A 1984 Bubble phenomena in sound fields: part two, *Ultrasonics* **22**, 115.
Prosperetti A 1991 The thermal behaviour of oscillating gas bubbles, *J Fluid Mech* **222**, 587.
Prosperetti A 1996 A new hypothesis on single-bubble sonoluminescence, *J Acoust Soc Am* **100**, 2677. (Abstract)
Prosperetti A 1997 A new mechanism for sonoluminescence, *J Acoust Soc Am* **101**, 2003.
Prosperetti A, Crum LA and Commander KW 1988 Nonlinear bubble dynamics, *J Acoust Soc Am* **83**, 502.
Prosperetti A and Hao Y 1999 Modelling of spherical gas bubble oscillations and sonoluminescence, *Phil Trans Roy Soc Lond A* **357**, 203.
Prosperetti A and Lezzi A 1986 Bubble dynamics in a compressible liquid. Part 1. First-order theory, *J Fluid Mech* **168**, 457.
Pryde JA 1992 *The Liquid State*, Hutchinson, London, 2nd edition, paperback.
Putterman SJ 1995 Sonoluminescence: Sound into light, *Sci Amer* **272** No 2, 32.
Putterman S, Evans PG, Vazquez G and Weninger KR 2001 Is there a simple theory of sonoluminescence? *Nature* **409**, 782; with reply by Hilgenfeldt S, Grossman S and Lohse D.
Putterman SJ and Weninger KR 2000 Sonoluminescence: How bubbles turn sound into light, *Annu Rev Fluid Mech* **32**, 445.
Rapaport DC 1980 The event scheduling problem in molecular dynamic simulation, *J Comput Phys* **34**, 184.
Rapaport DC 1997 *The Art of Molecular Dynamics Simulation*, Cambridge University Press.
Rayleigh Lord 1917 On the pressure developed in a liquid during collapse of a spherical cavity, *Phil Mag* **34**, 94.
Reddy SP 1985 *Phenomena Induced by Intermolecular Interactions*, ed Birnbaum, Plenum p 129.
Rickayzen G and Powles JG 2002 A collapsing bubble in a fluid: a statistical mechanical approach, *Molecular Physics* **100**, 3823.
Roberts PH and Wu CC 1996 Structure and stability of a spherical implosion, *Phys Lett A* **213**, 59.
Roberts PH and Wu CC 2003 The Shock-Wave Theory of Sonoluminescence, pages 1–27 in *Shock Focussing Effect in Medical Science and Sonoluminescence*, Srivasta RC, Leutloff D, Takayama K and Grönig H (eds), Springer.
Ruuth SJ, Putterman S and Merriman B 2002 Molecular dynamics of the response of a gas to a spherical piston: Implications for sonoluminescence, *Phys Rev E* **66**, 036310.
Rybicki GB and Lightman AP 1979 *Radiative Processes in Astrophysics*, Wiley Interscience, New York.
Saksena TK and Nyborg WL 1970 Sonoluminescence from stable cavitation, *J Chem Phys* **53**, 1722.
Schwinger J 1992a Casimir energy for dielectrics, *Proc Natl Acad Sci USA* **89**, 4091.
Schwinger J 1992b Casimir energy for dielectrics: Spherical geometry, *Proc Natl Acad Sci USA* **89**, 11118.
Schwinger J 1993a Casimir light: A glimpse, *Proc Natl Acad Sci USA* **90**, 958.
Schwinger J 1993b Casimir light: The source, *Proc Natl Acad Sci USA* **90**, 2105.
Schwinger J 1993c Casimir light: Photon pairs, *Proc Natl Acad Sci USA* **90**, 4505.
Schwinger J 1993d Casimir light: Pieces of the action, *Proc Natl Acad Sci USA* **90**, 7285.
Schwinger J 1994 Casimir light: Field pressure, *Proc Natl Acad Sci USA* **91**, 6473.
Sehgal CM and Verrall RE 1982 A review of the electrical hypothesis of sonoluminescence, *Ultrasonics* **20**, 37.
Seigel R and Howell JR 1972 *Thermal Radiation Heat Transfer*, McGraw-Hill.

Sochard S, Wilhelm AM and Delmas H 1997 Modelling of free radicals production in a collapsing gas-vapour bubble, *Ultrasonics Sonochemistry* **4**, 77.
von Sonntag C, Mark G, Tauber A and Schuchmann H 1999 OH Radical formation and dosimetry in the sonolysis of aqueous solutions, *Advances in Sonochemistry* Vol 5, 109, JAI Press.
Spinrad H 1964 Molecular hydrogen lines in the spectra of cool stars, *Astrophys J* **140**, 1639.
Srinivasan D and Holroyd LV 1961 Optical spectrum of the sonoluminescence emitted by cavitated water, *J Appl Phys* **32**, 446.
Stepanovsky YuP and Sergeeva GG 2000 Sonoluminescence as a physical vacuum excitation, cond-mat/0002434.
Storey BD and Szeri AJ 2000 Water vapour, sonoluminescence and sonochemistry, *Proc Roy Soc Lond A* **456**, 1685.
Suslick KS ed 1988 *Ultrasound: Its Chemical, Physical and Biological Effects*, VCH Publishers, New York p 230.
Suslick KS 1990 Sonochemistry, *Science* **247**, 1439.
Suslick KS and Crum LA 1997 *Sonochemistry and Sonoluminescence, Encyclopedia of Acoustics* ed Crocker MJ, John Wiley, pp 271–281.
Suslick KS, Doktycz SJ and Flint EB 1990 On the origin of sonoluminescence and sonochemistry, *Ultrasonics* **28**, 280.
Suslick KS, Hammerton DA and Cline RE 1986 The sonochemical hot spot, *J Am Chem Soc* **108**, 5641.
Taylor KJ and Jarman PD 1970 The spectra of sonoluminescence, *Australian J Phys* **23**, 319.
Taylor RL and Caledonia G 1969 Experimental determination of the cross-sections for neutral bremsstrahlung I. Ne, Ar AND Xe, *J Quant Spectrosc Radiat Transfer* **9**, 657.
Toegel R, Gompf B, Pecha R and Lohse D 2000 Does water vapor prevent upscaling sonoluminescence? *Phys Rev Lett* **85**, 3165.
Trentalange S and Pandey S 1996 Bose–Einstein correlations and sonoluminescence, *J Acoust Soc Am* **99**, 2439.
Trilling L 1952 The collapse and rebound of a gas bubble, *J Appl Phys* **23**, 14.
Tsiklauri D 1997 Two-component radiation model of the sonoluminescing bubble, *Phys Rev E* **56**, R6245.
Unnikrishnan CS and Mukhopadhyay S 1996 Comment on "Sonoluminescence as quantum vacuum radiation" (Eberlein (1996b)) and Reply by Eberlein, *Phys Rev Lett* **77**, 4690.
Vaughan PW and Leeman S 1986 Some comments on mechanisms of sonoluminescence, *Acustica* **59**, 279.
Vaughan PW and Leeman S 1987 Sonoluminescence: A new light on cavitation, Proceedings of Ultrasonics International 1987, Butterworth, London, page 297.
Vazquez G, Camara C, Putterman S and Weninger H 2001 Sonoluminescence: nature's smallest blackbody, *Opt Lett* **26**, 575.
Vazquez G, Camara C, Putterman SJ and Weninger H 2002 Blackbody spectra for sonoluminescing hydrogen bubbles, *Phys Rev Lett* **88**, 197402.
Verrall R and Sehgal C 1987 Sonoluminescence, *Ultrasonics* **25**, 29.
Visser M, Liberati S, Belgiorna F and Sciama DW 1999 Sonoluminescence: Bogolubov coefficients for the QED vacuum of a time-dependent dielectric bubble, *Phys Rev Lett* **83**, 678.
Vuong VQ and Szeri AJ 1996 Sonoluminescence and diffusive transport, *Phys Fluids* **8**, 2354.
Vuong VQ, Szeri AJ and Young DA 1999 Shock formation within sonoluminescence bubbles, *Phys Fluids* **11**, 10.

Walton AJ 1977 Triboluminescence, *Adv Phys* **26**, 887.
Weninger KR, Camara CG and Putterman SJ 2000b Observation of bubble dynamics within luminescent cavitation clouds: Sonoluminescence at the nano-scale, *Phys Rev E* **63**, 016310.
Weninger KR, Evans PG and Putterman SJ 2000a Time correlated single photon Mie scattering from a sonoluminescing bubble, *Phys Rev E* **61**, 1020.
Weninger KR, Putterman SJ and Barber BP 1996 Angular correlations in sonoluminescence: Diagnostic for the sphericity of a collapsing bubble, *Phys Rev E* **54**, R2205.
Weyl WA 1951 Surface structure of water and some of its physical and clinical manifestations, *J Colloid Sci* **6**, 389.
Weyl WA and Marboe EC 1949 Some mechano-chemical properties of water, *Research* **2**, 19.
Whitham GB 1974 *Linear and Nonlinear Waves*, Wiley-Interscience, New York.
Willison JR 1998 Sonoluminescence: Proton-tunneling radiation, *Phys Rev Lett* **81**, 5430.
Wu CC and Roberts PH 1993 Shock-wave propagation in a sonoluminescing gas bubble, *Phys Rev Lett* **70**, 3424.
Wu CC and Roberts PH 1994 A model of sonoluminescence, *Proc Roy Soc A* **445**, 323.
Wu CC and Roberts PH 1996 Structure and stability of a spherical shock wave in a van der Waals gas, *Q J Mech Appl Math* **49**, 501.
Xiao C, Heyes DM and Powles JG 2002 The collapsing bubble in a liquid molecular dynamics simulations, *Molecular Physics* **100**, 3451.
Xu N, Wang L and Hu X 1998 Numerical study of electronic impact and radiation in sonoluminescence, *Phys Rev E* **57**, 1615.
Xu N, Wang L and Hu X 1999 Extreme electrostatic phenomena in a single sonoluminescing bubble, *Phys Rev Lett* **83**, 2441.
Xu N, Wang L and Hu X 2000 Bremsstrahlung of nitrogen and noble gases in single-bubble sonoluminescence, *Phys Rev E* **61**, 2611.
Yano FB and Koonin SE 1977 Determining pion source parameters in relativistic heavy-ion collisions, *Phys Lett* **78B**, 556.
Yasui K 1995 Effects of thermal conduction on bubble dynamics near the sonoluminescence threshold, *J Acoust Soc Am* **98**, 2772.
Yasui K 1996 Variation of liquid temperature at bubble wall near the sonoluminescence threshold, *J Phys Soc Japan* **65**, 2830.
Yasui K 1997a Chemical reactions in a sonoluminescing bubble, *J Phys Soc Japan* **66**, 2911.
Yasui K 1997b Alternative model of single-bubble sonoluminescence, *Phys Rev E* **56**, 6750.
Yasui K 1998a Effect of surfactants on single-bubble sonoluminescence, *Phys Rev E* **58**, 4560.
Yasui K 1998b Effect of non-equilibrium evaporation and condensation on bubble dynamics near the sonoluminescence threshold, *Ultrasonics* **36**, 575.
Yasui K 1998c Single-bubble dynamics in liquid nitrogen, *Phys Rev E* **58**, 471.
Yasui K 1999a Mechanism of single-bubble sonoluminescence, *Phys Rev E* **60**, 1754.
Yasui K 1999b Effect of a magnetic field on sonoluminescence, *Phys Rev E* **60**, 1759.
Yasui K 1999c Single-bubble sonoluminescence from hydrogen, *J Chem Phys* **111**, 5384.
Yasui K 1999d Single-bubble and multibubble sonoluminescence, *Phys Rev Lett* **83**, 4297.
Yasui K 2000 Mechanism of single-bubble sonoluminescence, 15th ISNA Göttingen 1999, *Am Inst Phys Conference Proc* Vol 524, p 437.
Yasui K 2001a Single-bubble sonoluminescence from noble gases, *Phys Rev E* **63**, 035301(R).
Yasui K 2001b Temperature in multibubble sonoluminescence, *J Chem Phys* **115**, 2893.
Yasui K 2001c Effect of liquid temperature on sonoluminescence, *Phys Rev E* **64**, 016310.

Yasui K 2002 Effect of volatile solutes on sonoluminescence, *J Chem Phys* **116**, 2945.
Yasui K 2003 Personal communication.
Yasui K, Tuziuti T, Iida Y and Mitome H 2003 Theoretical study of the ambient-pressure dependence of sonochemical reactions, *J Chem Phys* **119**, 346.
Young FR 1976 Sonoluminescence from water containing dissolved gases, *J Acoust Soc Am* **60**, 100.
Yuan L, Cheng HY, Chu M-C and Leung PT 1998 Physical parameters affecting sonoluminescence: A self-consistent hydrodynamic study, *Phys Rev E* **57**, 4265.
Zel'dovich YB and Raizer YP 1966 *Physics of Shock Waves and High-Temperature Hydrodynamic Phenomena*, The Dover Publications edition, published in 2002, is a republication in one volume of the two-volume work originally published in English in 1966 and 1967 by the Academic Press. It has 916 pages and is priced at £20.

**CHAPTER FIVE**

# Conclusions

Sonoluminescence is the process
whereby a phonon is converted
into a photon.

*Professor Alan Purvis,*
*University of Durham, UK*

A little huff, a little blow,
And up the magic bubbles go,
Some large, some small,
Some not lasting long at all.

*Anonymous*

## 5.1 PRELIMINARIES

Sonoluminescence was discovered 70 years ago by Frenzel and Schultes (1934). This was light from clouds of bubbles, or multibubble sonoluminescence. After World War II progress was made as mainly described in Chapter 2 with multibubble sonoluminescence. In 1990, after earlier discoveries (see Sec. 3.1), Gaitan and his PhD supervisor Crum (Gaitain et al. (1992)) made the breakthrough which has guided sonoluminescence research ever since. Gaitan and Crum found that they could *stably* trap a 5 μm bubble at the velocity node of a standing sound wave in a glass flask. They could then control the pressure, temperature, contents and other parameters of the bubble gas that was producing the sonoluminescence.

As an indicator of how research on sonoluminescence has accelerated I have 400 articles in the 56 years from 1934 to 1990, and no less than 500 articles in the 11 years from 1990 to 2001 and the pace shows no signs of slackening, witness the 15 papers on sonoluminescence at the International Congress on Acoustics in Rome 2001. There has been an avalanche of papers on sonoluminescence, especially from the groups of Putterman (Barber et al. (1997) and Lohse (Brenner et al. (2002)). The mystery of where the light comes from has proved a fertile field for PhD students all over the world.

Much effort is being given to the measurement of bubble dynamics with lasers and streak cameras. Also to the detection of the elusive sonoluminescence spectra.

Two major questions still remain to be completely answered:

- What exactly is going on in the bubble?
- How exactly is the light produced?

Experimental work would be simpler if a rare gas were used. The light is bright for Xe and Ar and we do not have the complications of dissociation, chemical reactions, solubility and so on, although we still have water vapor from the surrounding water. Theoretical work would be easier if He were studied as this is a very compact stable atom. However although it produces the hottest radiation, it also produces the dimmest (Greenland (1999)) so that comparison with experiment would be difficult. Xe, or Ar which is readily available, would be more suitable. In 1989 I wrote in my book *Cavitation* (Young (1989, 1999)), "My own view is that sonoluminescence results from the heating of the cavity contents during the compression phase of the oscillating bubble. This causes excitation of the gas in the cavity, thereby promoting the formation and subsequent recombinations of excited species, with the emission of light." There is at present no completely satisfactory theorem of sonoluminescence which explains all experimental observations.

Sonoluminescence has been with us for 69 years, and I have lived with it for 42 years after first encountering Jarman (1959a,b) at Imperial College.

Gaitan's discovery was a major surprise, and I think that sonoluminescence has more surprises in store for us.

## 5.2 USES

A direct use for sonoluminescence seems unlikely for two reasons. First the difficulty of setting up and fine tuning the water cell, transducer and detection system. Second, the fact that only a fraction, about one millionth (see Sec. 4.7.3), of the energy in a collapsing bubble ends up as light. Still, remember that in its early days, the *laser* was called a discovery without a use.

However, sonoluminescence may be of *use indirectly*. Ahmad et al. (1987) made a study of the ultrasonic debridement (removal of tissue) of root canals in teeth. They wished to investigate the mechanism and relative efficiency of hand held files and ultrasonic scalers for cleaning. To detect whether cavitation was a cleaning mechanism they mounted the ultrasonic file in a solution of 2.5% sodium hypochlorite and used *sonoluminescence* as a detector of cavitation. The sonoluminescence was detected by an image intensifier tube, amplifier and television video recorder. The experiment was then repeated using a sickle-shaped ultrasonic scaler. No light was detected from the oscillating file. With the scaler, however, very bright light was immediately observed around the tip, Fig. 5.1. Walmsley et al. (1990) give a study of cavitational activity on the root surface of teeth during ultrasonic scaling. An image intensifier was again used, and sonoluminescence and probably tribo-luminescence was observed.

Grieser and Ashokkumar (2001) have recently shown that sonoluminescence can be used to excite fluorescent molecules to emit light themselves, often at much higher intensities, a process the authors called *sonophotoluminescence*, see §2.2.7. This form of emission, due to photoexcitation, changes in the same way as a sonoluminescent signal, with variations in solution content and solution concentration, and thus provides an accurate amplified version of the sonoluminescence signal.

Figure 5.1 Emission of light from scaler. L, light; T, tip of scaler (original magnification ×7). (Ahmad et al. (1987).)

Since sonochemistry is produced by cavitation, and cavitation often produces sonoluminescence, sonoluminescence can often be used as a probe for examining the properties of bubbles in solution that are exposed to sound. For an account of sonochemistry, see Young (1999) page 353.

Beckett and Hua (2001) measured the correlation between the sonoluminescence and sonochemical reaction yield, in the case of 1,4-dioxane decomposition. 1,4-Dioxane is an organic pollutant, and its sonochemical degradation is of interest to environmental scientists. To increase the yield, Beckett and Hua (2001) sparged the solution with an argon/oxygen mixture from 0% argon to 100% argon. Figure 5.2 shows the results for the optimum frequency for sonoluminescence and chemical activity of 358 kHz. Clearly, sonoluminescence can be used as an indicator for sonochemistry reactions yields. Sonoluminescence can also be used for studying the extreme conditions obtained within cavitation bubbles. Flint and Suslick (1991) analyzed the multibubble sonoluminescence spectra from *silicone oil* (polydimethylsiloxane) under a continuous Ar sparge at 0°C.

The observed emission came from the Swan band transitions, $d^3\Pi_g - a^3\Pi_u$, in the excited state $C_2$, which was modeled with synthetic spectra as a function of rotational and vibrational temperatures. From a comparison of synthetic and observed spectra, the effective cavitation temperature was found to be $5075 \pm 156$ K, claiming a rather high degree of accuracy. Maximum pressures in the bubble can also be measured. Also, changes in the light intensity can be used to infer changes in chemical reaction rates. This is best done using single bubble sonoluminescence, as with multibubble sonoluminescence the presence of a large number of bubbles is difficult to deal with. With a single bubble the dynamics is highly controllable and repeatable. The radius of the bubble can be measured during the acoustic cycle by light scattering methods, thus allowing study of the effects of

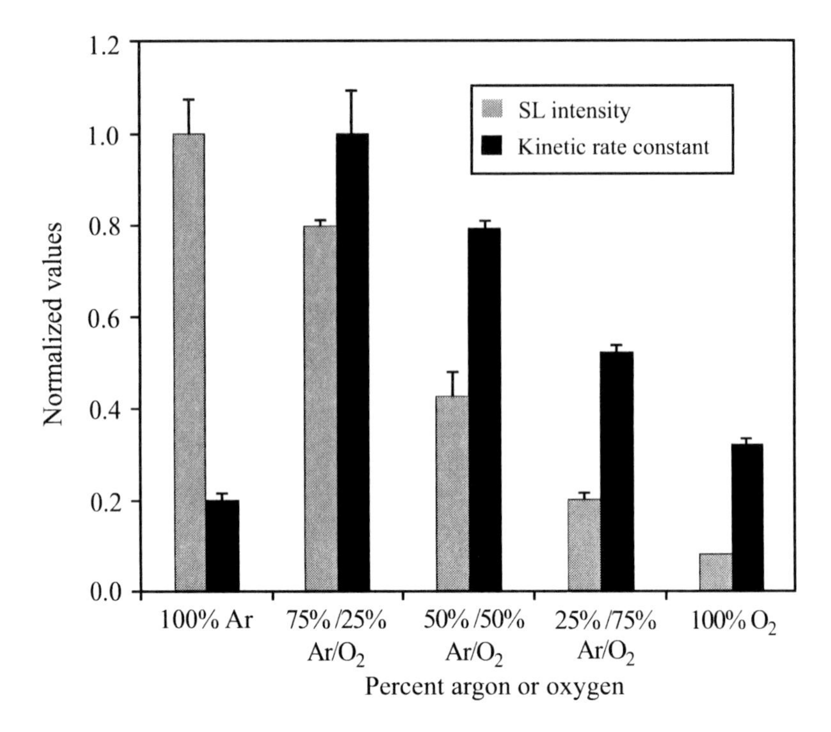


Figure 5.2 Relative SL intensities and 1,4-dioxane pseudo-first-order decomposition rate constants at various saturation gas ratios. All normalized SL intensities and decomposition rate constants are given as a fraction of the highest value of all gas ratios. Frequency = 358 kHz. Error bars represent one standard deviation. (Beckett and Hua (2001).)

various parameters, e.g. sound pressure amplitude, nature of the gas in the bubble etc, on cavitation and sonoluminescence.

Lohse (2002) discusses how a sonoluminescing single bubble can be viewed not as a light bulb, but as a high-temperature, high-pressure, *micro-reactor*. As the extreme conditions inside the bubble are adjustable through external parameters, such as forcing pressure or water temperature, so the bubble can be considered as a controlled high-temperature reaction chamber, filled with any chosen gas, offering opportunities to measure reaction rates in extreme temperature and pressure regimes. This has application to *sonochemistry* (Suslick (1990)).

Ogi et al. (2002) have recently reported a possible use of single bubble sonoluminescence for water purification and treatment. They use the intense ultraviolet light in sonoluminescence to activate a $TiO_2$ photocatalyst to decompose organic compounds in water.

## 5.3 FUTURE WORK

As regards *future work* much attention has been focused on the theories of sonoluminescence. Clues are from the spectra and flash widths of sonoluminescence. Weninger et al. (2000) discuss three equilibrium models of the spectrum: (1) black body (see Sec. 4.6.1), (2) bremsstrahlung (see Sec. 4.6.2), both in the weak and strong ionization limits, and (3) pressure-broadened atomic or molecular bound-state emissions of a nonionized, 5,000 K gas (see Sec. 2.2.7, Flint and Suslick (1991)). Weninger et al. (2000) state that the universal structure of the continuum for different rare gases suggests black body if the hot spot is opaque and bremsstrahlung if it is sufficiently small to be transparent. They suggest that it remains to be seen whether the more detailed formulation of process (3) can explain (a) the similarity of the spectra observed and (b) the means whereby a temperature of $\frac{1}{2}$ eV yields a thermal spectrum with a peak beyond 6 eV.

Vazquez et al. (2001) (see Sec. 4.6.1) conclude that combined measurements of spectral irradiance, Mie scattering, and flash width (as determined by time-correlated single-photon counting) suggest that sonoluminescence from hydrogen and rare gas bubbles is radiation from a *black body* ranging from 6,000 K ($H_2$) to 20,000 K (He) and a surface of emission whose radius $R_e$ ranges from 0.1 μm (He) to 0.4 μm (Xe). Thus although sonoluminescence is emitted from a *surface whose radius is smaller than the wavelength of light* and the photon-matter free path, its spectrum can match that of a black body (for 200 nm $< \lambda <$ 800 nm and $R_e > \sim 0.1$ μm. They state that whether black body behavior extends out into the Rayleigh–Jeans limit of large $\lambda$ is a question for future experiments. As an equilibrated black body radiates only from its *surface*, the extent to which the bubble's contents are stressed may be unrepresented in the observed spectrum. The spatial correlations and mode counting that would characterize such a small black body could constitute a new regime for the application of statistical optics. Inability to reconcile the long photon mean free path with the smallness of the hot spot suggests the need for new physics in the modeling of sonoluminescence. Vazquez et al. (2002) point out that the radius of emission $R_e$ of the black body is consistent with the radius of the collapsed bubble, but it remains unmeasured. The Hanbury–Brown and Twiss correlations should provide a route to this parameter (Trentalange and Pandey (1996)).

Weninger et al. (2001) discuss the measurement of the time dependence of the radius $R(t)$ of the collapsing bubble from which sonoluminescence originates. With Mie light scattering, care should be taken to collect light from a large solid angle, such as 30°–80° as in Weninger et al. (2000) in which case the intensity of light scattered is within 20% of $R^2$ for bubbles bigger than 0.6 μm. Weninger et al. (2001) suggest that the index of refraction inside the bubble be measured with an optical probe (such as Thomson scattering) that is sensitive to ionization. They note that a determination of $R(t)$ will help resolve the existence of shock waves or other energy focusing effects inside the bubble.

Ruuth et al. (2002) presented a major paper on molecular dynamics simulation of the response of a gas to a spherical piston (see Sec. 4.16). They modeled the interior of a collapsing rare gas bubble as a hard sphere gas driven by a spherical piston boundary moving according to the Rayleigh–Plesset equation. They investigated the energy focusing by molecular dynamics to answer the detailed questions of whether there is shock formation within the bubble, whether there is plasma formation, and what peak temperatures are achieved during the collapse.

Ruuth et al. (2002) refer to the existing theories of sonoluminescence e.g. Putterman and Weninger (2000), Hilgenfeldt et al. (1999) and Moss et al. (1999), which interpret the light emission as being due to thermal bremsstrahlung from a transparent plasma. These views are confounded by the observation that in many cases the sonoluminescence is accurately matched by a black body spectrum, which implies an opaque emitter (Vazquez et al. (2001). Ruuth et al. point out that another shortcoming of the weak ionization theory of sonoluminescence is that Xe should emit 1000 times more strongly than He (Putterman et al. (2001)). This factor of 1000 contrasts strongly with the experimentally observed factor of about 4 for single bubble sonoluminescence (Vazquez et al. (2002)). Note that Young (1976) obtained an experimental factor of about 100 for multibubble sonoluminescence.

Ruuth et al. (2002) also discuss ionization effects. Near the minimum radius of the bubble, collisions may become sufficiently energetic to ionize the gas atoms. Ionization exerts a very strong cooling effect on the gas, since on the order of 10 eV of thermal energy is removed from the gas by each ionization event. Indeed, if such energy losses are not included, xenon simulations can reach temperatures of more than $10^6$ K, while the inclusion of ionization cooling brings these peak temperatures down substantially. They account for ionization, but do not allow for subsequent electron–ion recombination to neutral atoms although this would be interesting to include at the next level of description. Particularly, as this could be an intriguing source of radiation as the hot spot decays.

For any molecular dynamics study an *event calendar* is needed. This calendar will store future events. As collisions and cell crossings occur, newly predicted collisions and cell crossings must be added to the calendar, and events that are no longer relevant must be removed. To manage the calendar efficiently, a binary tree data structure is used. Ruuth et al. (2002) plan to make a detailed study of the theoretical performance of the tree structure. Their molecular dynamics calculations indicated that extreme energy focusing occurred within a bubble, which in some cases is driven by a *shocklike* compression in the gas. Peak temperatures ranged from 40,000 K for He to 500,000 K for Xe. These were accompanied by high levels of ionization during the final collapse, and the formation of a transient, high-density plasma seemed quite likely. The simulations simply treated the bubble wall as a piston moving in with a prescribed velocity.

Ruuth et al. (2002) make the following seven suggestions for future work.

(1) They recommend coupling the internal molecular dynamics to the wall velocity to obtain a self-consistent bubble motion and internal dynamics. This could be done by coupling to Euler or Navier–Stokes models for the surrounding fluid. This may be particularly important for accurately computing the dynamics through the point of minimum radius.

(2) Another important area for future research would be adding *water vapor* into the bubble interior (see Sec. 2.2.4). This provides an *important cooling mechanism* because of the reduced polytropic exponent and because of the endothermic water dissociation (Brenner et al. (2002)). Because of the lower temperatures, less light should result.

(3) They also suggest that other bubble collapse geometries could be considered, and these may have different energy focusing characteristics. For example, one could consider a nonspherical collapse, hemispherical bubbles collapsing on a solid surface, or consider collapse geometries appropriate for bubble jetting scenarios. Similarly, one could see if special collapse profiles could be used to reach much higher internal temperatures, and otherwise explore the extremes of the energy focusing potential. Perhaps a mode could even be found in which small amounts of *deuterium–deuterium fusion* could be induced, assuming there is deuterium gas in the bubble (see Sec. 3.12).

(4) Including additional atomic physics such as atomic excitation, rotational and vibrational degrees of freedom (needed for non-rare gases or water vapor), and electron–ion recombination would all allow for more accurate energy accounting, and may also be directly related to light emission mechanisms.

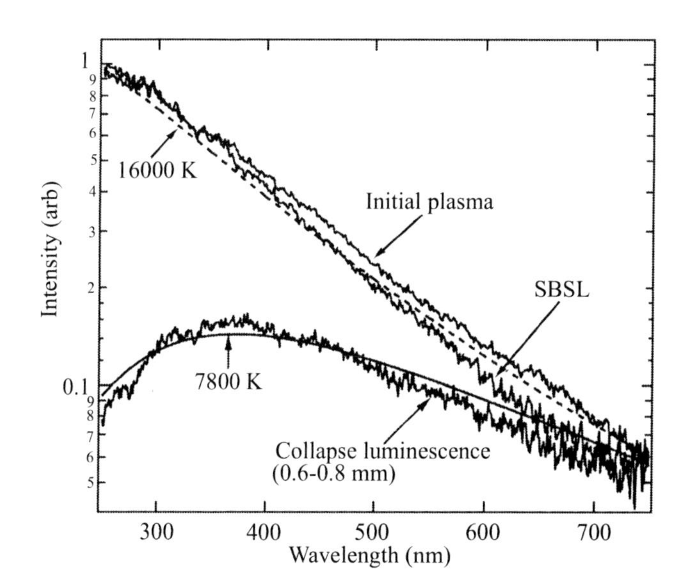


Figure 5.3 Spectral intensity versus wavelength for the luminescence from small bubbles (0.6–0.8 mm maximum radius), from the initial laser-induced plasma, and from an SBSL cell positioned at the point where the laser bubbles are created. Intensities have been scaled arbitrarily between the different curves. (Baghdassarian et al. (2001).)

(5) Another major direction would be to include electric field effects into the simulation. This would potentially allow simulation of the ions and electrons produced, which may have an important effect on the dynamics when sufficient ionization occurs. Moreover, by including atomic excitation, this approach would allow for direct simulation of the light emitting processes. With these effects included, an extremely detailed picture of the sonoluminescence phenomena could be laid out.

(6) It would be of great interest to investigate where less costly continuum models and Monte Carlo simulations are appropriate for studying sonoluminescence and to develop techniques for coupling these methods to detailed molecular dynamics simulations near the light emitting hot spot, in order to produce more complete models with greater predictive validity.

(7) There is also a great deal to explain *experimentally*. One example relevant to this study is that it would be useful to measure flash width as a function of ambient bubble radius (or, in practice, intensity and frequency of the driving sound field), for comparison with the scaling predictions of molecular dynamics and other models.

Black body radiation is also found in plasmas that are generated at the focus of a strong laser in water (Baghdassarian et al. (1999, 2001)). They scaled up the bubble size to 2 mm, in order to compress more gas and study consequently larger hot spots. The lower jagged curve in Fig. 5.3 shows the spectrum of laser-induced bubbles from a Nd:YAG laser producing 6 ns pulses with a maximum energy of 0.6 J at 1064 nm, for bubbles of 0.6–0.8 mm. The solid curve is a black body fit at 7800 K. A curious result (Baghdassarian et al. (1999)) is that the luminescence is independent of whether the water contains ambient dissolved air, is completely degassed, or is filled with 150 torr partial pressure of argon gas or 1 bar (760 torr) of xenon gas.

The ionization of the *water* by the YAG pulse creates an initial *plasma* 50–80 μm in radius that emits a bright pulse of light. The pulse has a duration of about 100 ns, with then a weak tail decaying over several μs. The spectrum of this pulse, plotted in Fig. 5.3, is labeled "initial plasma" and shows an intensity rising smoothly into the ultraviolet with no visible atomic lines. This compares well with a black body spectrum of 16,000 K (the dashed curve in Fig. 5.3).

To compare this luminescence with single bubble sonoluminescence (SBSL) Baghdassarian et al. (2001) mounted an acoustical SBSL cell at 33 kHz into the sample region of the setup, and aligned to center the acoustically trapped bubble to the same point where the laser created the plasma. Figure 5.3 shows the spectrum resulting from integrating several seconds of the SBSL pulses. This appears to have nearly the same wavelength dependence as that of the initial plasma i.e. that of a black body at 16,000 K. This supports a conclusion that SBSL is at least approximated by black body emission from a weakly ionized thermal plasma. As the bubble size is increased, a molecular band from $OH^{\bullet}$ becomes apparent in the spectrum of the laser-created bubbles. Baghdassarian et al. (2001) show that this provides a connection to multibubble sonoluminescence (Matula et al. (1995)) and to SBSL (Young et al. (2001)) where the same band is observed. The appearance of the $OH^{\bullet}$ band may be related to surface instabilities in the larger bubbles in the vicinity of the collapse point. Further work will be needed to understand the details of the plasma creation process, and how the bubble wall instabilities might affect it.

## REFERENCES


Ahmad M, Pitt Ford TR and Crum LA 1987 Ultrasonic debridement of root canals: An insight into the mechanics involved, *J Endodontics* **13**, 93.

Baghdassarian O, Chu H, Tabbert B and Williams GA 2001 Spectrum of luminescence from laser-created bubbles in water, *Phys Rev Lett* **86**, 4943.

Baghdassarian O, Tabbert B and Williams GA 1999 Luminescence characteristics of laser-induced bubbles in water, *Phys Rev Lett* **83**, 2437.

Barber BP, Hiller RA, Löfstedt R, Putterman SJ and Weninger KR 1997 Defining the unknowns of sonoluminescence, *Phys Rep* **281**, 65.

Beckett MA and Hua I 2001 Impact of ultrasonic frequency on aqueous sonoluminescence and sonochemistry, *J Phys Chem A* **105**, 3796.

Brenner MP, Hilgenfeldt S and Lohse D 2002 *Rev Mod Phys* **74**, 425.

Flint EB and Suslick KS 1991 The temperature of cavitation, *Science* **253**, 1397.

Frenzel H and Schultes H 1934 Luminescenz im ultraschallbeschickten Wasser, *Zeit Phys Chem* **B27**, 421.

Gaitan DF, Crum LA, Church CC and Roy RA 1992 Sonoluminescence and bubble dynamics for a single, stable, cavitation bubble, *J Acoust Soc Am* **91**, 3166.

Greenland PT 1999 Sonoluminescence, *Contemporary Physics* **40**,11.

Grieser F and Ashokkumar M 2001 The effect of surface active solutes on bubbles exposed to ultrasound, *Adv Colloid and Interface Science* 89–90, 423.

Hilgenfeldt S, Grossmann S and Lohse D 1999 A simple explanation of light emission in sonoluminescence, *Nature* **398**, 402.

Jarman PD 1959a Sonoluminescence, PhD thesis, Imperial College, University of London.

Jarman PD 1959b Measurement of sonoluminescence from pure liquids and some aqueous solutions, *Proc Phys Soc* **73**, 628.
Lohse D 2002 Inside a micro-reactor, *Nature* **418**, 381.
Matula TJ, Roy RA, Mourad PD, McNamara III WB and Suslick KS 1995 Comparison of multibubble and single-bubble sonoluminescence spectra, *Phys Rev Lett* **75**, 2602.
Moss WC, Young DA, Harte JA, Levatin JL, Rozsnyai BF, Zimmerman GB and Zimmerman IH 1999 Computed optical emissions from a sonoluminescing bubble, *Phys Rev E* **59**, 2986.
Ogi H, Hirao M and Shimoyama M 2002 Activation of $TiO_2$ photocatalyst by single-bubble sonoluminescence for water treatment, *Ultrasonics* **40**, 649.
Putterman SJ, Evans PG, Vasquez G and Weninger K 2001 Is there a simple theory of sonoluminescence? *Nature* **409**, 782.
Putterman SJ and Weninger KR 2000 Sonoluminescence: How bubbles turn sound into light, *Annu Rev Fluid Mech* **32**, 445.
Ruuth SJ, Putterman S and Merriman B 2002 Molecular dynamics simulation of the response of a gas to a spherical piston: Implications for sonoluminescence, *Phys Rev E* **66**, 036310.
Suslick KS 1990 Sonochemistry, *Science* **247**, 1439.
Trentalange S and Pandey SU 1996 Bose–Einstein correlations and sonoluminescence, *J Acoust Soc Am* **99**, 2439.
Vazquez G, Camara C, Putterman SS and Weninger K 2001 Sonoluminescence: Nature's smallest blackbody, *Opt Lett* **26**, 575.
Vazquez G, Camara C, Putterman SJ and Weninger K 2002 Blackbody spectra for sonoluminescing hydrogen bubbles, *Phys Rev Lett* **88**,197402.
Walmsley AD, Walsh TF, Laird WRE et al. 1990 Effects of cavitational activity on the root surface of the teeth during ultrasonic scaling, *J Clin Periodont* **17**, 306.
Weninger KR, Camara CG and Putterman SJ 2000 Observation of bubble dynamics within luminescent cavitation clouds: Sonoluminescence at the nano-scale, *Phys Rev E* **63**, 016310.
Weninger KR, Evans PG and Putterman SJ 2001 Comment on "Mie scattering from a sonoluminescing bubble with high spatial and temporal resolution", *Phys Rev E* **64**, 038301.
Young FR 1976 Sonoluminescence from water containing dissolved gases, *J Acoust Soc Am* **60**, 100.
Young FR 1989 Cavitation, McGraw-Hill; 1999 Cavitation, Imperial College Press.
Young JB, Nelson JA and Kang W 2001 Line emission in single-bubble sonoluminescence, *Phys Rev E* **86**,2673.

# Appendix

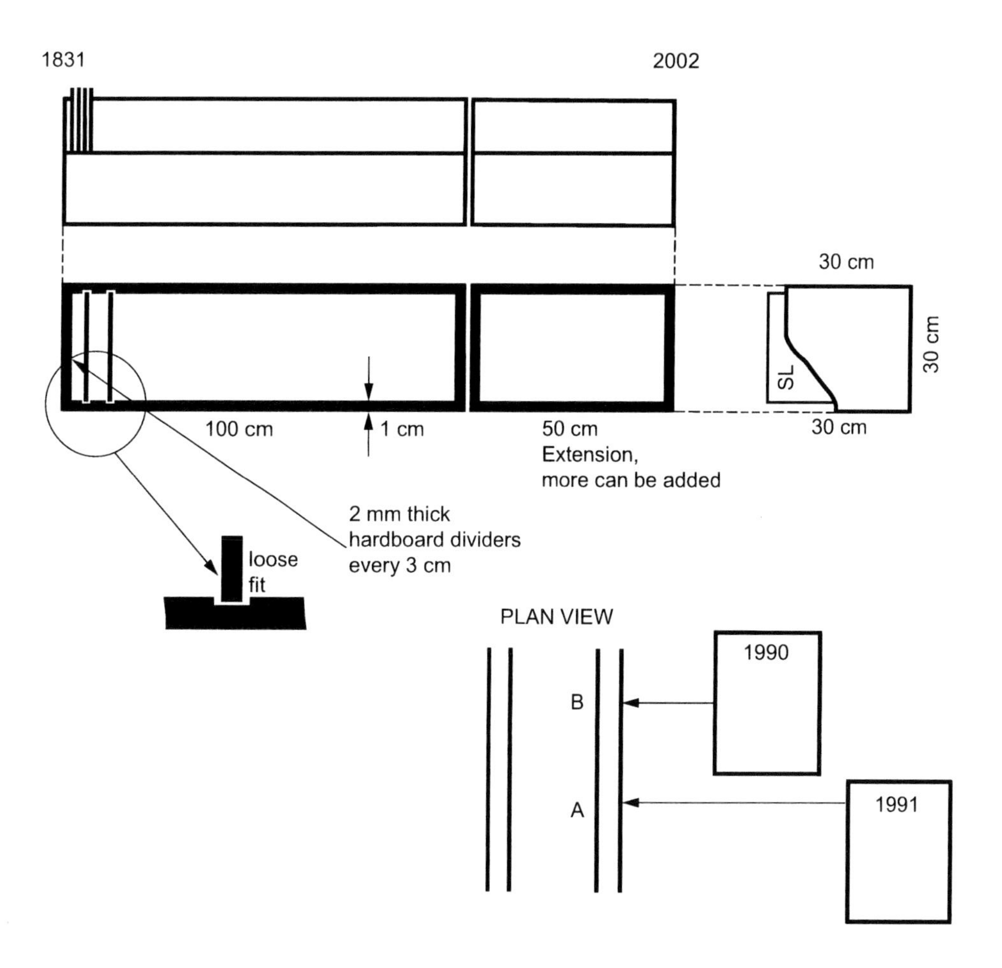


Figure A.1 Filing rack designed by the author's carpenter, Gordon Nicholas. If the filing rack is mounted by the right hand side of the writer's desk, he can retrieve any article very quickly. Will hold 800 articles. The year sheets 1990 and 1991 go in at B and A, and project 4 cm above the articles. All 1990 articles are arranged alphabetically between A and B, starting with Apfel at A.

Figure A.2 Filing rack.

# Subject Index

# Author Index